Data Analysis and Visualization in Genomics and Proteomics

Data Analysis and Visualization in Genomics and Proteomics

Editors
Francisco Azuaje
University of Ulster at Jordanstown, UK

and

Joaquín Dopazo
Spanish Cancer National Centre (CNIO), Madrid, Spain

John Wiley & Sons, Ltd

Telephone (+44) 1243 779777

Email (for orders and customer service enquiries): cs-books@wiley.co.uk
Visit our Home Page on www.wileyeurope.com or www.wiley.com

Other Wiley Editorial Offices

John Wiley & Sons Inc., 111 River Street, Hoboken, NJ 07030, USA

Jossey-Bass, 989 Market Street, San Francisco, CA 94103-1741, USA

Wiley-VCH Verlag GmbH, Boschstr. 12, D-69469 Weinheim, Germany

John Wiley & Sons Australia Ltd, 33 Park Road, Milton, Queensland 4064, Australia

John Wiley & Sons (Asia) Pte Ltd, 2 Clementi Loop #02-01, Jin Xing Distripark, Singapore 129809

John Wiley & Sons Canada Ltd, 22 Worcester Road, Etobicoke, Ontario, Canada M9W 1L1

Wiley also publishes its books in a variety of electronic formats. Some content that appears in print may not be available in electronic books.

Library of Congress Cataloging-in-Publication Data

British Library Cataloguing in Publication Data

A catalogue record for this book is available from the British Library

ISBN 0-470-09439-7

Typeset in 10.5/13pt Times by Thomson Press (India) Limited, New Delhi
Printed and bound in Great Britain by Antony Rowe Ltd., Chippenham, Wiltshire
This book is printed on acid-free paper responsibly manufactured from sustainable forestry in which at least two trees are planted for each one used for paper production.

Contents

Preface **xi**

List of Contributors **xiii**

SECTION I INTRODUCTION – DATA DIVERSITY AND INTEGRATION **1**

1 Integrative Data Analysis and Visualization: Introduction to Critical Problems, Goals and Challenges **3**

Francisco Azuaje and Joaquín Dopazo

1.1 Data Analysis and Visualization: An Integrative Approach 3
1.2 Critical Design and Implementation Factors 5
1.3 Overview of Contributions 8
References 9

2 Biological Databases: Infrastructure, Content and Integration **11**

Allyson L. Williams, Paul J. Kersey, Manuela Pruess and Rolf Apweiler

2.1 Introduction 11
2.2 Data Integration 12
2.3 Review of Molecular Biology Databases 17
2.4 Conclusion 23
References 26

3 Data and Predictive Model Integration: an Overview of Key Concepts, Problems and Solutions **29**

Francisco Azuaje, Joaquín Dopazo and Haiying Wang

3.1 Integrative Data Analysis and Visualization: Motivation and Approaches 29
3.2 Integrating Informational Views and Complexity for Understanding Function 31
3.3 Integrating Data Analysis Techniques for Supporting Functional Analysis 34
3.4 Final Remarks 36
References 38

SECTION II INTEGRATIVE DATA MINING AND VISUALIZATION – EMPHASIS ON COMBINATION OF MULTIPLE DATA TYPES 41

4 Applications of Text Mining in Molecular Biology, from Name Recognition to Protein Interaction Maps 43

Martin Krallinger and Alfonso Valencia

4.1 Introduction 44
4.2 Introduction to Text Mining and NLP 45
4.3 Databases and Resources for Biomedical Text Mining 47
4.4 Text Mining and Protein–Protein Interactions 50
4.5 Other Text-Mining Applications in Genomics 55
4.6 The Future of NLP in Biomedicine 56
Acknowledgements 56
References 56

5 Protein Interaction Prediction by Integrating Genomic Features and Protein Interaction Network Analysis 61

Long J. Lu, Yu Xia, Haiyuan Yu, Alexander Rives, Haoxin Lu, Falk Schubert and Mark Gerstein

5.1 Introduction 62
5.2 Genomic Features in Protein Interaction Predictions 63
5.3 Machine Learning on Protein–Protein Interactions 67
5.4 The Missing Value Problem 73
5.5 Network Analysis of Protein Interactions 75
5.6 Discussion 79
References 80

6 Integration of Genomic and Phenotypic Data 83

Amanda Clare

6.1 Phenotype 83
6.2 Forward Genetics and QTL Analysis 85
6.3 Reverse Genetics 87
6.4 Prediction of Phenotype from Other Sources of Data 88
6.5 Integrating Phenotype Data with Systems Biology 90
6.6 Integration of Phenotype Data in Databases 93
6.7 Conclusions 95
References 95

7 Ontologies and Functional Genomics 99

Fátima Al-Shahrour and Joaquín Dopazo

7.1 Information Mining in Genome-Wide Functional Analysis 99
7.2 Sources of Information: Free Text Versus Curated Repositories 100
7.3 Bio-Ontologies and the Gene Ontology in Functional Genomics 101
7.4 Using GO to Translate the Results of Functional Genomic Experiments into Biological Knowledge 103

7.5 Statistical Approaches to Test Significant Biological Differences 104
7.6 Using FatiGO to Find Significant Functional Associations in Clusters of Genes 106
7.7 Other Tools 107
7.8 Examples of Functional Analysis of Clusters of Genes 108
7.9 Future Prospects 110
References 110

8 The *C. elegans* Interactome: its Generation and Visualization 113

Alban Chesnau and Claude Sardet

8.1 Introduction 113
8.2 The ORFeome: the first step toward the interactome of *C. elegans* 116
8.3 Large-Scale High-Throughput Yeast Two-Hybrid Screens to Map the *C. elegans* Protein–Protein Interaction (Interactome) Network: Technical Aspects 118
8.4 Visualization and Topology of Protein–Protein Interaction Networks 121
8.5 Cross-Talk Between the *C. elegans* Interactome and other Large-Scale Genomics and Post-Genomics Data Sets 123
8.6 Conclusion: From Interactions to Therapies 129
References 130

SECTION III INTEGRATIVE DATA MINING AND VISUALIZATION – EMPHASIS ON COMBINATION OF MULTIPLE PREDICTION MODELS AND METHODS 135

9 Integrated Approaches for Bioinformatic Data Analysis and Visualization – Challenges, Opportunities and New Solutions 137

Steve R. Pettifer, James R. Sinnott and Teresa K. Attwood

9.1 Introduction 137
9.2 Sequence Analysis Methods and Databases 139
9.3 A View Through a Portal 141
9.4 Problems with Monolithic Approaches: One Size Does Not Fit All 142
9.5 A Toolkit View 143
9.6 Challenges and Opportunities 145
9.7 Extending the Desktop Metaphor 147
9.8 Conclusions 151
Acknowledgements 151
References 152

10 Advances in Cluster Analysis of Microarray Data 153

Qizheng Sheng, Yves Moreau, Frank De Smet, Kathleen Marchal and Bart De Moor

10.1 Introduction 153
10.2 Some Preliminaries 155
10.3 Hierarchical Clustering 157
10.4 *k*-Means Clustering 159

10.5 Self-Organizing Maps 159
10.6 A Wish List for Clustering Algorithms 160
10.7 The Self-Organizing Tree Algorithm 161
10.8 Quality-Based Clustering Algorithms 162
10.9 Mixture Models 163
10.10 Biclustering Algorithms 166
10.11 Assessing Cluster Quality 168
10.12 Open Horizons 170
References 171

11 Unsupervised Machine Learning to Support Functional Characterization of Genes: Emphasis on Cluster Description and Class Discovery 175
Olga G. Troyanskaya
11.1 Functional Genomics: Goals and Data Sources 175
11.2 Functional Annotation by Unsupervised Analysis of Gene Expression Microarray Data 177
11.3 Integration of Diverse Functional Data For Accurate Gene Function Prediction 179
11.4 MAGIC – General Probabilistic Integration of Diverse Genomic Data 180
11.5 Conclusion 188
References 189

12 Supervised Methods with Genomic Data: a Review and Cautionary View 193
Ramón Díaz-Uriarte
12.1 Chapter Objectives 193
12.2 Class Prediction and Class Comparison 194
12.3 Class Comparison: Finding/Ranking Differentially Expressed Genes 194
12.4 Class Prediction and Prognostic Prediction 198
12.5 ROC Curves for Evaluating Predictors and Differential Expression 201
12.6 Caveats and Admonitions 203
12.7 Final Note: Source Code Should be Available 209
Acknowledgements 210
References 210

13 A Guide to the Literature on Inferring Genetic Networks by Probabilistic Graphical Models 215
Pedro Larrañaga, Iñaki Inza and Jose L. Flores
13.1 Introduction 215
13.2 Genetic Networks 216
13.3 Probabilistic Graphical Models 218
13.4 Inferring Genetic Networks by Means of Probabilistic Graphical Models 229
13.5 Conclusions 234
Acknowledgements 235
References 235

14 Integrative Models for the Prediction and Understanding of Protein Structure Patterns **239**

Inge Jonassen

14.1 Introduction 239
14.2 Structure Prediction 241
14.3 Classifications of Structures 244
14.4 Comparing Protein Structures 246
14.5 Methods for the Discovery of Structure Motifs 249
14.6 Discussion and Conclusions 252
References 254

Index **257**

Preface

The sciences do not try to explain, they hardly even try to interpret, they mainly make models. By a model is meant a mathematical construct which, with the addition of certain verbal interpretations describes observed phenomena. The justification of such a mathematical construct is solely and precisely that it is expected to work.

John von Neumann (1903–1957)

These ambiguities, redundancies, and deficiencies recall those attributed by Dr. Franz Kuhn to a certain Chinese encyclopaedia entitled Celestial Emporium of Benevolent Knowledge. On those remote pages it is written that animals are divided into (a) those that belong to the Emperor, (b) embalmed ones, (c) those that are trained, (d) suckling pigs, (e) mermaids, (f) fabulous ones, (g) stray dogs, (h) those that are included in this classification, (i) those that tremble as if they were mad, (j) innum erable ones, (k) those drawn with a very fine camel's hair brush, (l) others, (m) those that have just broken a flower vase, (n) those that resemble flies from a distance.

Jorge Luis Borges (1899–1986)
The analytical language of John Wilkins. In *Other Inquisitions* (1937–1952). University of Texas Press, 1984.

One of the central goals in biological sciences is to develop predictive models for the analysis and visualization of information. However, the analysis and visualization of biological data patterns have traditionally been approached as independent problems. Until now, biological data analysis has emphasized the automation aspects of tools and relatively little attention has been given to the integration and visualization of information and models.

One fundamental question for the development of a systems biology approach is how to build prediction models able to identify and combine multiple, relevant information resources in order to provide scientists with more meaningful results.

Unsatisfactory answers exist in part because scientists deal with incomplete, inaccurate data and in part because we have not fully exploited the advantages of integrating data analysis and visualization models. Moreover, given the vast amounts of data generated by high-throughput technologies, there is a risk of identifying spurious associations between genes and functional properties owing to a lack of an adequate understanding of these data and analysis tools.

This book aims to provide scientists and students with the basis for the development and application of integrative computational methods to analyse and understand biological data on a systemic scale. We have adopted a fairly broad definition for the areas of *genomics* and *proteomics*, which also comprises a wider spectrum of '*omic*' approaches required for the understanding of the functions of genes and their products. This book will also be of interest to advanced undergraduate or graduate students and researchers in the area of bioinformatics and life sciences with a fairly limited background in data mining, statistics or machine learning. Similarly, it will be useful for computer scientists interested in supporting the development of applications for systems biology.

This book places emphasis on the processing of multiple data and knowledge resources, and the combination of different models and systems. Our goal is to address existing limitations, new requirements and solutions, by providing a comprehensive description of some of the most relevant and recent techniques and applications.

Above all, we have made a significant effort in selecting the content of these contributions, which has allowed us to achieve a unity and continuity of concepts and topics relevant to information analysis, visualization and integration. But clearly, a single book cannot do justice to all aspects, problems and applications of data analysis and visualization approaches to systems biology. However, this book covers fundamental design, application and evaluation principles, which may be adapted to related systems biology problems. Furthermore, these contributions reflect significant advances and emerging solutions for integrative data analysis and visualization. We hope that this book will demonstrate the advantages and opportunities offered by integrative bioinformatic approaches.

We are proud to present chapters from internationally recognized scientists working in prestigious research teams in the areas of biological sciences, bioinformatics and computer science. We thank them for their contributions and continuous motivation to support this project.

The European Science Foundation Programme on Integrated Approaches for Functional Genomics deserves acknowledgement for supporting workshops and research visits that led to many discussions and collaboration relevant to the production of this book.

We are grateful to our Publishing Editor, Joan Marsh, for her continuing encouragement and guidance during the proposal and production phases. We thank her Publishing Assistant, Andrea Baier, for diligently supporting the production process.

Francisco Azuaje and Joaquin Dopazo

Jordanstown and Madrid

October 2004

List of Contributors

Fátima Al-Shahrour, Bioinformatics Unit, Centro Nacional de Investigaciones Oncológicas, Melchor Fernandez Almagro 3, E-28039 Madrid, Spain

Rolf Apweiler, EMBL Outstation – Hinxton, European Bioinformatics Institute, Wellcome Trust Genome Campus, Hinxton, Cambridge CB10 1SD, UK

Terri K. Attwood, School of Biological Sciences, 2.205, Stopford Building, The University of Manchester, Oxford Road, Manchester M13 9PT, UK

Francisco Azuaje, School of Computing and Mathematics, University of Ulster at Jordanstown, BT37 0QB, Co. Antrim, Northern Ireland, UK

Alban Chesnau, Institute de Génétique Moléculaire, Centre National de la Recherche Scientifique, IFR 122, 1919 Route de Mende, 34293 Montpellier Cedex 5, France

Amanda Clare, Department of Computer Science, University of Wales, Penglais, Aberystwyth SY23 3DB, UK

Bart De Moor, Department of Electrical Engineering, ESAT-SCD, K.U. Leuven, Kasteelpark Arenberg 10, 3001 Leuven-Heverlee, Belgium

Frank De Smet, Department of Electrical Engineering, ESAT-SCD, K.U. Leuven, Kasteelpark Arenberg 10, 3001 Leuven-Heverlee, Belgium

Ramón Díaz-Uriarte, Bioinformatics Unit, Centrol Nacional de Investigaciones Oncológicas, Melchor Fernández Almagro 3, E-28039 Madrid, Spain.

Joaquín Dopazo, Bioinformatics Unit, Centrol Nacional de Investigaciones Oncológicas, Melchor Fernández Almagro 3, E-28039 Madrid, Spain

Jose L. Flores, Department of Computer Science, University of Mondragon, Larranña 16, E-20560 Oñati, Spain

Mark Gerstein, Department of Molecular Biophysics and Biochemistry, Yale University, Bass Center, 266 Whitney Avenue, P.O. Box 208114, New Haven, CT 06520-8114, USA

Iñaki Inza, Department of Computer Science and Artificial Intelligence, University of the Basque Country, P.O. Box 649, E-20080 Donostia, Spain

Inge Jonassen, Department of Informatics and Computational Biology Unit, Bergen Centre for Computational Science, University of Bergen, HIB, N-5020 Bergen, Norway

Paul J. Kersey, EMBL Outstation – Hinxton, European Bioinformatics Institute, Wellcome Trust Genome Campus, Hinxton, Cambridge CB10 1SD, UK

Martin Krallinger, Protein Design Group (PDG), National Biotechnology Center (CNB), Campus Universidad Autónoma (UAM), C/Darwin, 3, Ctra. de Colmenar Viejo Km 15,500, Cantoblanco, E-28049 Madrid, Spain

Pedro Larrañaga, Department of Computer Science and Artificial Intelligence, University of the Basque Country, P.O. Box 649, E-20080 Donostia, Spain

Haoxin Lu, Department of Molecular Biophysics and Biochemistry, Yale University, Bass Center, 266 Whitney Avenue, P.O. Box 208114, New Haven, CT 06520-8114, USA

Long J. Lu, Department of Molecular Biophysics and Biochemistry, Yale University, Bass Center, 266 Whitney Avenue, P.O. Box 208114, New Haven, CT 06520-8114, USA

Kathleen Marchal, Department of Electrical Engineering, ESAT-SCD, K.U. Leuven, Kasteelpark Arenberg 10, 3001 Leuven-Heverlee, Belgium

Yves Moreau, Department of Electrical Engineering, ESAT-SCD, K.U. Leuven, Kasteelpark Arenberg 10, 3001 Leuven-Heverlee, Belgium

S. R. Pettifer, Department of Computer Science, University of Manchester, Kilburn Building, Oxford Road, Manchester M13 9PT, UK

Manuela Pruess, EMBL Outstation – Hinxton, European Bioinformatics Institute, Wellcome Trust Genome Campus, Hinxton, Cambridge CB10 1SD, UK

Alexander Rives, Institute of Systems Biology, 1441 North 34th Street, Seattle, WA 98103, USA

Claude Sardet, Institut de Génétique Moléculaire, Centre National de la Recherche Scientifique, UMR5535, 1919 Route de Mende, 34293 Montpellier Cedex 5, France

Falk Schubert, Department of Computer Sciences, Yale University, 51 Prospect Street, New Haven, CT 06520, USA

Qizheng Sheng, Department of Electrical Engineering, ESAT-SCD, K.U. Leuven, Kasteelpark Arenberg 10, 3001 Leuven-Heverlee, Belgium

J. R. Sinnott, Room 2.102, School of Computer Science, Kilburn Building. The University of Manchester, Manchester M13 9PL, UK

Olga G. Troyanskaya, Department of Computer Science and Lewis-Sigler Institute for Integrative Genomics, Princeton University, 35 Olden Street, Princeton, NJ 08544, USA

Alfonso Valencia, Protein Design Group, CNB-CSIC, Centro Nacional de Biotechnologia, Cantoblanco, E-28049 Madrid, Spain

Haying Wang, School of Computing and Mathematics, University of Ulster at Jordanstown, BT37 0QB, Co. Antrim, Northern Ireland, UK

Allyson L. Williams, EMBL Outstation – Hinxton, European Bioinformatics Institute, Wellcome Trust Genome Campus, Hinxton, Cambridge CB10 1SD, UK

Yu Xia, Department of Molecular Biophysics and Biochemistry, Yale University, Bass Center, 266 Whitney Avenue, P.O. Box 208114, New Haven, CT 06520-8114, USA

Haiyuan Yu, Department of Molecular Biophysics and Biochemistry, Yale University, Bass Center, 266 Whitney Avenue, P.O. Box 208114, New Haven, CT 06520-8114, USA

I Introduction – Data Diversity and Integration

1

Integrative Data Analysis and Visualization: Introduction to Critical Problems, Goals and Challenges

Francisco Azuaje and **Joaquin Dopazo**

Abstract

This chapter introduces fundamental concepts and problems approached in this book. A rationale for the application of integrative data analysis and visualization approaches is presented. Critical design, implementation and evaluation factors are discussed. The chapter identifies barriers and opportunities for the development of more robust and meaningful methods. It concludes with an overview of the content of the book.

Keywords

biological data analysis, data visualization, integrative data analysis, functional genomics, systems biology, design principles

1.1 Data Analysis and Visualization: An Integrative Approach

With the popularization of high-throughput technologies, and the consequent enormous accumulation of biological data, the development of a systems biology era will depend on the generation of predictive models and their capacity to identify and combine multiple information resources. Such data, knowledge and models are associated with different levels of biological organization. Thus, it is fundamental

Data Analysis and Visualization in Genomics and Proteomics Edited by Francisco Azuaje and Joaquin Dopazo

to improve the understanding of how to integrate biological information, which is complex, heterogeneous and geographically distributed.

The *analysis* (including discovery) and *visualization* of relevant biological data patterns have traditionally been approached as independent computational problems. Until now biological data analysis has placed emphasis on the automation aspects of tools, and relatively little attention has been given to the integration and visualization of information and models, probably due to the relative simplicity of pre-genomic data. However, in the post-genomic era it is very convenient that these tasks complement each other in order to achieve higher integration and understanding levels.

This book provides scientists and students with the basis for the development and application of integrative computational methods to exchange and analyse biological data on a systemic scale. It emphasizes the processing of multiple data and knowledge resources, and the combination of different models and systems. One important goal is to address existing limitations, new requirements and solutions by providing comprehensive descriptions of techniques and applications. It covers different data analysis and visualization problems and techniques for studying the roles of genes and proteins at a system level. Thus, we have adopted a fairly broad definition for the areas of *genomics* and *proteomics*, which also comprises a wider spectrum of *omic* approaches required for the understanding of the *functions* of genes and their products.

Emphasis is placed on *integrative* biological and computational approaches. Such an integrative framework refers to the study of biological systems based on the combination of data, knowledge and predictive models originating from different sources. It brings together informational views and knowledge relevant to or originating from diverse organizational, functional modules.

Data analysis comprises systems and tools for identifying, organizing and interpreting relevant biological patterns in databases as well as for asking functional questions in a whole-genome context. Typical functional data analysis tasks include classification, gene selection or their use in predictors for microarray data, the prediction of protein interactions etc.

Data visualization covers the design of techniques and tools for formulating, browsing and displaying prediction outcomes and complex database queries. It also covers the automated description and validation of data analysis outcomes.

Biological data analysis and visualization have traditionally been approached as independent problems. Relatively little attention has been given to the integration and visualization of information and models. However, the integration of these areas facilitates a deeper understanding of problems at a systemic level.

Traditional data analysis and visualization lack key capabilities required for the development of a system biology paradigm. For instance, biological information visualization has typically consisted of the representation and display of information associated with lists of genes or proteins. Graphical tools have been implemented to visualize more complex information, such as metabolic pathways and genetic networks. Recently, more complex tools, such as *Ensembl* (Birney *et al.*, 2003), have integrated different types of information, e.g. genomic, functional, polymorphisms etc., on a genome-wide context. Other tools, such as *GEPAS* (Herrero *et al.*, 2004), integrate gene expression data as well as genomic and functional information for predictive analysis. Nevertheless, even state-of-the-art tools still lack the elements necessary to achieve a meaningful, robust integration and interpretation of multiple data and knowledge sources.

This book aims to present recent and significant advances in data analysis and visualization that can support system biology approaches. It will discuss key design, application and evaluation principles. It will address the combination of different types of biological data and knowledge resource, as well as prediction models and analysis tools. From a computational point of view it will demonstrate (a) how data analysis techniques can facilitate more comprehensive, user-friendly data visualization tasks and (b) how data visualization methods may make data analysis a more meaningful and biologically relevant process. This book will describe how this synergy may support integrative approaches to functional genomics.

1.2 Critical Design and Implementation Factors

This section briefly discusses important data analysis problems that are directly or partially addressed by some of the subsequent chapters.

Over the past eight years a substantial collection of data analysis and prediction methods for functional genomics has been reported. Among the many papers published in journals and conference proceedings, perhaps only a minority perform rigorous comparative assessment against well established and previously tested methodologies. Moreover, it is essential to provide more scientifically sound problem formulations and justifications. This is especially critical when adopting methodologies involving, for example, assumptions about the statistical independence between predictive attributes or the interpretation of statistical significance.

Such technical shortcomings and the need to promote health and wealth through innovation represent strong reasons for the development of shared, best practices for data analysis applications in functional genomics. This book includes contributions addressing one or more of these critical factors for different computational and experimental problems. They describe approaches, assess solutions and critically discuss their advantages and limitations.

Supervised and unsupervised classification applications are typical, fundamental tasks in functional genomics. One of the most challenging questions is not whether there are techniques available for different problems, but rather which 'specific' technique(s) should be applied and 'when' to apply them. Therefore, data analysis models must be evaluated to detect and control unreliable data analysis conditions, inconsistencies and irrelevance. A well known scheme for supervised classification is to generate indicators of accuracy and precision. However, it is essential to estimate the significance of the differences between prediction outcomes originating from different models. It is not uncommon to find studies published in recognized journals and conferences, which claim prediction quality differences, that do not provide evidence of statistical significance given the data available and the models under comparison. Chapters 5 and 12 are particularly relevant to understand these problems.

The lack of adequate evaluation methods also negatively affects clustering-based studies (see Chapters 7, 10 and 11). Such studies must provide quality indicators to measure the significance of the obtained clusters, for example in terms of their compactness and separation. Another important factor is to report statistical evidence to support the choice of a particular number of clusters. Furthermore, in annotation-based analyses it is essential to apply tools to determine the functional classes (such as gene ontology terms) that are significantly enriched in a given cluster (see Chapter 7).

Predictive *generalization* is the ability to correctly make predictions (such as classification) on data unseen during the model implementation process (sometimes referred to as training or learning). Effective and meaningful predictive data analysis studies should aim to build models able to generalize. It is usually accepted that a model will be able to achieve this property if its architecture and learning parameters have been properly selected. It is also critical to ensure that enough training data is available to build the prediction model. However, such a condition is difficult to satisfy due to resource limitations. This is a key feature exhibited, for instance, by a significant number of gene expression analyses. With a small set of training data, a prediction model may not be able to accurately represent the data under analysis. Similarly, a small test dataset may contribute to an unreliable prediction quality assessment. The problems of building prediction models based on small datasets and the estimation of their predictive quality deserve a more careful consideration in functional genomics. Model over-fitting is a significant problem for designing effective and reliable prediction models. One simple way to determine that a prediction model, M, is over-fitting a training dataset consists of identifying a model M', which exhibits both higher training prediction and lower test prediction errors in relation to M. This problem is of course directly linked to the prediction generalization problem discussed above. Thus, an over-fitted model is not able to make accurate predictions on unseen data. Several predictive quality assessment and data sampling techniques are commonly applied to address this problem. For example, the prediction performance obtained on a *validation dataset* may be used to estimate when a neural network training process should be stopped to improve generalization. Over-fitting basically indicates that a prediction learning process was not correctly conducted due to factors such as an inadequate selection of training data and/or learning parameters. The former factor is commonly a consequence of the availability of small datasets. It is crucial to identify factors, experimental conditions and constraints that contribute to over-fitting in several prediction applications for functional genomics. This type of study may provide guidelines to make well-informed decisions on the selection of prediction models. Solutions may be identified not only by looking into these constraints, but also by clearly distinguishing between prediction goals. A key goal is to apply models, architectures and learning parameters that provide both accurate and robust representation of the data under consideration. Further research is needed to understand how to adapt and combine prediction methods to avoid over-fitting problems in the presence of small or skewed data problems.

Feature selection is another important problem relevant to predictive data analysis and visualization. The problem of selecting the most relevant features for a classification problem has been typically addressed by implementing *filter* and *wrapper* approaches. Filter-based methods consist of statistical tests to detect features that are significantly differentiated among classes. Wrapper approaches select relevant features as part of the optimization of a classification problem, i.e. they are embedded into the classification learning process. Wrapper methods commonly outperform filter methods in terms of prediction accuracy. However, key limitations have been widely studied. One such limitation is the *instability problem*. In this problem variable, inconsistent feature subsets may be selected even for small variations in the training datasets and classification architecture. Moreover, wrapper methods are more computationally expensive. Instability may not represent a critical problem if the main objective of the feature selection task is to optimize prediction performance, such as classification accuracy. Nevertheless, deeper investigations are required if the goal is to assess biological relevance of features, such as the discovery of potential biomarkers. Further research is necessary to design methods capable of identifying robust and meaningful feature relevance. These problems are relevant to the techniques and applications presented in Chapters 5, 6, 12 and 13.

The area of functional genomics present novel and complex challenges, which may require a redefinition of conceptions and principles traditionally applied to areas such as engineering or clinical decision support systems. For example, one important notion is that significant, meaningful feature selection can be achieved through both the reduction and maximization of feature *redundancy* and *diversity* respectively. Therefore, crucial questions that deserve deeper discussions are the following. Can feature similarity (or correlation) be associated with redundancy or irrelevance? Does feature diversity guarantee the generation of biologically meaningful results? Is feature diversity a synonym of relevance? Sound answers will of course depend on how concepts such as feature relevance, diversity, similarity and redundancy are defined in both computational and biological contexts.

Data mining and knowledge discovery consist of several, iterative and interactive analysis tasks, which may require the application of heterogeneous and distributed tools. Moreover, a particular analysis and visualization outcome may represent only a component in a series of processing steps based on different software and hardware platforms. Therefore, the development of system- and application-independent schemes for representing analysis results is important to support more efficient, reliable and transparent information analysis and exchange. It may allow a more structured and consistent representation of results originating from large-scale studies, involving for example several visualization techniques, data clustering and statistical significance tests. Such representation schemes may also include metadata or other analysis content descriptors. They may facilitate not only the reproducibility of results, but also the implementation of subsequent analyses and inter-operation of visualization systems (Chapter 9). Another important goal is to allow their integration with other data and information resources. Advances mainly oriented to the data generation problem, such as the *MicroArray Gene Expression Markup Language* (MAGE-ML), may offer useful guidance to develop methods for the representation and exchange of predictive data analysis and visualization results.

1.3 Overview of Contributions

The remainder of the book comprises 13 chapters. The next two chapters overview key concepts and resources for data analysis and visualization. The second part of the book focuses on systems and applications based on the combination of multiple types of data. The third part highlights the combination of different data analysis and visualization predictive models.

Chapter 2 provides a survey of current techniques in data integration as well as an overview of some of the most important databases. Problems derived from the enormous complexity of biological data and from the heterogeneity of data sources in the context of data integration and data visualization are discussed.

Chapter 3 overviews fundamental concepts, requirements and approaches to (a) integrative data analysis and visualization approaches with an emphasis on the processing of multiple data types or resources and (b) integrative data analysis and visualization approaches with an emphasis on the combination of multiple predictive models and analysis techniques. It also illustrates problems in which both methodologies can be successfully applied, and discusses design and application factors.

Chapter 4 introduces different methodologies for text mining and their current status, possibilities and limitations as well as their relation with the corresponding areas of molecular biology, with particular focus on the analysis of protein interaction networks.

Chapter 5 introduces a probabilistic model that integrates multiple information sources for the prediction of protein interactions. It presents an overview of genomic sources and machine learning methods, and explains important network analysis and visualization techniques.

Chapter 6 focuses on the representation and use of genome-scale phenotypic data, which in combination with other molecular and bioinformatic data open new possibilities for understanding and modelling the emergent complex properties of the cell. Quantitative trait locus (QTL) analysis, reverse genetics and phenotype prediction in the new post-genomics scenario are discussed.

Chapter 7 overviews the use of bio-ontologies in the context of functional genomics with special emphasis on the most used ones: The Gene Ontology. Important statistical issues related to high-throughput methodologies, such as the high occurrence of false or spurious associations between groups of genes and functional terms when the proper analysis is not performed, are also discussed.

Chapter 8 discusses data resources and techniques for generating and visualizing interactome networks with an emphasis on the interactome of *C. elegans*. It overviews technical aspects of the large-scale high-throughput yeast two-hybrid approach, topological and functional properties of the interactome network of *C. elegans* and their relationships with other sources such as expression data.

Chapter 9 reviews some of the limitations exhibited by traditional data management and visualization tools. It introduces *UTOPIA*, a project in which re-usable software components are being built and integrated closely with the familiar desktop environment to make easy-to-use visualization tools for the field of bioinformatics.

Chapter 10 reviews fundamental approaches and applications to data clustering. It focuses on requirements and recent advances for gene expression analysis. This contribution discusses crucial design and application problems in interpreting, integrating and evaluating results.

Chapter 11 introduces an integrative, unsupervised analysis framework for microarray data. It stresses the importance of implementing integrated analysis of heterogeneous biological data for supporting gene function prediction. It explains how multiple clustering models may be combined to improve predictive quality. It focuses on the design, application and evaluation of a knowledge-based tool that integrates probabilistic, predictive evidence originating from different sources.

Chapter 12 reviews well-known supervised methods to address questions about differential expression of genes and class prediction from gene expression data. Problems that limit the potential of supervised methods are analysed. It places special stress on key problems such as the inadequate validation of error rates, the non-rigorous selection of data sets and the failure to recognize observational studies and include needed covariates.

Chapter 13 presents an overview of probabilistic graphical models for inferring genetic networks. Different types of probabilistic graphical models are introduced and methods for learning these models from data are presented. The application of such models for modelling molecular networks at different complexity levels is discussed.

Chapter 14 introduces key approaches to the analysis, prediction and comparison of protein structures. For example, it stresses the application of a method that detects local patterns in large sets of structures. This chapter illustrates how advanced approaches may not only complement traditional methods, but also provide alternative, meaningful views of the prediction problems.

References

Birney, E. and Ensembl Team (2003) Ensembl: a genome infrastructure. *Cold Spring Harb Symp Quant Biol*, **68**, 213–215.

Herrero, J., Vaquerizas, J. M., Al-Shahrour, F., Conde, L., Mateos, A., Diaz-Uriarte, J. S. and Dopazo, J. (2004) New challenges in gene expression data analysis and the extended GEPAS. *Nucleic Acids Res*, **32** (web server issue): W485–W491.

2

Biological Databases: Infrastructure, Content and Integration

Allyson L. Williams, Paul J. Kersey, Manuela Pruess and **Rolf Apweiler**

Abstract

Biological databases store information on many currently studied systems including nucleotide and amino acid sequences, regulatory pathways, gene expression and molecular interactions. Determining which resource to search is often not straightforward: a single-database query, while simple from a user's perspective, is often not as informative as drawing data from multiple resources. Since it is unfeasible to assemble details for all biological experiments within a single resource, data integration is a powerful option for providing simultaneous user access to many resources as well as increasing the efficiency of user queries. This chapter provides a survey of current techniques in data integration as well as an overview of some of the most important individual databases.

Keywords

data integration, data warehousing, distributed annotation server (DAS), biological databases, genome annotation, protein classification, automated annotation, sequence clustering

2.1 Introduction

The exponential growth of experimental molecular biology in recent decades has been accompanied by growth in the number and size of databases interpreting and describing the results of such experiments. In particular, the development of

Data Analysis and Visualization in Genomics and Proteomics Edited by Francisco Azuaje and Joaquin Dopazo

automated technologies capable of determining the complete sequence of an entire genome and related high-throughput techniques in the fields of transcriptomics and proteomics have contributed to a dramatic growth in data. While all of these databases strive for complete coverage within their chosen scope, the domain of interest for some users transcends individual resources. This may reflect the user's wish to combine different types of information, or the inability of a single resource to fully contain the details of every relevant experiment. Additionally, large databases with broad domains tend to offer less detailed information than smaller, more specialized, resources, with the result that data from many resources may need to be combined to provide a complete picture. This chapter provides a survey of current techniques in data integration and an overview of some of the most important individual resources. A list of web sites for these as well as other selected databases is available at the end of the chapter in Table 2.2.

2.2 Data Integration

Much of the value of molecular biology resources is as part of an interconnected network of related databases. Many maintain cross-references to other databases, frequently through manual curation. These cross-references provide the basic platform for more advanced data integration strategies that have to address additional problems, including (a) the establishment of the identity of common objects and concepts, (b) the integration of data described in different formats, (c) the resolution of conflicts between different resources, (d) data synchronization and (e) the presentation of a unified view. The resolution of specific conflicts and the development of unified views rely on domain expertise and the needs of the user community. However, some of the other issues can be addressed through generic approaches such as standard identifiers, naming conventions, controlled vocabularies, adoption of standards for data representation and exchange, and the use of data warehousing technologies.

Identification of common database objects and concepts

Many generic data integration systems assume that individual entities and concepts have common definitions and a shared identifier space. In practice, different identifiers are often used for a single entity, and the concepts in different resources may be non-coincident or undefined. For example, a protein identifier in the *EMBL/GenBank/DDBJ* nucleotide sequence database (Benson *et al.*, 2004; Kulikova *et al.*, 2004; Miyazaki *et al.*, 2004) represents one protein-coding nucleotide sequence in a single submission to the database. If the same sequence had been submitted many times, there would be several identifiers for the same protein. An accession number in the *UniProt Knowledgebase* (Apweiler *et al.*, 2004), by contrast, is a protein identifier not necessarily restricted to a single submission or sequence. Identical translations from different genes within a species, or alternative sequences derived from the same gene, are merged into the same record. Such semantic differences need to be understood before devising an integration strategy.

Using standard names for biological entities significantly helps the merging of data with different identifier spaces. Many of the eukaryotic model organism databases enjoy *de facto* recognition from the scientific community for their right to define 'official' names for biological entities such as genes. These groups take their lead from expert committees such as the International Union of Biochemistry and Molecular Biology and the International Union of Pure and Applied Chemistry (IUBMB/IUPAC, 2004). Collaborations often result in approved gene names from one species used in naming orthologues from other species.

Recently, there has been a major effort to supplement the use of standard names with standard annotation vocabularies. The approach pioneered with *Gene Ontology* (GO) (Harris *et al.*, 2004), a controlled vocabulary for the annotation of gene products, has proved a successful and flexible template. Features of GO include a well defined domain, a commitment to provide a definition for each term, an open model for development through which many partners can collaboratively contribute to vocabulary development and the arrangement of terms in a directed acyclic graph (DAG). A DAG is a hierarchical data structure that allows the expression of complex relationships between terms. The hierarchical relationships make it possible to integrate annotations with different degrees of specificity using common parent terms, while the use of a graph rather than a tree structure makes it possible to express overlapping concepts without creating redundant terms. The power of this approach has led to the widespread adoption of GO by many resources, facilitating the integration of annotation and encouraging the development of many similar projects in other domains. A number of these projects can be accessed through the *Open Biological Ontologies* website (OBO, 2004).

Integration of data in different formats

In addition to nomenclature and semantics, data integration requires the resolution of differences in syntax, as resources may describe the same data in different formats. Even where a single data type is studied, specialized tools are often needed to access data from different sources. This problem is magnified with the development of high-throughput transcriptomics and proteomics techniques: potentially, there are as many data formats as there are equipment manufacturers. One successful approach for dealing with this problem has been pioneered by the Microarray Gene Expression Data (MGED) Society, a consortium of data producers, public databases and equipment manufacturers (MGED, 2004). The MGED Society has created the Minimal Information About a Microarray Experiment (MIAME) standard, which defines the information needed to describe a microarray experiment (Brazma *et al.*,

2001). As such, MIAME is a semantic standardization, but has led to the development of a syntactic standard for writing MIAME-compliant information, MicroArray Gene Expression Mark-up Language (MAGE-ML), to serve as a data exchange and integration format (Spellman *et al.*, 2002). Central to the success of this approach has been (a) the use of Extensible Markup Language (XML) (W3C Consortium, 2004), an open standard that does not tie users to particular database vendors, (b) the concentration on a minimal set of information to maximize the chances of agreement between partners, (c) the use of controlled vocabularies within the standard wherever possible and (d) the adoption of the standard by most of the key participants within this domain. Similar developments are currently underway in various fields of genomics (where the *Generic Model Organism Database* Project (Stein *et al.*, 2002), a consortium of model organism databases, is defining a universal database schema) and proteomics (where controlled vocabularies and data exchange standards are being developed under the auspices of the Human Proteomics Organisation Proteomics Standards Initiative (HUPO PSI) (Hermjakob *et al.*, 2004a)).

DAS: integration of annotation on a common reference sequence

Frequently, molecular biology annotation is assigned to regions of nucleic acid or protein sequences. Such annotation can be reliably integrated, provided data producers agree on the sequence and a co-ordinate system for describing locations. The Distributed Annotation Server (DAS) protocol facilitates this by defining a lightweight exchange format for sequence annotation data (DAS, 2004). A DAS system has three principal components: a reference sequence server, annotation servers that serve annotation for a given sequence and clients that retrieve data from the annotation servers. DAS has been designed to enable individual data producers to serve data easily, with the client performing the integration. The standard format makes it possible to write highly configurable client applications (typically graphical genome browsers) that can be re-used to integrate any compliant data. A further advantage is that anyone running a DAS client makes their own policy decisions on which servers to query for annotation, making it possible to produce different integrated views of the same reference sequence.

Data warehousing technologies

In spite of the emergence of common exchange formats, there is no standard technology used in the production of molecular biology databases. DAS is a powerful technology but is dependent on a simple data model, a standard representation of data according to this model and an agreement by data producers on a common reference sequence. Integration of more complex and irregular data into a system where users can query all data, regardless of source, requires some database-specific knowledge,

Table 2.1 Characteristics of different technical approaches to data integration

	SRS	EnsMart	DiscoveryLink	Grid
Warehouse or distributed resource?	Centralized (with gateways to external resources)	Centralized (with support for query chaining)	Distributed	Distributed
Update frequency	Periodic	Periodic	Instantaneous	Instantaneous
'Data transformation' layer	Flat file parsers	Flexible	APIs/wrappers for each individual resource	Web service descriptions
Query engine	Flat file indexing	RDBMS	Separately located from individual data resources	Flexible

and can be supported by the use of scalable, generic data warehousing technologies. A data warehouse is a database designed to hold secondary data derived from (potentially many) primary sources, in a schema designed to optimize the performance of expected queries rather than to protect the integrity of the data: the warehouse is periodically updated from the primary sources, but not synchronized between updates. Examples of systems employing data warehousing techniques in molecular biology include the *Sequence Retrieval System* (SRS) (Etzold, Ulyanov, and Argos, 1996) and *EnsMart* (Kasprzyk *et al.*, 2004). *DiscoveryLink* (Hass *et al.*, 2001) and *Grid* technologies (Foster, 2003) employ related strategies in an attempt to overcome the disadvantages of the warehousing approach. An overview of these different approaches is given in Table 2.1.

The development of a resource that supports integrative querying typically requires (a) the definition of a data model, (b) the creation of software to extract information from the source databases and to fit it to the model, (c) the definition of a query interface and (d) the implementation of an efficient querying mechanism. The creation of a data model is difficult due to the size of the molecular biology domain and the likelihood that changes in the content of an individual resource may require revision of the unified model. One approach is to model the expected query structure rather than the underlying domain, which increases the efficiency of data retrieval. In SRS, the structure of records from source databases is defined in individual parsers written for each plain text formatted resource: each identified portion of a record is indexed and can be specified as a criterion in selection and display, with no deeper semantic analysis. Common to all parsers is the identification of cross-references between records, which SRS uses to support cross-querying between the underlying databases. The approach of SRS is lightweight and scalable: the European Bioinformatics Institute (EMBL-EBI) successfully maintains over 200 cross-referenced databases in their public SRS server (Zdobnov *et al.*, 2002). Though the system does not allow for the semantic interpretation of data or the resolution of conflicts, it does provide an integrated view of data already present in primary resources.

EnsMart offers similar functionality to SRS, but implemented in a relational database management system. EnsMart provides generic support for efficient querying of database schemas that fit certain design patterns; to take advantage of the functionality of EnsMart, warehouse designers must write and maintain the code required to transform their own data to fit the EnsMart model. An additional feature of EnsMart is support for query chaining between distinct warehouses in separate locations, where those databases share the use of a common identifier set or vocabulary.

A major problem with data warehousing is synchronization. For example, synchronization problems arise when two resources cross-reference different versions of a third, and grow when a warehouse is constructed from specific releases of data from different sources. This task is liable to be computationally intensive, and during the interval between successive builds recent updates are not available in the warehouse. The DiscoveryLink system offers an alternative to warehousing. A central query engine communicates with distributed resources to dynamically integrate data when a request is made. The individual resources provide the query engine with a descriptor, which the engine uses to determine the locations of the requested data items and from which resources each may be most efficiently retrieved. As with SRS, the system depends on the usage of a common system of identifiers and nomenclature. The benefit of this approach is that updates in source databases are instantly available without rebuilding a static warehouse from scratch. Additionally, the user is not required to know about the structure and content of individual resources: the mapping between query terms and source databases is defined in the resource descriptors. However, the performance of individual queries may be reduced because of the need to dynamically fetch and integrate data.

Some of the principles applied in DiscoveryLink mirror the ideas behind the development of Grid. Grid has been proposed as a next-generation infrastructure to support and enable the collaboration of people and resources through scalable computation and data management systems. Under this model, service providers describe available resources in a common format. These descriptions of resources are utilized when a middleware layer, contacted by the user, converts a resource-neutral query into a request for information from specific sites. The failure of any one site is covered by the existence of others offering the same data. For example, a query to the Grid would specify the sequence and structure of a protein, but not the database or service provider. Grid technologies have been successfully used in large-scale computing projects, but it is not clear whether they will be able to support generic public access to diverse resources. Grid requires synchronization of data between providers and the existence of a common terminology to describe the data and services they offer. The successful future use of Grid as a transparent tool for accessing molecular biology data will therefore require a prior solution to many of the current problems in data integration.

2.3 Review of Molecular Biology Databases

A representative set of molecular biology databases is described below, grouped into divisions broadly coinciding with their defined scope. The complex task of integrating multiple resources is evidenced in the large number of databases available. Providing a general introduction to biological resources demonstrates this complexity and summarizes the vast amount of information available to researchers. While there is not room to discuss every database, useful database links, including and extending those detailed in this section, can be found in Table 2.2 at the end of the chapter.

Bibliographic databases

Bibliographic databases contain summary information taken from a variety of sources including journals, conference reports, books and patents. Some such databases specialize in biology and medicine. Including over 14 million references and 4500 journals, *PubMed* is one of the largest databases of life science abstracts with *MEDLINE*, a bibliographic database of over 11 million records, as its main component (NCBI, 2002). A query interface is available both at the National Center for Biotechnology Information (NCBI) and at EMBL-EBI. *BIOSIS Previews* (BP), one of seven bibliographic databases provided by BIOSIS, contains 13 million records from 1969 to the present and has a scope similar to that of PubMed (BIOSIS, 2004b). With over 4000 journals and other sources shared by both BP and PubMed, they are similar in scope but still retain significant numbers of unique records (BIOSIS, 2004a).

Taxonomy databases

Taxonomy databases store information on organism classification, data necessary for completion of most biological database records. The *NCBI Taxonomy* database contains over 160 000 taxonomic nodes and draws data from a variety of resources (Wheeler *et al.*, 2000). It stores information on living, extinct, known, and unknown organisms. The database is widely cross-referenced by other molecular biology databases including those maintained at the NCBI, the EMBL/GenBank/DDBJ nucleotide sequence database and UniProt. *NEWT* is the UniProt taxonomy database and includes the NCBI Taxonomy, species specific to UniProt not yet part of the NCBI Taxonomy and curated external links (Phan *et al.*, 2003).

Sequence databases

The International Nucleotide Sequence Database (INSD) Collaboration provides a nucleotide sequence repository for the public domain, and is a joint effort of three partners in Europe, Asia and the Americas: EMBL-EBI, DNA Data Bank of Japan (DDBJ) and NCBI. The three organizations synchronize their data every 24 hours. Many types of sequence are stored in *EMBL/GenBank/DDBJ* records, including individual genes, whole genomes, RNA, third-party annotation, expressed sequence tags, high-throughput cDNAs and synthetic sequences. Large-scale genomic sequencing has led to the exponential growth of this repository, which contains over 39 million records and 65 billion nucleotides. Due to its completeness and standing as a primary data provider, EMBL/GenBank/DDBJ is the initial source for many molecular biology databases.

RefSeq is a collection of nucleic acid and protein sequences derived from organisms with completely deciphered genomes. It is based on data derived from EMBL/GenBank/DDBJ and supplemented by additional sets of curated or predicted data in organisms of particular scientific interest (Pruitt and Maglott, 2001). *Genome Reviews* offers standardized representations of the genomes of over 190 organisms with completely sequenced genomes, importing annotation from the UniProt Knowledgebase and other sources into records derived from EMBL/GenBank/DDBJ. Release 1.0 holds data on over 170 complete genomes.

UniProt, the Universal Protein Resource, is a comprehensive catalogue of data on protein sequence and function, maintained through a collaboration of the Swiss Institute of Bioinformatics (SIB), EMBL-EBI and the Protein Information Resource (PIR). UniProt consists of three layers: the Knowledgebase (*UniProt*), the Archive (*UniParc*) and the non-redundant databases (*UniRef*). UniParc is a repository for all protein sequences, providing a mechanism by which the historical association of database records and protein sequences can be tracked. It is non-redundant at the level of sequence identity, but may contain semantic redundancies. All reported sequences are represented in UniParc, while records later found to be incorrect are excluded from the UniProt Knowledgebase, an automatically and manually annotated protein database drawn mainly from EMBL/GenBank/DDBJ coding sequences and directly sequenced proteins. The Knowledgebase consists of two parts: *UniProt/Swiss-Prot*, manually annotated with information extracted from literature and curator-evaluated computational analysis, and *UniProt/TrEMBL*, an automatically annotated section containing records awaiting full manual annotation. UniProt contains cross-references to more than 50 databases, making it a hub of biomolecular information.

The *ImMunoGeneTics* (IMGT) Project maintains databases on immunoglobulins, T cell receptors, major histocompatibility complex (MHC) and related proteins of the immune system of human and other vertebrate species. EMBL/GenBank/DDBJ entries fitting these categories are retrieved and annotated to a high standard. *IMGT/LIGM* (Laboratoire d'ImmunoGénétique Moléculaire) holds immunoglobulin and T cell receptor records for many species, while *IMGT/HLA* (human leukocyte antigen) is a specialized database for human MHC sequences (Robinson *et al.*, 2003). The Immuno Polymorphism Database (IPD) Project maintains the *IPD-MHC* database. This database is complementary to IMGT/HLA, and contains MHC sequences from other vertebrates including dogs, cats and many species of apes and monkeys (IPD, 2004).

Gene databases

The Human Genome Organization (HUGO) is an international organisation created to promote the study of the human genome (HUGO, 2004). As part of HUGO, the Human Gene Nomenclature Committee (HGNC) maintains *Genew*, a database of approved human gene names and symbols (Wain *et al.*, 2002). Genew contains over 19 000 records and, based on current estimates of the total number of human genes, has roughly another 12 000 to name. This database of human genes is used by many others, including UniProt, ensuring common nomenclature across all human data. Similar gene-centric databases include the *Mouse Genome Database* (MGD) (Bult *et al.*, 2004), *FlyBase* (The FlyBase Consortium, 2003), the *Rat Genome Database* (RGD) (Twigger *et al.*, 2002) and *RatMap* (RatMap, 2004). *Mendelian Inheritance in Man* (MIM) (McKusick, 1998), currently in its 13th edition, and its online version *OMIM*, is a resource describing human genes and genetic disorders. OMIM contains over 15 000 entries and maintains a gene map of the cytogenetic locations of genes as well as a morbid map containing a list of diseases and their locations (OMIM, 2004).

Databases of automatically predicted genomic annotation

Genomic databases often store the sequence for the entire set of chromosomes of a given organism, as well as high-level manual and automated gene annotation. Manual annotation of genomes is slow and can take years to complete, therefore automatically annotated whole genome databases are useful in providing a 'best guess' of complete gene sets. *Ensembl*, a joint project of EMBL-EBI and the Wellcome Trust Sanger Institute, provides automatically generated annotation of raw genomic sequence data for many eukaryotic genomes including human, mouse and fruit fly (Birney *et al.*, 2004). The University of California – Santa Cruz (UCSC) *Genome Browser* is similar, providing access to the UCSC draft genomic sequences (Kent *et al.*, 2002). The NCBI's *Map Viewer* displays RefSeq data and provides maps for a variety of species (NCBI, 2004). Most represented species have at a minimum a genetic, sequence and radiation hybrid map.

Clustering databases

Sequence similarity is an important indicator of sequence function. Clustering databases can reduce the time spent searching for relevant sequence matches, generally by pre-computing sequence similarities and then grouping similar sequences. The *CluSTr* database automatically classifies UniProt sequences into groups of related proteins based on analysis of all pairwise sequence comparisons using the Smith-Waterman algorithm (Kriventseva, Servant and Apweiler, 2003). The storage of clusters at different levels of similarity enables biologically meaningful clusters to be selected. There are over 100 proteomes in CluSTr, with over 130 million sequence similarities and 1 million clusters. *UniGene* is a database of automatically clustered EMBL/GenBank/DDBJ sequences (NCBI, 1996). The goal of UniGene is to present one cluster per gene and it currently contains clusters from almost 50 species. In contrast, the database of *Clusters of Orthologous Groups* (COGs) clusters proteins according to phylogenetic lineages, with the intent of presenting ancient conserved domains (Tatusov *et al.*, 2001). The two main objectives of *UniRef* are to facilitate sequence merging in UniProt and to allow faster and more informative sequence similarity searches (Apweiler *et al.*, 2004). UniRef is composed of UniRef100, UniRef90 and UniRef50, which store representative non-redundant sets at the named similarity levels. The *International Protein Index* (IPI) offers non-redundant protein sets for human, mouse and rat, derived from the UniProt, Ensembl and RefSeq databases (Kersey *et al.*, 2004). IPI clusters source entries using extant annotation and sequence similarity to compact the raw data without merging similar but biologically distinct sequences.

Protein classification databases

CATH (Orengo, Pearl and Thornton, 2003) is a hierarchical domain classification database for protein structures taken from the *worldwide Protein Data Bank* (wwPDB). The four levels of the hierarchy that give the database its name are Class, Architecture, Topology, and Homologous superfamily. Classes are determined through secondary structure composition. Architectures classify the shape of protein using the orientation of secondary structures. Topologies are based on shape and connectivity between secondary structures, and homologous structures are grouped if there is enough evidence to theorize a common ancestor. *InterPro* is an integrated resource of protein families, domains and functional sites with data drawn from *PROSITE*, *Pfam*, *PRINTS*, *ProDom*, *SMART* (Simple Modular Architecture Research Tool), *TIGRFAMs*, PIR *SuperFamily* and *SUPERFAMILY* (Mulder *et al.*, 2003). Annotators manually curate InterPro, adding general abstracts and cross-references to databases such as GO, UniProt, CATH and SCOP. As of InterPro release 8.0, 93 per cent of UniProt/Swiss-Prot entries have InterPro cross-references.

Some databases use information from protein classification resources to provide a perspective on completed genomes and proteomes. *Integr8* uses protein classifications derived from InterPro combined with CluSTr groups, GO annotations and known structural information to provide information on the composition of complete proteomes (Pruess *et al.*, 2003). *STRING*, the Search Tool for the Retrieval of Interacting Genes/Proteins, presents protein classifications in their genomic context (von Mering *et al.*, 2003).

Structure databases

The *worldwide Protein Data Bank* (wwPDB), begun in 1972, contains over 24 400 protein structures (Berman, Henrick and Nakamura, 2003). It is a collaboration of the Research Collaboratory for Structural Bioinformatics (RCSB), the Macromolecular Structural Database (MSD-EBI) and the Protein Data Bank of Japan (PDBj). The majority of protein structures in the database are from x-ray crystallography, solution nuclear magnetic resonance (NMR) experiments and theoretical modelling. The first two methods are empirical, experimental methods and therefore are more reliable than theoretical modelling, which often involves matching a sequence against the experimentally determined structure of a similar sequence. The *Cambridge Structural Database* (CSD) stores almost 300 000 records of small organic molecules and metal–organic compounds, with no polypeptide or polysaccharide larger than 24 units (Allen, 2002). Most structures were identified using either x-ray or neutron diffraction. The *RESID* database of amino acid modifications describes smaller molecules than those in CSD (Garavelli, 2003). It includes entries for the 23 encoded alpha-amino acids together with over 300 predicted or observed co- or post-translational modifications. In addition to structural information, each record includes systematic and alternative names, atomic formulae and masses, enzyme activities generating the modifications and UniProt feature table annotations.

Expression databases

Microarray experiments provide a method for gathering gene expression and transcription information, creating large amounts of data that expression databases store and organize. *ArrayExpress* is a public repository for experimental microarray data, queryable via experiment, array or protocol (Brazma *et al.*, 2003). It uses the standard annotation format (MIAME) and data storage format (MAGE-ML) created by the MGED Society. There are over 140 experiments, 170 protocols and 800 arrays stored in ArrayExpress. The *Stanford Microarray Database*, containing over 3500 public two-colour experiments, contains more data than any other microarray database (Gollub *et al.*, 2003). Though it does not store or provide its data in MIAME format, future plans include moving to this standard.

2D-PAGE databases

Two-dimensional polyacrylamide gel electrophoresis (2D-PAGE) and sodium dodecyl sulfate PAGE (SDS-PAGE) experiments distribute proteins in a gel based on molecular weight, providing protein expression data. *SWISS-2DPAGE* stores the results of such experiments and adds a variety of cross-references to other

2D-PAGE databases and to UniProt (Hoogland *et al.*, 2000). A SWISS-2DPAGE entry also contains images of the gels and textual information such as physiology, mapping procedures, experimental data and references. Release 17.2 holds over 1200 protein entries and 36 maps. The human and mouse 2D-PAGE databases at the *Danish Centre for Human Genome Research* are intended to aid functional genome analysis in health and disease. The information from each gel is stored as its own database, accessible through a interactive image of the gel itself (Celis and Østergaard, 2004).

Interaction databases

Interaction databases model a variety of interactions between proteins, RNA, DNA and many other compounds, storing information on how molecules and systems interrelate. *IntAct* is an open source protein interaction database and analysis system. It holds interaction data, maintains annotation standards and provides search and analysis software (Hermjakob *et al.*, 2004b). There are over 27 000 proteins and 36 000 interactions, searchable and viewable using an interactive graphical web application of protein networks (Hermjakob *et al.*, 2004a). The *Biomolecular Interaction Network Database* (BIND) is compiled from data submissions and manually annotated interactions taken from peer-reviewed journal articles, and holds over 35 000 sequences and 90 000 interactions (Bader, Betel and Hogue, 2003). Each BIND record represents an interaction between biological objects such as proteins, DNA, RNA and ligands, which can be combined to form molecular pathways or complexes. The *Database of Interacting Proteins* (DIP) contains both manual and automated annotation of over 40 000 experimentally determined protein–protein interactions (Salwinski *et al.*, 2004). In addition to obtaining information from journal articles, DIP has added about 2000 entries through analysis of protein complexes present in wwPDB.

Enzyme databases

The *Integrated relational Enzyme* database (IntEnz) (Fleischmann *et al.*, 2004) was created under the auspices of the Nomenclature Committee (NC) of the IUBMB. The goal of IntEnz is to incorporate data from the NC-IUBMB Enzyme Classification list, the Enzyme Nomenclature database (*ENZYME*) (Bairoch, 2000) and the *Braunschweig Enzyme Database* (BRENDA) of enzyme function (Schomburg *et al.*, 2004). ENZYME contains records for every enzyme with an EC number. Each record stores recommended and alternative names, catalytic activity, cofactors, disease information and cross-references with UniProt. BRENDA provides similar records, with a breakdown by species for reactions, activities, cofactors, inhibitors and substrates.

Pathway databases

The *Kyoto Encyclopedia of Genes and Genomes* (KEGG) provides a variety of databases dealing with genes, proteins, chemical reactions and pathways (Kanehisa *et al.*, 2004). The *KEGG Pathway Database* contains data on metabolic pathways, regulatory pathways and molecular complexes. Each record is a manually drawn reference pathway diagram whose nodes are molecules relevant to the network type. For instance, metabolic networks use enzymes as nodes, while gene regulatory network nodes are transcription factors and target products.

2.4 Conclusion

The large number of available biological databases may seem overwhelming to many, and a thorough search for information on a gene requires the use of many disparate resources. Such a search might start with EMBL/GenBank/DDBJ for reference data on the nucleotide sequence. Microarray databases such as ArrayExpress would be useful in showing any available data on the expression of the gene. If the gene codes for a protein, then CluSTr and UniRef will help identify UniProt proteins with similar sequences. A search of InterPro will help classify the protein into a specific family and show any probable domains. The use of a 2D-PAGE database such as SWISS-2DPAGE may provide direct information on the expression of the protein. If it does not code for a protein then a search of the wide variety of non-coding RNA databases may yield more information. Fortunately, many of these databases contain cross-references to ease progression from one source of information to the next. Additionally, there are many useful integration tools and databases to help novice and experienced users alike. Integrated databases provide (a) quick, one-stop access to a variety of different types of information, (b) a base for more detailed searches, (c) a place for small or specialty databases to gain exposure to a wide variety of users and (d) an opportunity for complementary databases to learn about and collaborate with each other. Integration requires that disparate groups provide their data in a manner that can be read and manipulated by the main coordinating database: Integr8, for instance, has 11 institutes contributing to the database. To permit such an integration of data from a variety of difference sources, data standards such as MIAME and those developed by the PSI are of crucial importance. Common data standards make both distributed annotation systems and data warehouses feasible, which in turn allows collaborators to work on a single project transparently from anywhere in the world. Improvements to data access require strong collaborations, cross-referencing and integration, if the amount of available data is not to overwhelm the user.

Table 2.2 URLs for useful biological databases

Database type	Database name	URL
Nomenclature/ ontology		
	IUBMB	http://www.iubmb.org
	GO	http://www.geneontology.org
	OBO	http://obo.sourceforge.net
	MGED	http://www.mged.org
	GMOD	http://www.gmod.org
	HUGO/HGNC	http://www.gene.ucl.ac.uk/nomenclature
	HUPO	http://www.hupo.org
	SOFG	http://www.sofg.org
Integrated		
	MSD	http://www.ebi.ac.uk/msd
	EMP	http://www.empproject.com
	MEROPS	http://merops.sanger.ac.uk
	Integr8	http://www.ebi.ac.uk/integr8
	GeneCards	http://bioinfo.weizmann.ac.il/cards
Bibliography		
	PubMed	http://www.ncbi.nlm.nih.gov/PubMed
	MEDLINE	http://www.ebi.ac.uk/Databases/MEDLINE
	BIOSIS	http://www.biosis.org
	Zoological Record	http://www.biosis.org/products/zr
	EMBASE	http://www.embase.com
	AGRICOLA	http://agricola.nal.usda.gov
	CAB Abstracts	http://www.cabi.org
Taxonomy		
	NCBI Taxonomy	http://www.ncbi.nlm.nih.gov
	NEWT	http://www.ebi.ac.uk/newt
	Species 2000	http://www.sp2000.org
	ITIS	http://www.itis.usda.gov
	WBD	http://www.eti.uva.nl
Sequence		
	EMBL	http://www.ebi.ac.uk/embl
	GenBank	http://www.ncbi.nlm.nih.gov/Genbank
	DDBJ	http://www.ddbj.nig.ac.jp
	RefSeq	http://www.ncbi.nlm.nih.gov/RefSeq
	Genome Reviews	http://www.ebi.ac.uk/GenomeReviews
	UniProt	http://www.uniprot.org
	UniProt/Swiss-Prot	http://www.expasy.org/sprot
	UniProt/TrEMBL	http://www.ebi.ac.uk/trembl
	IMGT Databases	http://www.ebi.ac.uk/imgt
	IPD-MHC	http://www.ebi.ac.uk/ipd/mhc
	Entrez Protein	http://www.ncbi.nlm.nih.gov/Entrez
	Parasite Genomes	http://www.ebi.ac.uk/parasites/parasite-genome.html
	MIPS–CYGD	http://mips.gsf.de/genre/proj/yeast
	GPCRDB	http://www.gpcr.org
	RDP	http://rdp.cme.msu.edu
	TRANSFAC	http://www.gene-regulation.com

Table 2.2 (*Continued*)

Database type	Database name	URL
	EPD	http://www.epd.isb-sib.ch
	HIVdb	http://hivdb.stanford.edu
	REBASE	http://rebase.neb.com
Gene		
	Genew	http://www.gene.ucl.ac.uk/nomenclature
	MGD	http://www.informatics.jax.org
	FlyBase	http://www.flybase.org
	RGD	http://rgd.mcw.edu
	RATMAP	http://www.ratmap.org
	MIM/OMIM	http://www.ncbi.nlm.nih.gov/omim
	GDB	http://www.gdb.org
	SGD	http://www.yeastgenome.org
	Gramene	http://www.gramene.org
	TAIR	http://www.arabidopsis.org
	MaizeGDB	http://www.maizegdb.org
	AceDB	http://www.acedb.org
	ZFIN	http://www.zfin.org
	CGSC	http://cgsc.biology.yale.edu
	WormBase	http://www.wormbase.org
Prediction of genomic annotation		
	Ensembl	http://www.ensembl.org
	Genome Browser	http://genome.ucsc.edu
	Map Viewer	http://www.ncbi.nlm.nih.gov/mapview
Clustering		
	ClusSTr	http://www.ebi.ac.uk/clustr
	UniGene	http://www.ncbi.nlm.nih.gov/UniGene
	COGs	http://www.ncbi.nlm.nih.gov/COG
	UniRef	http://www.ebi.ac.uk/uniref
	IPI	http://www.ebi.ac.uk/IPI
	SYSTERS	http://systers.molgen.mpg.de
Protein classification		
	CATH	http://www.biochem.ucl.ac.uk/bsm/cath
	InterPro	http://www.ebi.ac.uk/interpro
	PROSITE	http://www.expasy.ch/prosite
	Pfam	http://www.sanger.ac.uk/Software/Pfam
	PRINTS	http://nmber.sbs.man.ac.uk/dbbrowser/PRINTS
	ProDom	http://prodes.toulouse.inra.fr/prodom.html
	SMART	http://smart.embl-heidelberg.de
	PIRSF	http://pir.georgetown.edu/iproclass
	SUPERFAMILY	http://supfam.mrc-lmb.cam.ac.uk/SUPERFAMILY
	TIGRFAMs	http://www.tigr.org/TIGRFAMs
	SCOP	http://scop.mrc-lmb.cam.ac.uk/scop
Structure		
	wwPDB	http://www.wwpdb.org
	CSD	http://www.ccdc.cam.ac.uk/products/csd
	RESID	http://www.ncifcrf.gov/RESID

Table 2.2 (*Continued*)

Database type	Database name	URL
	NDB	http://ndbserver.rutgers.edu
	DSSP	http://www.cmbi.kun.nl/gv/dssp
	HSSP	http://www.cmbi.kun.nl/gv/hssp
Expression		
	ArrayExpress	http://www.ebi.ac.uk/arrayexpress
	SMD	http://genome-www5.stanford.edu
	CGAP	http://cgap.nci.nih.gov
	GEO	http://www.ncbi.nlm.nih.gov/geo
2D-PAGE		
	SWISS-2DPAGE	http://www.expasy.org/ch2d
	DCHGR	http://proteomics.cancer.dk
Interaction		
	IntAct	http://www.ebi.ac.uk/intact
	BIND	http://bind.ca
	DIP	http://dip.doe-mbi.ucla.edu
	LIGAND	http://www.genome.ad.jp/ligand
Enzyme		
	IntEnz	http://www.ebi.ac.uk/intenz
	ENZYME	http://www.expasy.org/enzyme
	BRENDA	http://www.brenda.uni-koeln.de
Pathway		
	KEGG	http://www.genome.ad.jp/kegg
	BioCyc/EcoCyc	http://www.biocyc.org

References

Allen, F. H. (2002) The Cambridge Structural Database: a quarter of a million crystal structures and rising. *Acta Crystallogr B*, **58**, 380–388.

Apweiler, R., Bairoch, A., Wu, C. H. *et al.* (2004) UniPort: the Universal Protein knowledgebase. *Nucleic Acids Res*, **32** (database issue), D115–D119.

Bader, G. D., Betel, D. and Hogue, C. W. (2003) BIND: the Biomolecular Interaction Network Database. *Nucleic Acids Res*, **31**, 248–250.

Bairoch, A. (2000) The ENZYME database in 2000. *Nucleic Acids Res*, **28**, 304–305.

Benson, D. A., Karsch-Mizrachi, I., Lipman, D. J., Ostell, J. and Wheeler, D. L. (2004) GenBank: update. *Nucleic Acids Res*, **32** (database issue), D23–D26.

Berman, H., Henrick, K. and Nakamura, H. (2003) Announcing the worldwide Protein Data Bank. *Nat Struct Biol*, **10**, 980.

BIOSIS (2004a) *BIOSIS Previews: Search Strategies*. Available from http://www.biosis.org/strategies [accessed 02/09/04].

BIOSIS (2004b) *BIOSIS Previews: the World's Most Comprehensive Reference Database in the Life Sciences*. Available from http://www.biosis.org/products/previews [accessed 02/09/04].

Birney, E., Andrews, D., Bevan, P. *et al.* (2004) Ensembl 2004. *Nucleic Acids Res*. **32** (database issue), D468–D470.

Brazma, A., Hingamp, P., Quackenbush, J. *et al.* (2001) Minimum information about a microarray experiment (MIAME) – toward standards for microarray data. *Nat Genet*, **29**, 365–371.

Brazma, A., Parkinson, H., Sarkans, U. *et al.* (2003) ArrayExpress – a public repository for microarray gene expression data at the EBI. *Nucleic Acids Res.* **31**, 68–71.

Bult, C. J., Blake, J. A., Richardson, J. E. *et al.* (2004) The Mouse Genome Database (MGD): integrating biology with the genome. *Nucleic Acids Res.* **32** (database issue), D476–D481.

Celis, J. E. and Østergaard, M. (2004) *Julio Celis Database*, Available from http://proteomics.cancer.dk [accessed 02/09/04].

DAS (2004) www.biodas.org, Available from http://biodas.org [accessed 02/09/04].

Etzold, T., Ulyanov, A. and Argos, P. (1996) SRS: information retrieval system for molecular biology data banks. *Methods Enzymol*, **266**, 114–128.

Fleischmann, A., Darsow, M., Degtyarenko, K. *et al.* (2004) IntEnz, the integrated relational enzyme database. *Nucleic Acids Res*, **32** (database issue), D434–D437.

The FlyBase Consortium (2003) The FlyBase database of the *Drosophila* genome projects and community literature. *Nucleic Acids Res*, **31**, 172–175.

Foster, I. (2003) The grid: computing without bounds. *Sci Am*, **288**, 78–85.

Garavelli, J. S. (2003) The RESID Database of Protein Modifications: 2003 developments. *Nucleic Acids Res*, **31**, 499–501.

Gollub, J., Ball, C. A., Binkley, G. *et al.* (2003) The Stanford Microarray Database: data access and quality assessment tools. *Nucleic Acids Res*, **31**, 94–96.

Harris, M. A., Clark, J., Ireland, A. *et al.* (2004) The Gene Ontology (GO) database and informatics resource. *Nucleic Acids Res*, **32** (database issue), D258–D261.

Hass, L. M., Schwarz, P. M., Kodali, P. *et al.* (2001) DiscoveryLink: A system for integrated access to life sciences data sources. *IBM Systems J*, **40**, 489.

Hermjakob, H., Montecchi-Palazzi, L., Bader, G. *et al.* (2004a) The HUPO PSI's molecular interaction format – a community standard fot the representation of protein interaction data. *Nat Biotechnol*, **22**, 177–183.

Hermjakob, H., Montecchi-Palazzi, L., Lewington, C. *et al.* (2004b) IntAct: an open source molecular interaction database. *Nucleic Acids Res*, **32** (database issue), D452–D455.

Hoogland, C., Sanchez, J. C., Tonella, L. *et al.* (2000) The 1999 SWISS-2DPAGE database update. *Nucleic Acids Res*, **28**, 286–288.

HUGO (2004) *General Information about HUGO*, Available from http://www.hugo–international.org/hugo/HUGO–mission–statement.htm [accessed 02/09/04].

IPD (2004) *IPD Database*, Available from http://www.ebi.ac.uk/ipd [accessed 02/09/04].

IUBMB/IUPAC (2004) *Biochemical Nomenclature Commitees*, Available from http://www.chem.qmul.ac.uk/iupac/jcbn [accessed 02/09/04].

Kanehisa, M., Goto, S., Kawashima, S., Okuno, Y. and Hattori, M. (2004) The KEGG resource for deciphering the genome. *Nucleic Acids Res*, **32** (database issue), D277–D280.

Kasprzyk, A., Keefe, D., Smedley, D. *et al.* (2004) EnsMart: a generic system for fast and flexible access to biological data. *Genome Res*, **14**, 160–169.

Kent, W. J., Sugnet, C. W., Furey, T. S. *et al.* (2002) The human genome browser at UCSC. *Genome Res*, **12**, 996–1006.

Kersey, P. J., Duarte, J., Williams, A. *et al.* (2004) The International Protein Index: an integrated database for proteomics experiments. *Proteomics*, **4**, 1985–1988.

Kriventseva, E. V., Servant, F. and Apweiler, R. (2003) Improvements to CluSTr: the database of SWISS-PROT+TrEMBL protein clusters. *Nucleic Acids Res*, **31**, 388–389.

Kulikova, T., Aldebert, P., Althorpe, N. *et al.* (2004) The EMBL Nucleotide Sequence Database. *Nucleic Acids Res*, **32** (database issue), D27–D30.

McKusick, V. A. (1998) *Mendelian Inheritance in Man. A Catalog of Human Genes and Genetic Disorders*. Johns Hopkins University Press, Baltimore, MD.

MGED (2004) MGED NETWORK: *Ontology Working Group (OWG)*, Available from http://mged.sourceforge.net/ontologies/index.php [accessed 02/09/04].

Miyazaki, S., Sugawara, H., Ikeo, K., Gojobori, T. and Tateno, Y. (2004) DDBJ in the stream of various biological data. *Nucleic Acids Res*, **32** (database issue), D31–D34.

Mulder, N. J., Apweiler, R., Attwood, T. K. *et al.* (2003) The InterPro Database, 2003 brings increased coverage and new features. *Nucleic Acids Res*, **31**, 315–318.

NCBI (1996) *NCBI News: August 1996*, Available from http://www.ncbi.nlm.nih.gov/Web/Newsltr/aug96.html#advance [accessed 02/09/04].

NCBI (2002) *What's the Difference Between MEDLINE®* and PubMed®? *Fact Sheet*, Available from http://www.nlm.nih.gov/pubs/factsheets/dif_med_pub.html [accessed 02/09/04].

NCBI (2004) *Entrez Map Viewer Help Document*, Available from http://www.ncbi.nlm.nih.gov/mapview/static/MapViewerHelp.html [accessed 02/09/04].

OBO (2004) *About OBO*, Available from http://obo.sourceforge.net [accessed 02/09/04].

OMIM (2004) *Online Mendelian Inheritance in Man, OMIM (TM).* Available from http://www.ncbi.nlm.nih.gov/omim/ [accessed 02/09/04].

Orengo, C. A., Pearl, F. M. and Thornton, J. M. (2003) The CATH domain structure database. *Methods Biochem Anal*, **44**, 249–271.

Phan, I. Q., Pilbout, S. F., Fleischmann, W. and Bairoch, A. (2003) NEWT, a new taxonomy portal. *Nucleic Acids Res*, **31**, 3822–3823.

Pruess, M., Fleischmann, W., Kanapin, A. *et al.* (2003) The Proteome Analysis database: a tool for the in silico analysis of whole proteomes. *Nucleic Acids Res*, **31**, 414–417.

Pruitt, K. D. and Maglott, D. R. (2001) RefSeq and LocusLink: NCBI gene-centered resources. *Nucleic Acids Res*, **29**, 137–140.

RatMap (2004) *RatMap: The Rat Genome Database*, Available from http://ratmap.gen.gu.se [accessed 02/09/04].

Robinson, J., Waller, M. J., Parham, P. *et al.* (2003) IMGT/HLA and IMGT/MHC: sequence databases for the study of the major histocompatibility complex. *Nucleic Acids Res*, **31**, 311–314.

Salwinski, L., Miller, C. S., Smith, A. J. *et al.* (2004) The Database of Interacting Proteins: 2004 update. *Nucleic Acids Res*, **32** (database issue), D449–D451.

Schomburg, I., Chang, A., Ebeling, C. *et al.* (2004) BRENDA, the enzyme database: updates and major new developments. *Nucleic Acids Res*, **32** (database issue), D431–D433.

Spellman, P. T., Miller, M., Stewart, J. *et al.* (2002) Design and implementation of microarray gene expression markup language (MAGE-ML) *Genome Biol*, **3**, RESEARCH0046.

Stein, L. D., Mungall, C., Shu, S. *et al.* (2002) The generic genome browser: a building block for a model organism system database. *Genome Res*, **12**, 1599–1610.

Tatusov, R. L., Natale, D. A., Garkavtsev, I. V. *et al.* (2001) The COG database: new developments in phylogenetic classification of proteins from complete genomes. *Nucleic Acids Res*, **29**, 22–28.

Twigger, S., Lu, J., Shimoyama, M. *et al.* (2002) Rat Genome Database (RGD): mapping disease onto the genome. *Nucleic Acids Res*, **30**, 125–128.

von Mering, C., Huynen, M., Jaeggi, D. *et al.* (2003) STRING: a database of predicted functional associations between proteins. *Nucleic Acids Res*, **31**, 258–261.

W3C Consortium (2004) *Extensible Markup Language (XML).* Available from http://www.w3c.org/XML [accessed 02/09/04].

Wain, H. M., Lush, M., Ducluzeau, F. and Povey, S. (2002) Genew: the human gene nomenclature database. *Nucleic Acids Res*, **30**, 169–171.

Wheeler, D. L., Chappey, C., Lash, A. E. *et al.* (2000) Database resources of the National Center for Biotechnology Information. *Nucleic Acids Res*, **28**, 10–14.

Zdobnov, E. M., Lopez, R., Apweiler, R. and Etzold, T. (2002) The EBI SRS server – new features. *Bioinformatics*, **18**, 1149–1150.

3

Data and Predictive Model Integration: an Overview of Key Concepts, Problems and Solutions

Francisco Azuaje, Joaquin Dopazo and **Haiying Wang**

Abstract

This chapter overviews the combination of different data sources and techniques for improving functional prediction. Key concepts, requirements and approaches are introduced. It discusses two main strategies: (a) integrative data analysis and visualization approaches with an emphasis on the processing of multiple data types or resources and (b) integrative data analysis and visualization approaches with an emphasis on the combination of multiple predictive models and analysis techniques. It also illustrates problems in which both methodologies can be successfully applied.

Keywords

integrative data mining, integrative data visualization, gene expression analysis, protein networks, functional prediction

3.1 Integrative Data Analysis and Visualization: Motivation and Approaches

The combination of multiple data sources is both a fundamental requirement and a goal for developing a large-scale and dynamic view of biological systems. Data originating from multiple levels of complexity and organization are interrelated to

Data Analysis and Visualization in Genomics and Proteomics Edited by Francisco Azuaje and Joaquin Dopazo

assess their functional predictive abilities. For instance, quantitative relationships between gene expression correlation and protein–protein interaction, and gene and protein expression correlation have been studied (Allocco, Kohane, and Butte, 2004). Typical questions addressed by such studies include the following. Is there a significant connection between highly expressed genes and highly expressed proteins? Is the expression correlation exhibited by a pair of genes significantly associated with the likelihood of finding their products in the same protein complex? These quantitative relationships support the design of prediction models to facilitate functional classification and interpretation. In a post-genomic scenario the possibility of answering functional questions on a one-gene-at-a-time basis is being abandoned in favour of a systemic approach. In such an approach the accuracy of individual predictions is sacrificed at the expense of a deeper knowledge of how the different system components interact among them to play different biological roles. Thus, systems biology provides a more complex, integrated view of function, which differs from the traditional, naive method of assigning a given activity or role to a single protein.

A massive collection of computational and statistical techniques is available to analyse and visualize different types of 'omic' information. The most important computational question is not whether there are options for a particular problem. Rather, bioinformaticians are becoming more concerned about questions such as *how to combine different techniques? When? Why?*

The combination of multiple prediction models is fundamental to address limitations and constraints exhibited by individual approaches. Moreover, their integration may improve the accuracy, reliability and understandability of prediction tasks under different experimental and statistical assumptions and conditions. For example, it has been demonstrated that the combination of multiple, diverse classification models may significantly outperform the prediction outcomes obtained from the application of individual classifiers (Kittler *et al.*, 1998). Thus, model diversity is a crucial factor to achieve multiple views of the same problem, reduce bias and improve the coverage of the prediction space. Diversity may be obtained not only through the application of multiple models, but also through the implementation of different methods for selecting data, features and prediction outcomes.

In general, two major computational categories of integrative data analysis and visualization approaches may be identified: (a) those approaches that place an emphasis on the processing of multiple data types and (b) those approaches that rely on the combination of multiple predictive models and analysis techniques. The first approach may of course apply multiple predictive computational models, but its main goal is to combine different types of biological data sets in order to improve a prediction task or to achieve a more complete, dynamic view of a biological problem. An example of this type of approach is the combination of expression, cellular localization and protein interaction data for the prediction of protein complex membership. Although the second approach may (or may not) process different types of data, its main objective is to implement different statistical and/or machine learning models to improve predictive quality. One example is the combination of several clustering algorithms, including neural networks, to improve accuracy and coverage in the functional characterization of genes based on microarray data.

This chapter discusses these two main data analysis and visualization problems by providing an overview of recent key investigations and applications for functional genomics. It also illustrates problems in which both methodologies can be successfully applied.

3.2 Integrating Informational Views and Complexity for Understanding Function

The organizational modules of the cell may be divided into several types of '*omic*' information. For example, the transcriptome refers to the set of information transcribed from coding sequences, which is defined by their expression patterns. The interactome specifies the existing interactions between molecules in the cell, including protein–protein and protein–DNA interactions. The reader is referred to the work of Ge, Walhout and Vidal (2003) for a discussion on the classification of 'omic' approaches.

Information originating from each 'omic' approach may be incomplete, incorrect or irrelevant. Their predictive quality and usefulness may be significantly compromised by the presence of several false negatives and false positives. Each data source offers a different, partial view of the functional roles of genes and proteins, but they may also generate overlapping views of the same problem. Therefore, their integration may provide the basis for more effective and meaningful functional predictors. Moreover, it may support the generation and validation of new hypotheses. For instance, if *method A* suggests that gene product *X* interacts with gene product *Y*, it would be then important to apply other methods to assess the relevance or validity of this interaction. Phenotypic information describing the essentiality of these genes together with their expression patterns may aid in the identification of their participation in common biological pathways or related functions. Thus, these putative roles may reflect the relevance of this interaction.

An integrative prediction process aims to exploit the existing quantitative relationships between different 'omic' data sets. These relationships may indicate the types of constraints and integration mechanisms that need to be defined. Thus, for instance, an important problem is to investigate how different data sets are statistically correlated. In some applications it is important to assess the significance of such relationships with respect to relationships detected from random data sets. Advances in this area include techniques to describe how gene expression correlation and interactome data are interrelated in *S. cerevisiae*. Several correlation measures, such as the *Pearson coefficient* and the *cosine distance*, may be used. A typical strategy consists of depicting the distribution of expression correlation values for interactome data sets, which may be compared with the distribution obtained from random protein pairs (Ge, Walhout and Vidal, 2003). These comparisons indicate, for example, that interacting proteins are more likely to be encoded by genes strongly correlated by their expression profiles (Jansen *et al.*, 2003). Another technique consists of plotting the likelihood of finding two proteins in the same protein complex as a function of their expression correlation coefficients (Jansen, Greenbaum and Gerstein, 2002). The validity of this methodology for detecting transcriptome–interactome relationships in multi-cellular organisms requires further investigation. For instance, it has been suggested that these relationships can be observed in *C. elegans*, at least for particular types of tissue (Walhout *et al.*, 2002).

This data visualization procedure may be easily extended to estimate other functional properties, such as the likelihood of finding pairs of genes regulated by a common transcription factor on the basis of their gene expression correlation. It has been shown that pairs of genes with significantly correlated expression patterns are much likelier to be bound by a common transcription factor in comparison to those pairs exhibiting weaker expression correlations (Allocco, Kohane and Butte, 2004).

Inter-relationships between interactome and phenome, transcriptome and translatome and transcriptome and phenome have also been studied (Ge, Walhout and Vidal, 2003). Such associations may motivate different interpretations, which sometimes may be specific to particular organisms or functional roles, but which may be reconciled and integrated to formulate hypotheses or to support the development of more effective prediction models (Ge, Walhout and Vidal, 2003). Figure 3.1 illustrates typical plots for visualizing potential significant relationships between different 'omic' properties.

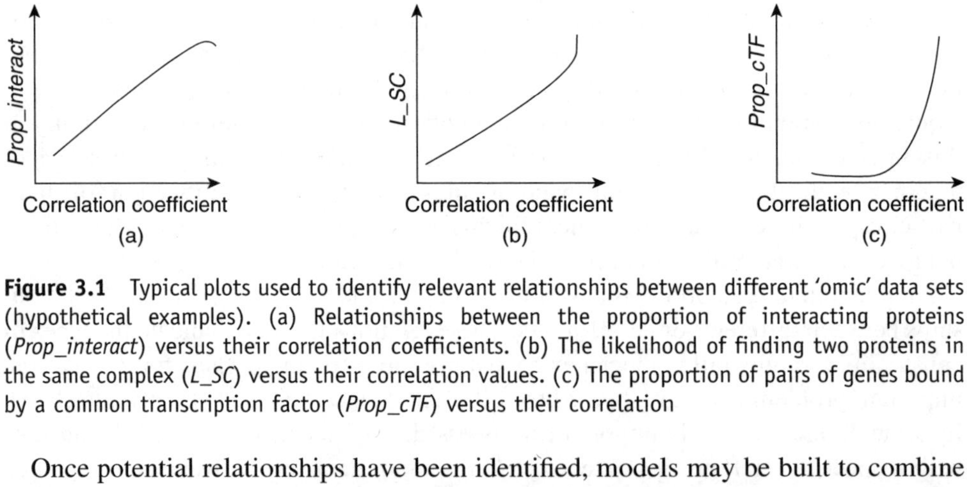

Figure 3.1 Typical plots used to identify relevant relationships between different 'omic' data sets (hypothetical examples). (a) Relationships between the proportion of interacting proteins (*Prop_interact*) versus their correlation coefficients. (b) The likelihood of finding two proteins in the same complex (*L_SC*) versus their correlation values. (c) The proportion of pairs of genes bound by a common transcription factor (*Prop_cTF*) versus their correlation

Once potential relationships have been identified, models may be built to combine evidence or prediction outcomes derived from different data sources. Several machine learning methods, such as decision trees and neural networks, may be applied to implement this task. For instance, integrative models based on Bayesian networks have been applied to predict protein–protein interactions in yeast. One recent advance (Jansen *et al.*, 2003) reported the integration of different types of experimental interaction data, functional annotations, mRNA expression and essentiality data to improve the identification of protein–protein interactions. One important advantage shown by probabilistic frameworks is that they provide an assessment of the predictive relevance and reliability of each integrated source. They are useful to deal with different types of data and missing values. Moreover, relationships between sources are expressed in terms of conditional probabilities, which in many applications facilitate the interpretation of results. One limitation is that these models often require the user to make strong assumptions about the independence of the information sources, which may not be easy to justify or accurate to generate reliable predictions.

Integrative data analysis approaches are also fundamental tools for refining or adapting other systemic models such as metabolic networks (Ideker *et al.*, 2001). In this case different types of data, such as mRNA expression, protein expression and physical interaction data, may be used to measure responses to systematic perturbations. Data clustering techniques and correlation visualization tools (including those discussed above) may be applied to summarize these responses and their associations with functional roles or processes.

One important problem that requires further research is the development of methods to visualize not only different information sources, but also multiple analysis outcomes. These techniques should support both interactive and iterative tasks. A key limitation, which was discussed in Chapter 1, is that the areas of data analysis (or data mining) and visualization have traditionally evolved as separate disciplines. Typical information visualization tools have been designed to process single data sources. Moreover, they have put emphasis on the problem of displaying final analysis outcomes, without providing more hierarchical, multi-resolution views of prediction processes. Thus, an integrative data visualization approach is necessary not only to complement integrative data analyses, but also to make them more meaningful.

Information visualization platforms currently available allow researchers to merge multiple data sources to highlight relevant relationships, such as those represented in regulatory networks (Baker *et al.*, 2002). Regulatory networks may be, for instance, displayed together with other types of information such as gene expression correlation and interaction information. Different experimental methods or relationships may be represented by using colour-coding schemes associated with the nodes and edges in the network.

Integrative visualization tools should provide multiple graphical and analytical views of other organizational levels or 'omic' sources, including pathways and functional annotations. The *VisAnt* platform is one such option (Hu *et al.*, 2004), in which metabolic data, gene homology, annotations and cross-referencing information of genes and proteins are integrated. One important challenge for this type of research is to support a flexible, open and integrated display of heterogeneous information sources and analysis outcomes. In this direction, the *Ensembl* (Birney *et al.*, 2004) project, which incorporates tools such as *EnsMart* (Kasprzyk *et al.*, 2004), allows user-friendly integration of different types of information in a genomic context, including cross-genome comparisons. Other relevant systems are the NCBI's Map Viewer and the UCSC genome browser, which also incorporate multi-source genomic information through web-based interfaces.

A fundamental condition to achieve an integrative data analysis and visualization paradigm is the ability to integrate diverse outcomes originating from the application of multiple prediction models.

3.3 Integrating Data Analysis Techniques for Supporting Functional Analysis

One important characteristic exhibited by the models introduced above is that they combine multiple data sources by mainly applying only one type of prediction model, such as a single classification technique. An alternative integrative prediction approach may also take advantage of the diversity of available prediction models and techniques. It has been demonstrated that different techniques can unveil various aspects of different types of data such as gene expression data (Leung and Cavalieri, 2003). The combination of diverse models can overcome the dependency on problem- or technique-specific solutions.

One such integrative approach is known as *Multisource Association of Genes by Integration of Clusters*; this was proposed by Troyanskaya and co-workers (Troyanskaya *et al.*, 2003). It applies probabilistic reasoning and unsupervised learning to integrate different types of large-scale data for functional prediction. The system has been tested on *S. cerevisiae* by combining multiple classification techniques based on microarray, physical and genetic interactions and transcription factor binding sites data. An assessment of functional prediction relevance in yeast has been performed by processing Gene Ontology annotations derived from the *S. cerevisiae* Genome Database. The inputs to the integrative probabilistic prediction framework may consist of clustering-driven predictions based on gene expression correlation and other functional relationships between pairs of gene products. This framework allows, for instance, the combination of classification outcomes generated by several clustering techniques such as *k*-means, self-organizing maps and hierarchical clustering. The system estimates the probability that a pair of gene products is functionally interrelated. Such a relationship is defined by their involvement in the same biological process, as defined by the Gene Ontology. This approach clearly demonstrates how an integrative approach may outperform single-source prediction techniques, such as models based solely on microarray data. Moreover, it highlights the advantages of combining multiple classification methods. Troyanskaya further discusses this integrative framework and its applications in Chapter 11.

Other authors, such as Wu *et al.* (2002), have shown the importance of applying multiple clustering methods to discover relevant biological patterns from gene expression data. This type of model aims to integrate classification outcomes originating from several clustering methods such as hierarchical clustering, *k*-means and self-organizing maps. One important assumption is that these methods may produce partially overlapping expression clusters. Multiple partitions may be obtained by running different clustering algorithms using several learning parameters or numbers of clusters. Without going into details, a functional class prediction derived from a clustering experiment may be associated with a probability value, *P*. It estimates the possibility that a cluster of genes was obtained by chance and allows assignment of a gene to multiple functional categories. Thus, integrative predictions are made on the basis of the minimum *P*-value exhibited by a category in a cluster. The computational predictions and experimental validation performed by Wu *et al.* further demonstrate the importance of integrating several machine learning and statistical methods to improve biological function predictions based on a single data source. One key advantage of combining multiple clustering-based prediction outcomes is that it allows the association of multiple, reliable functional predictions with a gene product based on a probabilistic framework. Clusters may be automatically linked to significant functional categories by processing a reference knowledge base, such as the Gene Ontology. The implementation of tools for automatically annotating clusters is a fundamental problem to achieve integrative data analysis goals. In Chapter 7, Al-Shahrour and Dopazo will discuss the problem of assigning significant functional classes to gene clusters based on Gene Ontology annotations. Figure 3.2 summarizes basic tasks required in a clustering-driven integrative framework for predicting functional classes.

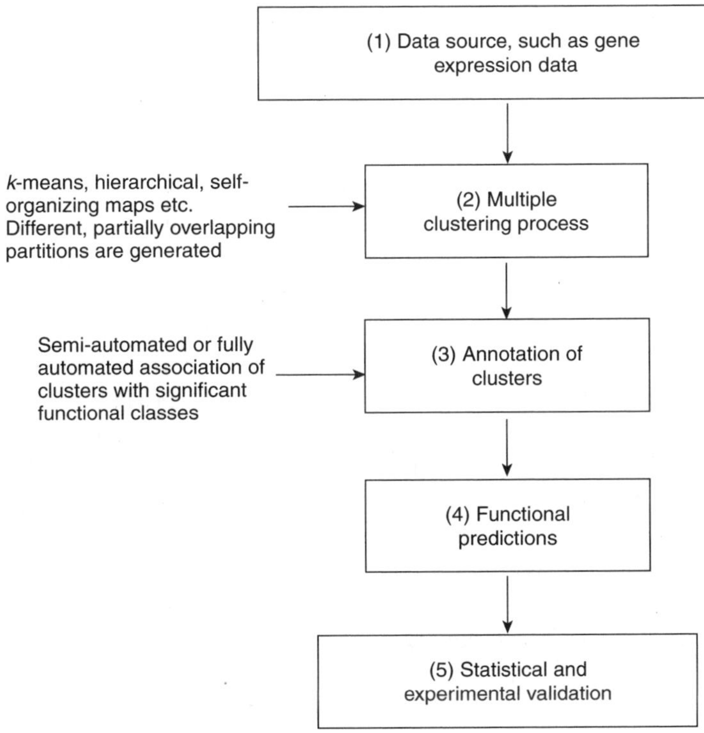

Figure 3.2 Clustering-based integrative prediction framework: basic tasks and tools. Different, partially overlapping partitions are generated by implementing different clustering techniques, based on different learning parameters and numbers of clusters. Probabilistic assessment of the significance of clusters in relation to functional categories is required for automatically labelling clusters and assigning classes to genes

In Chapter 10 Sheng and co-workers review several clustering techniques and methods for assessing the statistical quality of clusters. Different statistical methods may be combined to support the evaluation of clusters in terms of their significance, consistency and validity. This is a problem that deserves more attention and investigation in order to improve the design and interpretation of functional genomics studies, especially those analyses based on gene expression clusters. For instance, the application of *null hypothesis tests* and *internal* and *external validity indices* may be applied to select relevant, significant partitions and clusters. The estimation of the 'correct number of clusters' represented in a dataset is a complex task, which may strongly influence the products of a predictive analysis process. These tests may be used for (a) providing evidence against the hypothesis 'there are no clusters in the data' (null hypothesis tests), (b) finding the optimal partition on the basis of several inter- and intra-cluster distances (internal validity indices) or (c) assessing the agreement between an experimental partition and a reference partition (external indices). The experimental partition is the partition under study, while the reference dataset may be a partition with *a priori* known cluster structure. Bolshakova and Azuaje (2003) have proposed strategies to integrate the outcomes originating from multiple cluster validity indicators, which may be used to generate more reliable and robust predictions about the correct number of clusters.

3.4 Final Remarks

The goal of integrative data analysis and visualization is not only to increase the accuracy and sensitivity of functional prediction tasks, but also to achieve better insights into the problems under consideration. Even when this type of approach has become of great importance in genomics and proteomics, the problem of combining a wide variety of information to form a coherent and consistent picture of functional prediction problems has lagged. Moreover, current advances combine different types of data, relying on the application of a single prediction model (Zhang *et al.*, 2004), which are often based on strong assumptions about the statistical independence or distribution of the data under study (Jansen *et al.*, 2003). To fully exploit integrative data and visualization there is a need to process data derived from different sources. Similarly, it is fundamental to combine diverse predictive views originating from multiple classifiers or prediction models. Furthermore, it is crucial to continue studying relationships between apparently unrelated data, which may provide the basis for novel prediction information sources and models to be integrated.

Sections 3.2 and 3.3 overviewed two key strategies to perform integrative data analysis and visualization in functional genomics. Within such an integrative framework it is also possible to define problems, methods and applications according to (a) the type of data integration and (b) the level at which predictive model integration is achieved.

According to the type of data integration, integrative approaches can be categorized as follows.

Redundant information integration approaches. These approaches process information provided from a group of sources that represent the same type of functional data, e.g. expression data, but with a different degree of accuracy or confidentiality. Applications may be based on the integration of replicated sources that measure similar properties, but which may be noisy, inaccurate or subject to statistical variations (Edwards *et al.*, 2002). They generally aim to reduce the overall uncertainty and increase the predictive accuracy.

Complementary information integration approaches. These approaches integrate information from sources that represent different variables or properties of the prediction problem under consideration. Complementary information integration aims at combining partial, incomplete and noisy information to get a global picture of the prediction problem domain. One typical example is the combination of expression and interaction data sets to predict protein complex membership. Multiple sources provide information that may not be perceived by using individual experimental methods.

According to the level in which information integration is performed, problems and applications can be categorized as follows.

Integration at the level of input representation. Information provided from the sources is fused before performing prediction or classification tasks. This process may be implemented by integrating in a unique input feature vector the attribute values that represent the different variables under study. For instance, Zhang *et al.* (2004) grouped several gene- and protein-pair properties into a single binary feature representation to predict co-complexed pairs in *S. cereivisiae* based on decision trees.

Integration at the level of feature pre-processing. In this case the product of different feature filtering or selection procedures applied to an information source is combined before performing a classification task. Based on a combination of several feature selection schemes, including *signal-to-noise ratio* and an *evolving classification function* technique, Goh, Song and Kasabov (2004) have recently introduced a hybrid feature selection method to improve classification of gene expression data. This study highlighted the advantages of a hybrid, integrative method for gene selection.

Integration at the level of classification. Information provided from different sources or prediction models is processed independently; their prediction outcomes are generated, and then integrated in order to make a final prediction about the functional problem under consideration. One example from this category is the integration of serial and parallel competitive classifiers such as ensembles of neural networks and decision trees (Tan and Gilbert, 2003; Hu and Yoo, 2004).

The application of integrative data analyses at the pre-processing and classification levels based on different types of functional data deserves further investigation. It may offer powerful tools not only to improve predictive quality (accuracy and coverage), but also to support the generation of more comprehensive studies at a systems level.

As a final caveat, it is important to remark that, while on the one hand the overabundance of data can fuel our understanding of biological phenomena, on the other hand one must not neglect the possibility of observing spurious associations between genes and functional properties due to pure chance. It is then necessary to establish rigorous frameworks for the analysis and validation of data-driven functional predictions at a genomic scale.

References

Allocco, D. J., Kohane, I. S. and Butte, A. J. (2004) Quantifying the relationship between co-expression, co-regulation and gene function. *BMC Bioinformatics*, **5** (18).

Baker, C. A. H., Carpendale, M. S. T., Prusinkiewicz, P. and Surette, M. G. (2002) GeneVis: visualisation tools for genetic regulatory network dynamics. In *Proceedings of 13th IEEE Visualisation 2002 Conference*, Boston, MA, 243–250.

Birney E., Andrews D., Bevan P. *et al.* (2004) Ensembl 2004. *Nucleic Acids Res*, **32**, D468-D470.

Bolshakova, N., and Azuaje, F. (2003) Cluster validation techniques for genome expression data. *Signal Processing*, **83** (4), 825–833.

Edwards, A. M., Kus, B., Jansen, R., Greenbaum, D., Greenblatt, J., and Gerstein, M. (2002) Bridging structural biology and genomics: assessing protein interaction data with known complexes. *Trends Genetics*, **18**, 529–536.

Ge, H., Walhout, A. J. M., and Vidal, M. (2003) Integrating 'omic' information: a bridge between genomics and system biology. *Trends Genetics*, **19** (10), 551–560.

Goh, L., Song, Q., and Kasabov, N. (2004) A novel feature selection method to improve classification of gene expression data. In *Proceedings of the Second Conference on Asia–Pacific Bioinformatics: Vol. 29*, 161–166, Dunedin, New Zealand.

Hu, X., and Yoo, I. (2004) Cluster ensemble and its applications in gene expression analysis. In *Proceedings of the Second Conference on Asia–Pacific Bioinformatics: Vol. 29.* 297–302, Dunedin, New Zealand.

Hu, Z., Mellor, J., Wu, J., and DeLisi, C. (2004) VisANT: an online visualisation and analysis tool for biological interaction data. *BMC Bioinformatics*, **5** (17).

Ideker, T., Thorsson, V., Ranish, J. A., Christmas, R., Buhler, J., Eng, J. K., Bumgarner, R., Goodlett, D. R., Aebersold, R., and Hood, L. (2001) Integrated genomic and proteomic analyses of a systematically perturbed metabolic network. *Science*, **292** (4), 929–934.

Jansen, R., Greenbaum, D., and Gerstein, M. (2002) Relating whole-genome expression data with protein–protein interactions. *Genome Research*, **12** (1), 37–46.

Jansen, R., Yu, H., Greenbaum, D., Kluger, Y., Krogan, N. J., Chung, S., Emili, A., Snyder, M., Greenblatt, J. F., and Gerstein, M. (2003) A Bayesian networks approach for predicting protein–protein interactions from genomic data. *Science*, **302** (17), 449–453.

Kasprzyk, A., Keefe, D., Smedley, D. *et al.* (2004) EnsMart – a generic system for fast and flexible access to biological data. *Genome Res*, **14**, 160–169.

Kittler, J., Hatef., M., Duin, R. P. W., and Matas, J. (1998) On combining classifiers. *IEEE Trans Pattern Anal Machine Intell*, **20** (3), 226–239.

Leung, Y. F. and Cavalieri, D. (2003) Fundamentals of cDNA microarray data analysis. *Trends Genetics*, **19** (11), 649–659.

Tan, A. C. and Gilbert, D. (2003) Ensemble machine learning on gene expression data for cancer classification. *Appl Bioinformatics*, **2** (Suppl. 3), S75–S83.

Troyanskaya, O. G., Dolinski, K., Owen, A. B., Altman, R. B., and Botstein, D. (2003) A Bayesian framework for combining heterogeneous data sources for gene function prediction (in *Saccharomyces cerevisiae*). *Proc Natl Acad Sci USA*, **100** (14), 8348–8353.

Walhout, A. J., Reboul, J., Shtanko, O. *et al.* (2002) Integrating interactome, phenome, and transcriptome mapping data for the *C. elegans* germline. *Curr Biol*, **12**, 1952–1958.

Wu, L. F., Hughes, T. R., Davierwala, A. P., Robinson, M. D., Stoughton, R., and Altschuler, S. J. (2002) Large-scale prediction of *Saccharomyces cerevisiae* gene function using overlapping transcriptional clusters. *Nat Genet*, **31**, 255–265.

Zhang, L. V., Wong, S. L., King, O. D., and Roth, F. P. (2004) Predicting co-complexed protein pairs using genomic and proteomic data integration. *BMC Bioinformatics*, **5** (38).

II

Integrative Data Mining and Visualization – Emphasis on Combination of Multiple Data Types

4

Applications of Text Mining in Molecular Biology, from Name Recognition to Protein Interaction Maps

Martin Krallinger and **Alfonso Valencia**

Abstract

The broad range of genome sequencing projects and large-scale experimental characterization techniques are responsible for the rapid accumulation of biological data. A range of data-mining techniques have been developed to handle this tremendous data overload. Experimentally produced data obtained by large-scale protein interaction experiments imply a large variability of information types. Information related to the protein and gene sequences, structures and functions, as well as interaction complexes and pathways, is produced with powerful experimental approaches and published in the form of scientific articles. The scientific text constitutes the primary source of functional information not only for single researchers, but also for the annotation of databases. To be able to handle all this information it is important to link gene and protein sequences with the available functional information deposited in the biomedical literature, where the experimental evidence is described in detail. The emerging field of biomedical text mining has provided the first generation of methods and tools to extract and analyze collections of genes, functions and interactions. Here we describe the current status, possibilities and limitations offered by those methods and their relation with the corresponding areas of molecular biology, with particular attention to the analysis of protein interaction networks.

Data Analysis and Visualization in Genomics and Proteomics Edited by Francisco Azuaje and Joaquin Dopazo

Keywords

text mining, information retrieval, information extraction, named entity recognition, natural language processing, protein–protein interaction, protein function

4.1 Introduction

The increasing amount of sequence information (Hayashizaki, 2003) and gene discoveries derived from genome projects as well as the fast growth of scientific literature (Hoffmann and Valencia, 2003) result in large data collections. Still the main source of information on protein and gene functions is the biomedical literature. Thus literature search not only constitutes a fundamental step at the planning stage of research, but is also essential for the interpretation of experimental results.

In the context of genome sequencing projects and high-throughput experiments there is also an increasing demand for extensive compilations of functional characterizations of individual genes. In particular, the growing field of large-scale protein interaction approaches has developed an intense demand for annotation and characterization associated with the construction of the first specific databases (see Section 4.4).

To extract and store functional descriptions contained in articles, a large number of annotation databases has been developed. These databases contain structured information relative to genes or proteins, where the corresponding annotations are manually extracted by a domain expert through literature study. Those annotations are translated automatically to other genes sharing a significant sequence similarity. This process involves a number of uncertainties, and has been recently evaluated by several groups (Rost, 2002; Todd, Orengo and Thornton, 2002; Osterman and Overbeek, 2003). Our group has estimated the number of potential errors in the translation of detailed annotations between genomes as at least 20 per cent (Devos and Valencia, 2001, 2000).

It has become clear that only direct experimental characterization and clear pointers to the corresponding literature will be able to unravel the errors introduced in databases by repeated annotation transference exercises. Roberts (2004) and Karp (2004) describe the need for direct experimental assessment of protein functions. The need for direct literature references and systematic annotation of function has become particularly obvious in the field of protein interactions. The newly created databases containing information on large-scale protein–protein interaction data are now one of the primary targets of these systematic annotation approaches. The most notorious collection of biomedical literature is PubMed, with over 12 million citations for biomedical articles (NCBI, 2004).

There is an increasing interest in combining information of structured databases such as annotation and genome databases with unstructured textual information in the form of scientific literature (Buckingham, 2004). Nevertheless, it is a cumbersome task to maintain and keep alive the links between experimental results published in scientific journals and the genes stored in annotation and interaction databases.

New proteins mentioned in the literature are not the only ones that have to be included in the annotation databases. The entries that are already contained in annotation databases have also to be revised periodically in order to keep the annotation up to date.

Considering the already existing data explosion in molecular biology, different initiatives to extract and maintain the links between gene sequences and functional information are being explored on the basis of the combination of text mining and natural language processing (NLP) techniques.

NLP is a branch of information science that deals with natural language information and free text analysis. NLP technologies have been developed to manage, process and index unstructured textual information. Text mining could be defined as the analysis and extraction of knowledge from large collections of free textual data by using automatic or semi-automatic systems.

The identification of protein interactions has become an area of particular interest, in part because it is related to the methodological advances in the large-scale experimental methods (Sali *et al.*, 2003), but also because detecting protein interactions in the literature is a problem particularly relevant to the current information extraction technologies.

4.2 Introduction to Text Mining and NLP

NLP research has undergone significant advances because of the range of machine readable textual information and dictionaries that have been compiled in recent years. In the case of biomedical research, large collections of abstracts of scientific articles are now electronically available. A short description of the basic text mining and statistical natural language processing topics is given in Table 4.1. Manning and Schuetze (1999) provide a general overview into this discipline.

Table 4.1 Important topics in text mining and NLP

Topics	Definitions
Information Extraction (IE)	deals with the recovery or extraction of relevant information or meaning from textual data: among the targets of information extraction are entities, relations and events.
Information Retrieval (IR)	or document retrieval, is a process that, given a query presented by the user, tries to recover all the relevant documents from a collection of documents.
Part of Speech (POS) tagging	process of labelling each word with its corresponding part of speech tag (e.g. noun, verb, adjective).
Corpus development	development of collections of documents or textual data.
Named entity recognition (NER)	identification of entity names (e.g. people, organizations, places) in text, and by extension also of genes and protein symbols.

Due to the complexity of biomedical vocabulary and writing styles, the identification of bioentities like genes, proteins or chemical compounds is considered to be the first crucial step in this area. The variety of expressions and synonyms which refer to the actual protein object in free text is currently the main obstacle for advancement in this field.

A variety of different methods has been implemented to tag proteins and genes within free text, such as ad hoc rule-based approaches, methods based on dictionaries of genes and pattern matching. Machine-learning techniques have also been trained to identify these bioentities.

A very interesting set of methods combines manually and automatically generated rules in different proportions. There is also a considerable number of hybrid techniques taking advantage of the strength of different strategies. Among the encountered difficulties are the identification of typographical variants or synonyms of gene names and the disambiguation of gene symbols that correspond to common words. Krallinger *et al.* (2004) provide a more detailed description of the difficulties encountered in protein tagging. Tanabe and Wilbur (2002) developed a system that used automatically generated rules based on part of speech (POS) information together with manually generated rules to tag gene and protein names. POS information was also used by the PROPER (PROtein Proper noun phrase Extracting Rules) system, presented by Fukuda *et al.* (1998), which does not rely on a dictionary of protein names, but rather uses surface characteristics such as certain symbols or capital letters to spot protein names. A large collection of machine-learning approaches has been used to extract protein names from running text. Kazama *et al.* (2002) for instance explored the use of support vector machines (SVMs) for biomedical named entity recognition.

Other major tasks in NLP are *information retrieval* (IR) and *information extraction* (IE). Information retrieval is concerned with the retrieval of textual information from document collections, e.g. all the documents relevant to a certain protein or disease. Very popular IR systems are those that perform automated searches for text in hypertext networked databases such as the Internet. This is the case of widely used applications such as the web search engine *Google*.

Most experimental biologists take advantage of information retrieval systems without being aware of the underlying NLP methodologies. For instance, when querying the PubMed database in search of certain scientific articles, an IR system is used to retrieve the desired documents. The information retrieval system available at the NCBI, which allows queries of the PubMed database through the web, is known as ENTREZ (Schuler *et al.*, 1996).

IE aims to identify semantic structures and other specific types of information within free text using strategies based on POS information, ontologies or the identification of common patterns. An example of information extraction is the identification of relations such as protein–protein interactions in abstracts. In order to achieve information extraction, dictionaries and thesauri that provide semantic classes and ontologies that provide structured collections of terms are both of great interest. Yandell and Majoros (2002) pointed out the use of hierarchically structured ontologies for query expansion, exploiting the semantic relationships among terms contained in the ontology and providing examples of the different classes of semantic classifications. For a detailed discussion of ontologies, refer to Chapter 7.

4.3 Databases and Resources for Biomedical Text Mining

Text-mining tools use a broad spectrum of resources: some of them consist of information stored in structured databases such as annotation databases, and some others are basically formed by collections of unstructured free text, such as scientific abstracts. Among the existing resources are general domain-independent tools, such as POS taggers or stemmers, and domain-specific resources, such as protein taggers or domain-specific thesauri and ontologies.

A common feature of biological data is the difficulty of interconnecting the different data sources. For example, to link terms contained in an ontology to free text is often only possible through complex inference processes based on domain specific background knowledge. It is the domain specificity and the complex syntactical and semantic characteristics of the biomedical text that make general-purpose NLP tools difficult to apply, and the domain-independent tools are often insufficient to cope with certain aspects of scientific publications. The coverage of biomedical dictionaries and other lexical resources is low regarding the terms that actually appear in free text.

This is due partially to the vast number of new terms introduced by researchers in their publications, as well as the use of typographical and lexical variants in running text. POS tagging with general text taggers in biomedical texts is rather error prone, and the grammatical use of words in scientific text may vary significantly. Moreover, most of the domain-specific words are unknown to the taggers. Also, term frequency information derived from a generic corpus, such as newspaper collections or the Internet, cannot be easily extrapolated to molecular biology texts. It is thus important to be aware of the limitations of generic tools.

Scientific article databases

Scientific literature is one of the most important information sources for biomedical research. In order to provide access to biomedical citations, the National Center for Biotechnology Information (NCBI) developed PubMed (Wheeler *et al.*, 2003), which stores citations electronically submitted by publishers.

The access to the bibliographic information is assured through MEDLINE (the bibliographic database) which contains over 12 million references to biomedical journal articles (June 2004), and provides abstracts for a considerable number of them. These abstracts are nowadays the primary data source for NLP research in the biomedical field. Although significant effort is made towards the development of text-mining techniques to handle full text articles, their limited access imposes a considerable hurdle. As Andrade and Valencia (1998) have already pointed out, the use of abstracts is nonetheless associated with several advantages. They are formed using generally short sentences with a reduced and non-ambiguous set of vocabulary, making them a valuable data set for information extraction techniques.

Genome and annotation databases

One of the main concerns of genome research is the proper interpretation of automatically (e.g. using bioinformatics methods) and experimentally obtained results. Thus structured databases containing information of collections of genes/proteins were developed not only for single organisms (genome databases) but also for proteins derived from different species, in well organized efforts such as SwissProt (see Table 4.2). All protein databases are now being unified under the

Table 4.2 Some of the available genome and annotation databases

Database name	Reference	URL
SwissProt	(Boeckmann *et al.*, 2003)	http://us.expasy.org/sprot
MIPS	(Mewes *et al.*, 2004)	http://mips.gsf.de
EXProt	(Ursing *et al.*, 2002)	http://www.cmbi.kun.nl/EXProt
FlyBase	(FlyBase Consortium, 2003)	http://ybase.bio.indiana.edu
SGD	(Dwight *et al.*, 2002)	http://www.yeastgenome.org
WormBase	(Harris *et al.*, 2004)	http://www.wormbase.org
RefSeq, LocusLink	(Pruitt and Maglott, 2001)	http://www.ncbi.nlm.nih.gov/
GeneBank	(Benson *et al.*, 2004)	http://www.ncbi.nlm.nih.gov/Genbank
EMBL-EBI	(EBI, 2004)	http://www.ebi.ac.uk/Databases/

name of UniProt (Leinonen *et al.*, 2002). The gene dictionaries stored within those databases are of tremendous importance for most of the gene-tagging tools aimed at linking the protein sequence and structured information of those entities with the functional information derived from free text where those genes are mentioned. The keywords and annotations manually associated with these proteins may serve as a gold standard for text-mining tools when trying to extract annotations or keyword associations. A collection of relevant databases is provided by the European Institute of Bioinformatics (EBI) and the European Molecular Biology Laboratory (EMBL). Cross-linking the different databases is often important to extract all the available knowledge about a given protein, e.g. definitions and synonyms. Table 4.2 provides a small fraction of the existing resources of genome and annotation databases with practical relevance for text mining purposes. SwissProt may provide keyword annotations of proteins, gene names and symbols, while LocusLink and GeneBank can be seen as integrative resources useful as extensive gene dictionaries. Genome databases such as SGD, FlyBase and WormBase are useful for text-mining approaches, which focus on organism-specific information extraction and retrieval. In Chapter 2 relevant biological databases are discussed in detail.

Bio-ontologies and text mining

Ontologies are static knowledge repositories that have been commonly used in information technology in order to classify semantically entities such as proteins, and are described extensively in Chapter 7. They are used to model the meaning of concepts within a certain domain. They also provide concepts with a semantic dimension and define the relations between them, and are especially useful for data interoperability.

Gene Ontology (GO) (Ashburner *et al.*, 2000) is currently the most extended ontology of molecular biology concepts related to gene products. It is built upon a set of controlled vocabulary, which describes gene products in terms of their associated molecular function, cellular component and biological process. GO has been used to annotate proteins using this set of controlled concepts. Camon *et al.* (2004) outlined some of the relevant aspects associated to the use of GO for protein annotation. GO is not the only ontology used within the molecular biology domain; the Open Biological Ontologies (OBO) initiative gathers together a collection of links to several ontologies used in the domain of biology. Although the use of structured vocabulary, e.g. ontologies, might have considerable advantages, it is also problematic in certain aspects for NLP tools, as they consist of controlled concepts that often do not correspond to natural language expressions. An analysis of the lexical properties of the gene ontology was conducted by McCray *et al.* (2002). They concluded that terms contained in GO are suitable in general for usage by NLP and text-mining methods.

Biomedicine and molecular biology lexicons

Thesauri comprise dictionary entries, their meanings and synonyms, and are thus a terminological knowledge source for text mining of biomedical literature in the form of vocabulary repositories. The Unified Medical Language System (UMLS) Metathesaurus (Bodenreider, 2004), is an extensive biomedical vocabulary collection. It has more than 2 million names for about 900 000 concepts (June 2004). Relations between the concepts are also provided, and it integrates Medical Subject Headings (MeSH), the NCBI taxonomy and GO. MeSH terms are composed of controlled vocabulary that is used to index PubMed articles. UMLS is specially suitable when performing terminology-based text-mining approaches. The UMLS has been used by several NLP methodologies. Srinivasan *et al.* (2002) for example analysed concepts contained in UMLS as to whether they are used in free text (PubMed abstracts and titles). They concluded that many UMLS concepts are encountered in the free text (around 34 per cent), but the coverage of concepts belonging to certain areas is rather low.

The GENIA corpus and the BioCreative contest data

The GENIA corpus (Kim *et al.*, 2003) consists of a semantically annotated collection of 2000 MEDLINE abstracts. These abstracts were annotated manually by domain experts according to a previously defined ontology (the GENIA ontology), containing a total of 100 000 annotations. The main goal of the GENIA project is to provide a gold standard and reference material for text-mining tools.

The Critical Assessment of Information Extraction in Biology (BioCreative) contest also provides a high-quality corpus for biomedical text mining. It contains data sets relevant for tools dealing with protein and gene name recognition and protein annotation using Gene Ontology terms (http//:www:pdg.cnb.uam.es/ BioLINK/ BioCreative.eval.html (Blaschke *et al.*, 2004) to appear in a special issue of BMC bioinformatics).

4.4 Text Mining of Protein–Protein Interactions

The extraction of protein–protein interactions is one of the main concerns within the biomedical text-mining community due to the experimental advances in protein interaction characterization. The most important experimental methods producing protein interaction data are summarized in Table 4.3. Among the experimental techniques used to study protein–protein interactions are protein arrays, mRNA microarrays, the yeast-two hybrid system and methods that produce structural data related to protein complexes such as nuclear magnetic resonance (NMR) and x-ray crystallography. Refer to Chapter 8 for details related to the visualization of interactomes.

Protein interaction data

The recent advances in large-scale protein interaction experiments have resulted in an information overload, which has been addressed mainly using bioinformatic methods. Databases responsible for storing interaction data in a structured way have also been developed. In case of the bioinformatic approaches there are three basic types of method to study protein interactions, namely genome-based methods, sequence-based methods and physical docking methods.

Table 4.3 Experimental methods for protein interaction characterization

Method	Description
Protein arrays	are based on immobilized protein fragments, proteins or antibodies which are displayed on a specially treated, gridlike surface. After treatment with selected samples containing certain proteins, patterns of binary interactions are obtained (Phizicky *et al.*, 2003).
mRNA expression microarrays	are used to measure the concentrations of mRNA of the corresponding proteins, which often correlate with a common functional control, due to participation in the same biological process, or because the interacting proteins are located on the same operon (Marcotte *et al.*, 1999).
Affinity purification associated with mass spectrometry	experiments are able to identify protein complexes and the individual components of those complexes (Aebersold and Mann, 2003).
X-ray crystallography, Nuclear magnetic resonance (NMR) spectroscopy and electron crystallography	are methods able to provide structural information on protein complexes at various resolutions. Fluorescence resonance energy transfer (FRET) and chemical cross-linking techniques might be used to study sub-unit contacts and to delimit the spatial proximity of the interacting sub-units.
Yeast two-hybrid system	is based on the modular properties of eukaryotic transcription factors (e.g. GAL4), which comprise two functional domains, one binding the DNA promoter sequences and a transcription activation domain. Each of these domains is fused to distinct proteins. If these proteins interact, the two transcription factor domains come into close spatial proximity and are thus able to activate the transcription of a reporter gene (Fields *et al.*, 1989).

Genome-based methods use information contained in the genome to extract potential interaction partners. For instance, the gene neighbourhood (genes that appear close in a given genome) may display a coordinated regulation due to functional interactions, especially in prokaryotes. Cases of gene fusion give clues about the functional interactions of the two proteins forming the fused gene. They are thus also an indicator for protein interactions if similar genes are not fused in other organisms.

Sequence-based methods explore features of the protein sequence indicating interactions. For these approaches phylogenetic characteristics of gene or protein sequences are often exploited. For instance, the comparative analysis of phylogenetic profiles often reveals functional relationships between two proteins, which may imply in some cases physical interactions between them. Phylogenetic information is also exploited by the mirror tree method. The detection of correlated mutations provides evidence for co-evolution, often found in interacting proteins. This feature was utilized by Pazos and Valencia (2002) for their *in silico* two-hybrid method. Finally, there are also physical docking algorithms used to predict protein interactions; they constitute an ambitious attempt to predict interactions of proteins forming complexes using structural information. So far, these methods seem only to be successful in cases where the structures of the interacting proteins are experimentally determined, e.g. through x-ray crystallography. The main difficulties encountered by these methods are the conformational changes displayed generally by proteins after binding of the interaction partners (Camacho *et al.*, 2003).

When automatically extracting protein interactions using information extraction methods, it is important to benchmark the results using a curated set of known or experimentally determined protein–protein interaction data. For other bioinformatics tools that analyze protein interactions such data sources are also crucial. Available datasets providing information for protein–protein interaction prediction are contained in Table 4.4. When using the existing resources for automatic protein interaction analysis and prediction it is important to keep in mind that there are a variety of interaction types and that experimental data is often heavily dependent on the kind of interaction analyzed. Interactions between proteins might be direct or through adapter proteins. There are transient and stable interactions forming complexes. Interactions also depend on the cellular conditions, meaning that some interactions are only encountered under certain specific circumstances.

Table 4.4 An overview of the available data sources relevant for the study of protein interactions

Database name	Reference	URL
BIND	(Bader, Betel and Hogue, 2003)	http://bind.ca
DIP	(Xenarios *et al.*, 2002)	http://dip.doe-mbi.ucla.edu
GRID	(Breitkreutz *et al.*, 2003)	http://biodata.mshri.on.ca/grid
HPID	(Han *et al.*, 2004)	http://www.hpid.org
HPRD	(Peri *et al.*, 2004)	http://www.hprd.org
IntAct	(Hermjakob *et al.*, 2004)	http:/www.ebi.ac.uk/intact
MINT	(Zanzoni *et al.*, 2002)	http://cbm.bio.uniroma2.it/mint
STRING	(vonMering *et al.*, 2003)	http://string.embl.de
ECID	(Juan and Valencia, Personal communication)	http://www.pdg.cnb.uam.es/ECID

Text mining and protein–protein interactions

Protein–protein interactions have been particularly attractive for the development of text-mining methods. The reasons are probably related to the possibility of discovering associations between objects quoted in the reduced space of scientific abstracts, even if co-occurrence may be related to very different types of relationship, e.g. homologous proteins derived from different organism sources.

Part of the attraction of protein interactions for text-mining research is also related to the possibility of developing interesting algorithms for the detection of relations between equivalent objects. In one of the first approaches Blaschke *et al.* (1999) addressed the problem by including interaction patterns (in addition to the two protein names), which should be relevant to define the relationship between the co-occurring proteins. Hoffmann and Valencia (2004) developed one of the most recent efforts to facilitate the access to the literature regarding protein interactions.

Blaschke *et al.* (1999) constructed a set of 14 carefully predefined words reflecting protein–protein interaction based on domain knowledge, thus avoiding the complexity of semantic analysis. Some of the verbs used by this system were activate, bind and suppress, among others. To deduce the direction of the interaction relative to the interaction partners, they also analyzed the order of the protein names, and their distance within the text segments. This system was applied to the *Drosophila* Pelle system and to the cell cycle control in *Drosophila*. The results showed that the system was effective in those cases of simple interaction types, which were often consistently repeated. In cases where the interactions were of complex nature (i.e. interaction information formulated in long sentences with complex grammatical structure), and the number of occurrences within free text was low, the interactions were more difficult to extract.

Blaschke and Valencia (2001b) presented a protein interaction discovery tool named SUISEKI (System for Information Extraction of Interactions). It is a hybrid statistical–computational linguistics method, as it performs syntactical analysis of phrases and statistical analysis of matched patterns. The patterns were characterized by the presence of at least two protein names related by certain frames. These frames attempt to capture the basic ways of expressing protein interactions in free text (Blaschke and Valencia, 2002). The initial set of frames was extracted manually by filtering large amounts of free text, including relations and sentences containing negations. The system also integrates visualization modules that allow the display and modification of the extracted interactions (see Figure 4.1). The accuracy of the interactions extracted with SUISEKI was assessed using a corpus of cell-cycle-related articles. The obtained results pointed out a strong correlation between the extracted interactions and their frequency in the text corpus. SUISEKI was additionally evaluated by its capacity of extracting the text corresponding to the interactions deposited in the DIP database (Blaschke and Valencia, 2001a; Xenarios *et al.*, 2002). Interestingly, for almost one-third of the interactions it was impossible to identify their origin because it was impossible for the system to find the correspondence between the protein names used in the database and in the abstracts. In other cases the information contained in the abstract was insufficient to identify interactions that were only described in the full text articles. There were also cases where SUISEKI failed because the set of frames included in that version of the system were not able to cope with certain grammatically complicated sentence structures. Other approaches

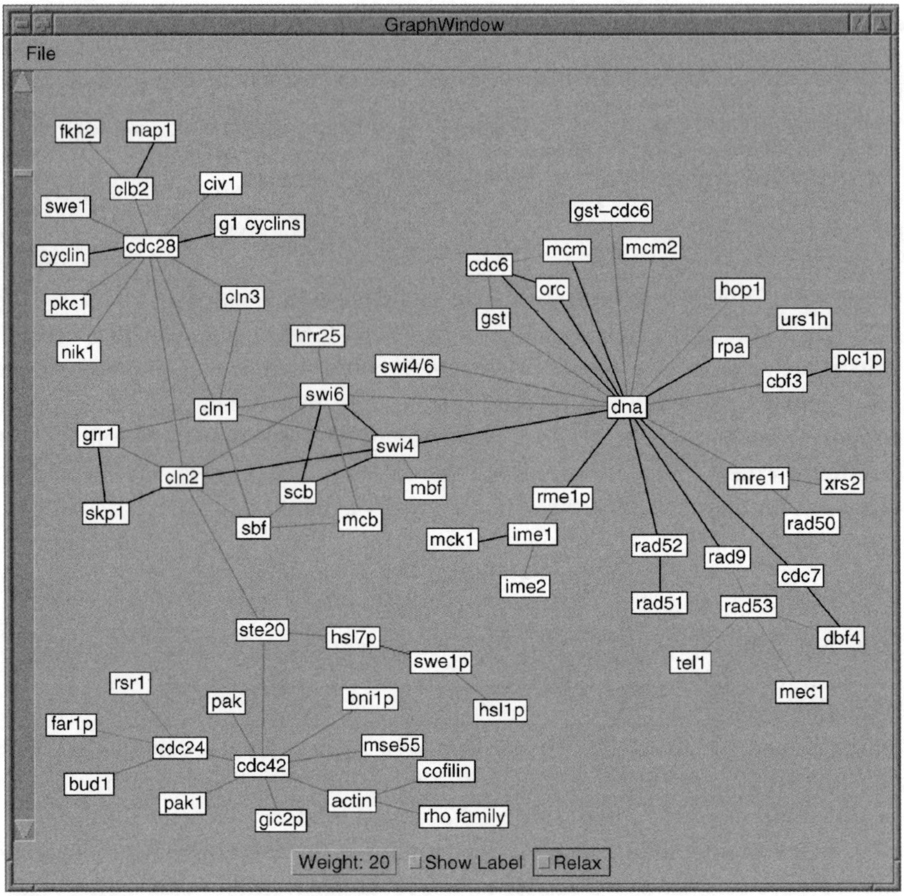

Figure 4.1 The analysis of the interaction network formed around the cell cycle corpus. This analysis provides a good practical example of the type of information provided by SUISEKI and the possibilities open to the analysis by human experts. The overview provided in this figure shows how the main protein components implicated in this biological system were detected automatically. In the upper left corner we see Cdc28, a key cyclin-dependent kinase (CDK) involved in cell cycle control. The activity of Cdc28 is controlled by the G1 cyclins (cln1, 2, 3) and the G2 cylins (clb1, 2, 3 and 4). The interaction between Cdc28 and the cyclins controls the activity of the transcription factors Sbf and Mbf. (Picture kindly provided by Christian Blaschke)

with philosophies similar to SUISEKI have been developed more recently. Ono *et al.* (2001) implemented a method that also used POS tagging and pattern matching to extract protein–protein interactions. It is important to keep in mind that although the precision of this type of frame-based approach tends to be better than that based on simple co-occurrence, the coverage is necessarily smaller. A different type of problem related to protein interactions has been addressed by Marcotte *et al.* (2001) with the construction of a tool to select abstracts containing information about interactions, as an assistant to the database annotation process. Later, Donaldson *et al.* (2003) constructed PreBIND and Textomy, an information extraction system that uses support vector machines (SVMs) to accomplish this task in support of the BIND database curators.

4.5 Other Text-Mining Applications in Genomics

One of the main possible applications of the information derived from text-mining techniques is to assist in the interpretation of high-throughput experimental approaches. A short survey of text-mining applications in different fields of genomics is provided in this section. Andrade and Valencia (1998) addressed the problem of automatic annotation of the function of protein families using textual information. They developed a system to extract automatically functionally relevant keywords from free text to describe biomolecules. Even if keywords are useful for human interpretation, they are often not suitable for data interoperability, and they contain little information regarding their respective relationships.

Sequence similarity is the base for identifying suitable structure templates for a given query sequence. In practice, the retrieved hits often display weak similarities that have been filtered manually using background knowledge. MacCallum *et al.* (2000) presented SAWTED (Structure Assignment With Text Description), a system that uses standard document comparison algorithms and information contained in text descriptions of SwissProt annotations to help in the identification of remote homologues.

Text-mining tools have also been constructed to classify proteins according to their sub-cellular location using abstracts or database records; for example, Nair and Rost (2002) exploited lexical information present in annotation database records to predict the sub-cellular location of proteins. It is in this context of protein function annotation where ontologies are necessary to maintain the coherence of the isolated terms that may be extracted from text. The GO has been explored by various text-mining approaches as an information source when trying to automatically extract annotation. Raychaudhuri *et al.* (2002) built a maximum entropy classifier to automatically associate GO terms with MEDLINE abstracts, and to annotate the corresponding genes. As sequence similarity often gives clues about functional similarity, the combination of sequence homology and text information was also explored in an attempt to annotate proteins automatically using GO (Xie *et al.* 2002).

To be useful in real word scenarios it is important to provide the appropriate tools to facilitate the interaction between expert database curators and the literature. Krallinger and Padron (2004) developed a method to extract passages containing evidence for the GO term corresponding to proteins. The results of the application of their procedure to a large dataset was evaluated by expert annotators of the GO database (at the BioCreative competition), concluding that the results were helpful as an initial attempt to highlight textual fragments for annotation extraction.

The other technology that has attracted the attention of the text-mining community is the analysis of the results of massive gene expression array experiments. The most obvious application is related to the interpretation of the functional connections between genes that display similar levels of gene expression (mRNA concentration). Raychaudhuri and Altman (2003) explored the functional coherence of gene clusters using textual data associated with the genes represented in the microarrays. Oliveros *et al.* (2000) analyzed the similarity of expression patterns of genes and their correlation with automatically extracted biological terms.

The GeneWays system integrates several text analysis modules to mine signal-transduction pathway data developed by Rzhetsky *et al.* (2004). It extracts relations between processes or substances and provides a relationship learner module.

4.6 The Future of NLP in Biomedicine

The growing interest in text-mining applications has led to a need to evaluate the performances of the various systems, and existing tools. Independent evaluations on a common dataset are essential to assess the state of the art of the various methods and to plan future developments. The Knowledge Discovery and Data Mining (KDD) Challenge Cup (Yeh *et al.*, 2003) focused on the evaluation of different approaches to determine whether a given article contains experimental evidence for gene products. To assess different protein and gene tagging tools, as well as to predict protein annotations the BioCreative contest was held (Blaschke *et al.*, 2004), to appear in a special issue of *BMC Bioinformatics*. Additionally, these experiments provide extremely valuable sets of annotated text, which can complement the few available marked sets.

Given the complex nature of human language, and especially of domain-specific expressions used in biology, text mining and NLP techniques still have a long way to go. Nevertheless, the current systems show sustained capacity to handle the large amount of accumulating data. With the improvements expected in the future, and the construction of ontologies for biology, more efficient knowledge discovery tools will emerge to extract additional knowledge and mine the data treasure contained in scientific articles.

Acknowledgements

The work of Krallinger was sponsored by the DOC scholarship programme of the Austrian Academy of Sciences. The Protein Design Group of Alfonso Valencia was supported by TEMBLOR (QLRT-2001-00015), ORIEL (IST-2001-32688) and BioSapiens (LHSG-CT-2003-503265) EC grants.

References

Aebersold, R. and Mann, M. (2003) Mass spectrometry-based proteomics. *Nature*, **422**, 198–207.

Andrade, M. and Valencia, A. (1998) Automatic extraction of keywords from scientific text: application to the knowledge domain of protein families. *Bioinformatics*, **14**, 600–607.

Ashburner, M. and Ball, C., Blake, J. *et al.* (2000) Gene ontology: tool for the unification of biology. The Gene Ontology Consortium. *Nat Genet*, **25**, 25–29.

Bader, G., Betel, D., and Hogue, C. (2003) BIND: the Biomolecular Interaction Network Database. *Nucleic Acids Res*, **31**, 248–250.

Benson, D. and Karsch-Mizrachi, I. *et al.* (2004) GenBank: update. *Nucleic Acids Res*, **32**, D23–D26.

Blaschke, C. and Andrade, M. A. *et al.* (1999) Automatic extraction of biological information from scientific text: protein–protein interactions. *Proc Int Conf Intell Syst Mol Biol*, 60–67.

Blaschke, C. and Andres Leon, E. *et al*, (2004) Evaluation of BioCreative assessment of task 2. *BMC Bioinformatics*.

Blaschke, C. and Valencia, A. (2001a) Can bibliographic pointers for known biological data be found automatically? Protein interactions as a case study. *Comp Funct Genom*, **2**, 196–206.

Blaschke, C. and Valencia, A. (2001b) The potential use of SUISEKI as a protein interaction discovery tool. *Genome Inform Ser Workshop Genome Inform*, **12**, 123–134.

Blaschke, C. and Valencia, A. (2002) The frame-based module of the Suiseki information extraction system. *IEEE Intelli. Sys*, **17**, 14–20.

Bodenreider, O. (2004) The Unified Medical Language System (UMLS): integrating biomedical terminology. *Nucleic Acids Res*, **32**, 267–270.

Boeckmann, B. *et al.* (2003) The SWISS-PROT protein knowledgebase and its supplement TrEMBL in 2003. *Nucleic Acids Res*, **31**, 365–370.

Breitkreutz, B. *et al.* (2003) The GRID: the General Repository for Interaction Datasets. *Genome Biol*, **4** (3), R23.

Buckingham, S. (2004) Bioinformatics: data's future shock. *Nature*, **428**, 774–777. Camacho, C. *et al.* (2003) Successful discrimination of protein interactions. *Proteins*, **52**, 92–97.

Camon, E. *et al.* (2004) The Gene Ontology Annotation (GOA) Database: sharing knowledge in Uniprot with Gene Ontology. *Nucleic Acids Res*, **32**, 262–266.

Devos, D. and Valencia, A. (2000) Practical limits of function prediction. *Proteins*, **41**, 98–107.

Devos, D. and Valencia, A. (2001) Intrinsic errors in genome annotation. *Trends Genet*, **17**, 429–431.

Donaldson, I. *et al.* (2003) PreBIND and Textomy: Mining the biomedical literature for protein–protein interactions using a support vector machine. *BMC Bioinformatics*, **4**, 11.

Dwight, S. *et al.* (2002) Saccharomyces Genome Database (SGD) provides secondary gene annotation using the Gene Ontology (GO). *Nucleic Acids Res*, **30**, 69–72.

EBI. (2004) Public web-based services from the European Bioinformat Institute. *Nucleic Acids Res*, **32**, W3–9.

Fields, S. *et al*, (1989) A novel genetic system to detect protein–protein interactions. *Nature*, **340**, 245–246.

FlyBase Consortium, 2003. The FlyBase database of the Drosophila genome projects and community literature. *Nucleic Acids Res*, **31**, 172–175.

Fukuda, K. *et al.* (1998) Toward information extraction: identifying protein names from biological papers. *Pac Symp Biocomput*, 707–718.

Han, K. *et al.* (2004). HPID: the human protein interaction database. *Bioinformatics*, **20**, 2466–2470.

Harris, T. *et al.* (2004) Worm-Base: a multi-species resource for nematode biology and genomics. *Nucleic Acids Res*, **32**, D411–D417.

Hayashizaki, Y. (2003) RIKEN mouse genome encyclopedia. *Mech Ageing Dev*, **124**, 93–102.

Hermjakob, H. *et al.* (2004) IntAct: an open source molecular interaction database. *Nucleic Acids Res*, **32**, D452–D455.

Hoffmann, R. and Valencia, A. (2003) Life cycles of successful genes. *Trends Genet*, **19**, 79–81.

Hoffmann, R. and Valencia, A. (2004) A gene network for navigating the literature. *Nat Genet*, **36**, 664.

Karp, P. (2004) *Enzyme Genomics Project Whitepaper*, 1–3.

Kazama, J. *et al.* (2002) Tuning support vector machines for biomedical named entity recognition. *Proc Natural Language Processing in the Biomedical Domain.*

Kim, J. *et al.* (2003) GENIA corpus: semantically annotated corpus for bio-textmining. *Bioinformatics*, **19**, i180–i182.

Krallinger, M. and Padron, M. (2004) Prediction of GO annotation by combining entity specific sentence sliding window profiles. *Proc. BioCreative Challenge Evaluation Workshop 2004.*

Krallinger, M. *et al.* (2004) Assessing the correlation between contextual patterns and biological entity tagging. *Proc. COLING 2004.*

Leinonen, R. *et al.* (2002) UniProt archive. *Bioinformatics*, **0**, 1911–1920.

MacCallum, R. *et al.* (2000) SAWTED: structure assignment with text description enhanced detection of remote homologues with automated SWISS-PROT annotation comparisons. *Bioinformatics*, **16**, 125–129.

Manning, C. and Schuetze, H. (1999) *Foundations of Statistical Natural Language Processing*. MIT Press, Cambridge, MA.

Marcotte, E. *et al.* (1999) A combined algorithm for genome-wide prediction of protein function. *Nature*, **402**, 25–26.

Marcotte, E. *et al.* (2001) Mining literature for protein–protein interactions. *Bioinformatics*, **17**, 259–363.

McCray, A. *et al.* (2002) The lexical properties of the gene ontology. *Proc AMIA Symp*, 504–508.

Mewes, H. *et al.* (2004) MIPS: analysis and annotation of proteins from whole genomes. *Nucleic Acids Res*. **32**, D41–D44.

Nair, R. and Rost, B. (2002) Inferring sub-cellular localization through automated lexical analysis. *Bioinformatics*, **8**, S78–S86.

NCBI (2004) PubMed. Available from http://www.ncbi.nlm.nih.gov/entrez/ [accessed: 07/10/04].

Oliveros, J. *et al.* (2000) Expression profiles and biological function. *Genome Inform SerWorkshop Genome Inform*, **11**, 106–117.

Ono, T. *et al.* (2001) Automated extraction of information on protein–protein interactions from the biological literature. *Bioinformatics*, **17**, 155–161.

Osterman, A. and Overbeek, R. (2003) Missing genes in metabolic pathways: a comparative genomics approach. *Curr Opin Chem Biol*, **7**, 238–251.

Pazos, F. and Valencia, A. (2002) In silico two-hybrid system for the selection of physically interacting protein pairs. *Proteins*, **47**, 219–227.

Peri, S. *et al.* (2004) Human protein reference database as a discovery resource for proteomics. *Nucleic Acids Res*, **32**, D497–D501.

Phizicky, E. and *et al.* (2003) Protein analysis on a proteomic scale. *Nature*, **422**, 208–215.

Pruitt, K. and Maglott, D. (2001) RefSeq and LocusLink: NCBI gene-centered resources. *Nucleic Acids Res*, **29**, 137–140.

Raychaudhuri, S. and Altman, R. (2003) A literature-based method for assessing the functional coherence of a gene group. *Bioinformatics*, **19**, 396–401.

Raychaudhuri, S. *et al.* (2002) Associating genes with gene ontology codes using a maximum entropy analysis of biomedical literature. *Genome Res*, **12**, 203–214.

Roberts, R. (2004) Identifying protein function—a call for community action. *PLOS Biology*, **2**, 0001–0002.

Rost, B. (2002) Enzyme function less conserved than anticipated. *J Mol Biol*, **318**, 595–608.

Rzhetsky, A. *et al.* (2004) GeneWays: a system for extracting, analyzing, visualizing, and integrating molecular pathway data. *J. Biomed Inform*. **37**, 43–53.

Sali, A. *et al.* (2003) From words to literature in structural proteomics. *Nature*, **422**, 216–225.

Schuler, G. *et al.* (1996) Entrez: molecular biology database and retrieval system. *Methods Enzymol*, **266**, 141–162.

Srinivasan, S. *et al.* (2002) Finding UMLS metathesaurus concepts in MEDLINE. *Proc AMIA Symp*, 727–731.

Tanabe, L. and Wilbur, W. (2002) Tagging gene and protein names in biomedical text. *Bioinformatics*, **18**, 1124–1132.

Todd, A., Orengo, C. and Thornton, J. (2002) Sequence and structural differences between enzyme and nonenzyme homologs. *Structure*, **10**, 1435–1451.

Ursing, B. *et al.* (2002) EXProt: a database for proteins with an experimentally verified function. *Nucleic Acids Res*, **30**, 50–51.

vonMering, C. *et al.* (2003) STRING: a database of predicted functional associations between proteins. *Nucleic Acids Res*, **31**, 258–261.

Wheeler, D. *et al.* (2003) Database resources of the National Center for Biotechnology. *Nucleic Acids Res*, **31**, 28–33.

Xenarios, I. *et al.* (2002) DIP, the Database of Interacting Proteins: a research tool for studying cellular networks of protein interactions. *Nucleic Acids Res*, **30**, 303–305.

Xie, H. *et al.* (2002) Large-scale protein annotation through gene ontology. *Genome Res.*, **12**, 785–794.

Yandell, M. and Majoros, W. (2002) Genomics and natural language processing. *Nat Rev Genet*, **3**, 601–610.

Yeh, A. *et al.* (2003) Evaluation of text data mining for database curation: lessons learned from the KDD Challenge Cup. *Bioinformatics*, **19**, 331–339.

Zanzoni, A. *et al.* (2002) MINT: a Molecular INTeraction database. *FEBS Lett*, **513**, 135–140.

5

Protein Interaction Prediction by Integrating Genomic Features and Protein Interaction Network Analysis

Long J. Lu[†], **Yu Xia**[†], **Haiyuan Yu**[†], **Alexander Rives**, **Haoxin Lu**, **Falk Schubert** and **Mark Gerstein**

Abstract

The recent explosion of genome-scale protein interaction screens has made it possible to study protein interactions on a level of interactome and networks. In this chapter, we begin with an introduction of a recent approach that probabilistically combines multiple information sources to predict protein interactions in yeast. Specifically, Section 5.2 describes the sources of genomic features. Section 5.3 provides a basic tutorial on machine-learning approaches and describes in detail the decision tree and naïve Bayesian network that have been used in above study. Section 5.4 discusses the missing value challenges in further development of our existing method. We then shift our attention to discuss protein–protein interactions in the context of networks in Section 5.5, where we present two important network analysis approaches: topology network analysis and modular network analysis. Finally we discuss advantages and key limitations of our method, and our vision of challenges in this area.

Keywords

protein–protein interactions, integration and prediction, Bayesian network, network topology, network modularity, network visualization

[†]These authors contribute equally to this chapter.

Data Analysis and Visualization in Genomics and Proteomics Edited by Francisco Azuaje and Joaquin Dopazo

5.1 Introduction

Protein–protein interactions are fundamental to cellular functions, and comprehensively identifying them is important towards systematically defining the biological role of proteins. New experimental and computational methods have produced a vast number of known or putative interactions catalogued in databases, such as MIPS (Mewes *et al.*, 2002), DIP (Xenarios *et al.*, 2002) and BIND (Bader *et al.*, 2001). Unfortunately, interaction datasets are often incomplete and contradictory (von Mering *et al.*, 2002, Edwards *et al.*, 2002). In the context of genome-wide analyses these inaccuracies are greatly magnified because the protein pairs that do not interact (negatives) far outnumber those that do (positives). For instance, in yeast the ~6000 proteins allow for ~18 million potential interactions, but the estimated number of actual interactions is below 100 000 (von Mering *et al.*, 2002; Bader and Hogue, 2002; Kumar and Snyder, 2002). Thus, even reliable techniques can generate many false positives when applied on a genomic scale. An analogy to this situation would be a diagnostic with a one per cent false-positive rate for a rare disease occurring in 0.1 per cent of the population, which would roughly produce one true positive for every 10 false ones. Consequently, when evaluating protein–protein interactions, one needs to integrate evidence from many different sources to reduce errors (Marcotte *et al.*, 1999; Jansen *et al.*, 2002).

In the era of post-genomic biology, it becomes particularly useful to think of cells as a complex network of interacting proteins (Eisenberg *et al.*, 2000; Hartwell *et al.*, 1999). Biology is increasingly moving from the study of the individual parts of the system separately to the study of the emergent properties of the entire system. Most biological functions are the result of the interactions of many different molecules. The challenge of systems biology is to develop models of biological functions that incorporate and elucidate this complexity.

This chapter has thus been written with the aim of introducing our recent development of a naïve Bayes approach in the protein complex membership prediction, and recent progress in the protein interaction network analysis. The first half of the chapter will provide a detailed description of the naïve Bayesian approach developed by Jansen *et al.* (2003) that probabilistically combines multiple information sources to predict protein interactions in yeast. We begin with Section 5.2, which describes the genomic features used in this approach. Section 5.3 provides a basic tutorial on machine-learning approaches and describes in detail the decision tree and naïve Bayes classifier that has been used in the above study. Section 5.4 discusses the missing value challenges in further development of our existing method. In the second half of the chapter, we shift our attention to discuss protein–protein interactions in the context of networks. Section 5.5 focuses on two important network analysis approaches: topology network analysis and modular network analysis (Rives and Galitski, 2003). A useful network visualization tool, TopNet (Yu *et al.*, 2004), will also be introduced. Finally, we will discuss advantages and limitations of our method, and our vision of challenges in this area.

5.2 Genomic Features in Protein Interaction Predictions

Jansen *et al.* (2003) recently showed how protein complexes can be predicted *de novo* with high confidence when multiple datasets are integrated, and demonstrated the application to yeast. These multiple datasets can be either noisy interaction datasets or circumstantial genomic evidence. The genomic data sources used in above study are the correlation of mRNA amounts in two expression datasets, two sets of information on biological function, and information about whether proteins are essential for survival (see below). Although none of these information sources are interaction *per se*, they contain information weakly associated with interaction: two subunits of the same protein complex often have co-regulated mRNA expression and similar biological functions and are more likely to be both essential or non-essential.

mRNA expression

Two sets of publicly available expression data – a time course of expression fluctuations during the yeast cell cycle and the Rosetta compendium, consisting of the expression profiles of 300 deletion mutants and cells under chemical treatments (Cho *et al.*, 1998; Hughes *et al.*, 2000) – have been used in the above study. These data are useful for the prediction of protein–protein interaction because proteins in the same complex are often co-expressed (Ge *et al.*, 2001). The Pearson correlation for each protein pair for both the Rosetta and cell cycle datasets indicates that these two datasets are strongly correlated. This problem can be circumvented by computing the first principal component of the vector of the two correlations. This first principal component is a stronger predictor of protein–protein interactions than either of the two expression correlation datasets by themselves. In order to perform Bayesian networks analysis (see Section 5.3), this first principal component of expression correlations is divided into 19 bins, and the overlap of each bin with the gold standard datasets (see below) is assessed (Table 5.1). The first column of Table 5.1 bears the name of the genomic feature and the number of bins we divide this feature into. The second column gives the number of protein pairs that this feature covers in the yeast interactome ($\sim$18 million pairs of proteins). The third column, which contains five subcolumns, shows the overlap between the genomic feature and the gold-standard (positive and negative) sets. The subcolumns positive (+) and negative (−) show how many protein pairs in the present bin of the genomic feature are among the protein pairs in the gold-standard positive set and negative set, respectively. The subcolumns sum(+) and sum(−) show the cumulative number of overlaps of the present and above bins. The subcolumn sum(+)/sum(−) is the ratio of sum(+) and sum(−). The next two columns are the conditional probabilities of the feature, and the last column is the likelihood ratio L, which is the ratio of the conditional probabilities in the two preceding columns. More details on the likelihood ratio are given in Section 5.3.

Table 5.1 Combining genomic features to predict protein–protein interactions in yeast

mRNA expression correlation		Number of protein pairs	Gold-standard overlap					P(Exp/+)	P(Exp/−)	Likelihood ratio (L)
			positive(+)	negative(−)	sum(+)	sum(−)	sum(+)/sum(−)			
Bins	0.9	678	16	45	16	45	0.36	2.10×10^{-3}	1.68×10^{-5}	124.9
	0.8	4 827	137	563	153	608	0.25	1.80×10^{-2}	2.10×10^{-4}	85.5
	0.7	17 626	530	2 117	683	2 725	0.25	6.96×10^{-2}	7.91×10^{-4}	88.0
	0.6	42 815	1073	5 597	1 756	8 322	0.21	1.41×10^{-1}	2.09×10^{-3}	67.4
	0.5	96 650	1089	14 459	2 845	22 781	0.12	1.43×10^{-1}	5.40×10^{-3}	26.5
	0.4	225 712	993	35 350	3 838	58 131	0.07	1.30×10^{-1}	1.32×10^{-2}	9.9
	0.3	529 268	1028	83 483	4 866	141 614	0.03	1.35×10^{-1}	3.12×10^{-2}	4.3
	0.2	1 200 331	870	183 356	5 736	324 970	0.02	1.14×10^{-1}	6.85×10^{-2}	1.7
	0.1	2 575 103	739	368 469	6 475	693 439	0.01	9.71×10^{-2}	1.38×10^{-1}	0.7
	0	9 363 627	894	1 244 477	7 369	1 937 916	0.00	1.17×10^{-1}	4.65×10^{-1}	0.3
	−0.1	2 753 735	164	408 562	7 533	2 346 478	0.00	2.15×10^{-2}	1.53×10^{-1}	0.1
	−0.2	1 241 907	63	203 663	7 596	2 550 141	0.00	8.27×10^{-3}	7.61×10^{-2}	0.1
	−0.3	484 524	13	84 957	7 609	2 635 098	0.00	1.71×10^{-3}	3.18×10^{-2}	0.1
	−0.4	160 234	3	28 870	7 612	2 663 968	0.00	3.94×10^{-4}	1.08×10^{-2}	0.0
	−0.5	48 852	2	8 091	7 614	2 672 059	0.00	2.63×10^{-4}	3.02×10^{-3}	0.1
	−0.6	17 423	N/A	2 134	7 614	2 674 193	0.00	0.00	7.98×10^{-4}	0.0
	−0.7	7 602	N/A	807	7 614	2 675 000	0.00	0.00	3.02×10^{-4}	0.0
	−0.8	2 147	N/A	261	7 614	2 675 261	0.00	0.00	9.76×10^{-5}	0.0
	−0.9	67	N/A	12	7 614	2 675 273	0.00	0.00	4.49×10^{-6}	0.0
Total number		18 773 128	7614	2 675 273	N/A	N/A	N/A	1.00	1.00	1.0

GO biological process similarity		Number of protein pairs	Gold-standard overlap					P(GO/+)	P(GO/−)	Likelihood ratio (L)
			positive(+)	negative(−)	sum(+)	sum(−)	sum(+)/sum(−)			
Bins	1–9	4 789	88	819	88	819	0.11	1.17×10^{-2}	1.27×10^{-3}	9.2
	10–99	20 467	555	3 315	643	4 134	0.16	7.38×10^{-2}	5.14×10^{-3}	14.4
	100–999	58 738	523	10 232	1166	14 366	0.08	6.95×10^{-2}	1.59×10^{-2}	4.4
	1000–9999	152 850	1003	28 225	2169	42 591	0.05	1.33×10^{-1}	4.38×10^{-2}	3.0
	10 000–∞	2 909 442	5351	602 434	7520	645 025	0.01	7.12×10^{-1}	9.34×10^{-1}	0.8
Total number		3 146 286	7520	645 025	N/A	N/A	N/A	1.00	1.00	1.0

MIPS functional similarity		Number of protein pairs	Gold-standard overlap					P(MIPS/+)	P(MIPS/−)	Likelihood Ratio (L)
			positive(+)	negative(−)	sum(+)	sum(−)	sum(+)/sum(−)			
Bins	1–9	6 584	171	1 094	171	1 094	0.16	2.12×10^{-2}	8.33×10^{-4}	25.5
	10–99	25 823	584	4 229	755	5 323	0.14	7.25×10^{-2}	3.22×10^{-3}	22.5
	100–999	88 548	688	13 011	1443	18 334	0.08	8.55×10^{-2}	9.91×10^{-3}	8.6
	1000–9999	255 096	6146	47 126	7589	65 460	0.12	7.63×10^{-1}	3.59×10^{-2}	21.3
	10 000–∞	5 785 754	462	1 248 119	8051	1 313 579	0.01	5.74×10^{-2}	9.50×10^{-1}	0.1
Total number		6 161 805	8051	1 313 579	N/A	N/A	N/A	1.00	1.00	1.0

Co-essentiality		Number of protein pairs	Gold-standard overlap					P(Ess/+)	P(Ess/−)	Likelihood ratio (L)
			positive(+)	negative(−)	sum(+)	sum(−)	sum(+)/sum(−)			
Bins	EE	384 126	1114	81 924	1114	81 924	0.014	5.18×10^{-1}	1.43×10^{-1}	3.6
	NE	2 767 812	624	285 487	1738	367 411	0.005	2.90×10^{-1}	4.98×10^{-1}	0.6
	NN	4 978 590	412	206 313	2150	573 724	0.004	1.92×10^{-1}	3.60×10^{-1}	0.5
Total number		8 130 528	2150	573 724	N/A	N/A	N/A	1.00	1.00	1.0

Biological functions

Interacting proteins often function in the same biological process (Schwikowski, Uetz and Fields, 2000; Vazquez *et al.*, 2003). This means that two proteins acting in the same biological process are more likely to interact than two proteins involved in different processes. In addition, proteins functioning in small, specific biological processes are more likely to interact than those functioning in large, general processes.

Two catalogues of functional information about proteins are collected from the MIPS functional catalogue – which is separate from the MIPS complexes catalogue – and the data on biological processes from Gene Ontology (GO) (Ashburner *et al.*, 2000). Most classification systems have the structure of a tree (e.g. MIPS) or a directed acyclic graph (DAG) (e.g. GO). Obviously, a pair of proteins should be very similar if there are only a few descendants of a given ancestor, whereas the similarity will be less significant if many proteins descend from it. Given two proteins that share a specific set of lowest common ancestor nodes in the classification structure, one can count the total number of protein pairs n that also have the exact same set of lowest common ancestors. This number is expected to be low for proteins that share a very detailed functional description, but very high for proteins that have no function in common. For instance, if a functional class contains only two proteins, then the count would yield $n=1$. On the other hand, if the root node is the lowest common ancestor of two proteins, n is on the order of the number of protein pairs contained in the classification.

The functional similarity between two proteins is thus quantified by the following procedure. First, two proteins of interest are assigned to a set of functional classes two proteins share, given one of the functional classification systems. Then the number of the $\sim$18 million protein pairs in yeast that share the exact same functional classes as the interested protein pairs is counted (yielding a count between 1 and $\sim$18 million). In general, the smaller this count, the more similar and specific is the functional description of the two proteins, while large counts indicate a very non-specific functional relationship between the proteins. Low counts (i.e. high functional similarity) are found to correlate with a higher chance of two proteins being in the same complex (Table 5.1).

Essentiality

Protein essentiality is also considered in the study (Mewes *et al.*, 2002). It should be more likely that both of two proteins in a complex are essential or non-essential, but not a mixture of these two attributes. This is because a deletion mutant of either one protein should by and large produce the same phenotype: they both impair the function of the same complex. Indeed, such a relationship is supported by the data (Table 5.1).

Finally, protein–protein interaction datasets generated by high-throughput experiments can also be seen as a special type of genomic feature.

Gold-standard datasets

The basic idea of how to integrate different sources of information is to assess each source of evidence for interactions by comparing it against samples of known positives and negatives, yielding a statistical reliability. Then, extrapolating genome-wide, the chance of possible interactions for every protein pair can be predicted by combining each independent evidence source according to its reliability. Thus, reliable reference datasets that serve as gold standards of positives (proteins that are in the same complex) and negatives (proteins that do not interact) are essential.

An ideal gold-standard dataset should satisfy the three following criteria: (1) independent from the data sources serving as evidence, (2) sufficiently large for reliable statistics and (3) free of systematic bias. It is important to note that different experimental methods carry with them different systematic errors – errors that cannot be corrected by repetition. Therefore, the gold-standard dataset should not be generated from a single experimental technique. Positive gold standards are extracted from the MIPS (Munich Information Center for Protein Sequences) complexes catalogue (version November 2001). It consists of a list of known protein complexes based on the data collected from the biomedical literature (most of these are derived from small-scale studies, in contrast to the high-throughput experimental interaction data). Only classes that are on the second level of MIPS complex code are considered. For instance, the MIPS class 'translation complexes' (500) contains the subclasses 'mitochondrial ribosome' (500.60), the 'cytoplasmic ribosome' (500.40) and a number of other subclasses related to translation-related complexes; we only considered pairs among proteins in those subclasses (500.*) as positives. Overall, this yielded a filtered set of 8250 protein pairs that are within the same complex.

A negative gold standard is harder to define, but essential for successful training. There is no direct information about which proteins do not interact. However, protein localization data provide indirect information if we assume that proteins in different compartments do not interact. A list of ~2.7 million protein pairs in different compartments are compiled from the current yeast localization data in which proteins are attributed to one of five compartments as has been done previously (Drawid and Gerstein, 2000). These compartments are the nucleus (N), mitochondria (M), cytoplasm (C), membrane (T for transmembrane), and secretory pathway (E for endoplasmic reticulum or extracellular).

5.3 Machine Learning on Protein–Protein Interactions

A wide spectrum of supervised methods can be applied to integrate genomic features in order to predict protein–protein interactions (see Chapter 12 for a revisit). Among them, machine-learning approaches, including simple unions and intersections of datasets, neural networks, decision trees, support-vector machines and Bayesian networks have been successfully applied to this goal. Below we try to elaborate basic concepts in machine learning, and provide a basic tutorial on how to employ decision trees and Bayesian networks in protein–protein interaction analysis.

According to Merriam-Webster's *Collegiate Dictionary*, learning is a process in which people 'gain knowledge or understanding of or skill in by study, instruction, or experience'. The key idea of 'learning' is to perform better based on past experience. Ever since computers were invented, people have tried to program them to learn (i.e. machine learning). Precisely, 'a computer program is said to *learn* from experience *E* with respect to some class of tasks *T* and performance measure *P*, if its performance at tasks in *T*, as measured by *P*, improves with experience *E*' (Mitchell, 1997). For example, researchers have used computer programs to recognize tumours based on biopsy results:

- task *T*, to determine whether an examinee has cancer or not
- performance measure *P*, percentage of correct predictions
- training experience *E*, biopsy results from cancer patients and normal people.

Let *X* and *Y* denote the sets of possible inputs and outputs. The learning algorithm needs to find the (approximate) target function *V* that takes each input $x_i \in X$ and gives the corresponding prediction $y_i \in Y$, i.e. output. If *Y* is a subset of the real numbers, we have a regression problem. Otherwise, we have a classification problem (binary or multiclass). In this case, the algorithm will determine whether a patient has cancer based on his/her biopsy result, which is a binary classification problem. There are, of course, other learning problems (e.g. reinforcement learning). Here, we are mainly interested in classification problems.

Why do we need machine learning? First, for many complicated problems, there is no known method to compute the accurate output from a set of inputs. Second, for other problems, computation according to known exact methods may be too expensive. In both cases, good approximate methods with reasonable amounts of computational demand are desired. Machine learning is a field in which computer algorithms are developed to learn approximate methods for solving different problems. Obviously, machine learning is a multidisciplinary field, including computer science, mathematics, statistics and so on.

Supervised learning versus unsupervised learning

Learning algorithms are usually divided into two categories: supervised and unsupervised. In supervised learning, a set of input/output example pairs is given, which is called the training set. The algorithms learn the approximate target function based on the training set. Once a new case comes in, the algorithms will calculate the output value based on the target function learned from the training set. By contrast, in unsupervised learning, a set of input values are provided, without the corresponding output. The learning task is to gain understanding of the process that generates input distribution. In this section, we will focus our discussion on supervised learning algorithms.

Decision trees

Decision tree learning is one of the most widely used algorithms to search for the best discrete-valued hypothesis (h) within H. Figure 5.1 illustrates a decision tree for the protein–protein interaction classification. Only the yeast protein pairs without missing values in genomic features and in gold-standard sets are considered. The decision tree tries to predict protein–protein interactions based on three genomic features using the ID3. S is a set of examples, in this case a collection of the protein pairs. E stands for entropy and G stands for information gain calculated according to the formula (5.1). Each diamond node is one attribute or genomic feature. This decision tree is constructed from the genomic features and gold-standard interactions described in Section 5.2. The learned decision tree classifies a new instance by going down the tree from the root to a certain leaf node. Each leaf node provides a classification for all instances within it. A test of a specific attribute is performed at each node, and each branch descending from that node corresponds to one of the possible values for this attribute.

The basic idea behind decision tree learning is to determine which attribute is the best classifier at a certain node to split the training examples. Many algorithms have

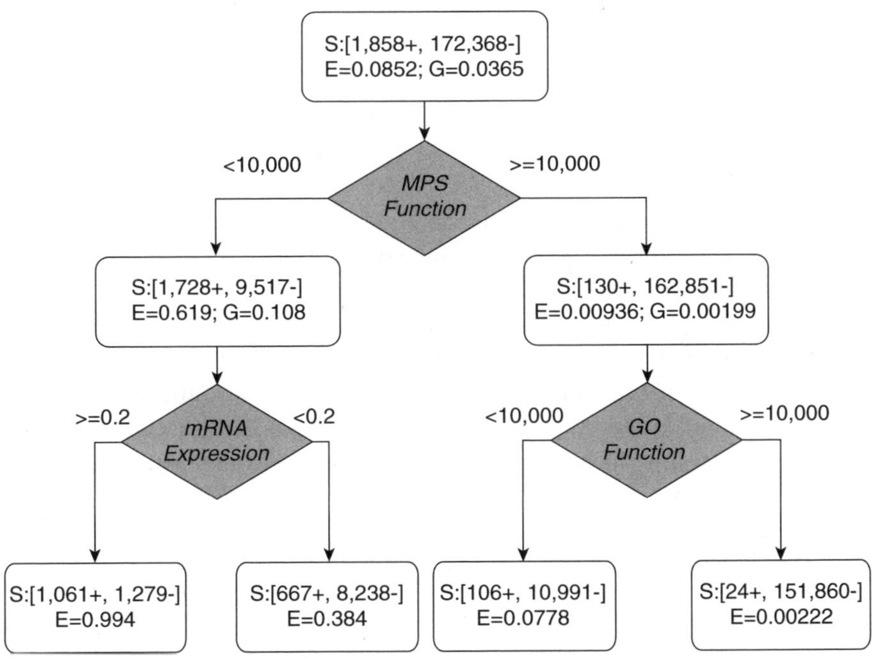

Figure 5.1 A typical decision tree

been developed to solve this problem, such as ID3, C4.5, ASSITANT and CART. Here, we focus on the ID3 algorithm (Quinlan, 1986).

In order to determine the best classifier, ID3 uses a statistical property, named *information gain*, to measure how well a given attribute separates the training examples with respect to the target classification. To define information gain, we need to introduce the concept of *entropy (E)* in information theory. Entropy is used to measure the impurity of a set of examples, which is calculated by the formula

$$E(S) \equiv -p_+ \log_2 p_+ - p_- \log_2 p_- \tag{5.1}$$

where S is a set of examples (positives and negatives) regarding some target concept, p_+ is the proportion of positives in S and p_- is the proportion of negatives in S. There are 1858 positives and 172368 negatives in the example shown in Figure 5.1. Therefore, the entropy is

$$E([1858_+, 172368_-]) = -\frac{1858}{174226}\log_2\left(\frac{1858}{174226}\right) - \frac{172368}{174226}\log_2\left(\frac{172368}{174226}\right) = 0.0852.$$

More generally, if the target concept can take on n different values (i.e. n different classes), the entropy of S relative to this n-wise classification is defined as

$$E(S) \equiv \sum_{i=1}^{n} -p_i \log_2 p_i \tag{5.2}$$

where p_i is the proportion of S belonging to class i.

Having defined entropy, we can now define information gain (G):

$$G(S, A) \equiv E(S) - \sum_{v \in \text{values}(A)} \frac{|S_v|}{|S|} E(S_v) \tag{5.3}$$

where A is an attribute associated with each instance. Here value (A) is the set of all possible values that A could take on, v is an element of values (A), and S_v is the set of instances whose value of A is v. Clearly, S_v is a subset of S.

The information gain measures the deduction of the impurity (entropy) of the training examples with respect to the target concept if they are split by a certain attribute. Therefore, the higher the information gain of an attribute, the better it classifies the examples. As a result, ID3 uses the value of the information gain to choose the best classifier at each node. For the training examples in Figure 5.1, the information gain values for all three attributes are

$$\begin{aligned} G(S, \text{MIPS function}) &= 0.0365 \\ G(S, \text{GO function}) &= 0.0216 \\ G(S, \text{mRNA expression}) &= 0.0088 \end{aligned}$$

The attribute 'MIPS function' has the highest value. Therefore, it is the root node in Figure 5.1. The same procedure is iterated for the child nodes, then the child nodes of these nodes, and so on. Each attribute can only be used once along each path. A path is terminated if either of the following two conditions is met: (1) all elements of the leaf node belong to the same class with respect to the target concept; (2) every attribute has appeared in the path. The whole tree is completed if all paths have been terminated. One complexity is that ID3 can only handle nominal attributes. If there are attributes with continuous values, such attributes could be used twice with different cut-offs along the same path.

Naïve Bayes classifier

Besides decision tree learning, another commonly used method is naïve Bayes learning, often called the naïve Bayes classifier. It is also applied to the kind of data in which each instance is associated with a set of nominal attributes. The naïve Bayes classifier assigns the most probable target value to the instance with the attribute values $\langle f_1, f_2, \ldots, f_n \rangle$:

$$h = \arg\max_{h_j \in H} P(h_j \mid f_1, f_2, \ldots, f_n) \tag{5.4}$$

Using the Bayes theorem, the formula can be rewritten as

$$h = \arg\max_{h_j \in H} \frac{P(f_1, f_2, \ldots, f_n \mid h_j) P(h_j)}{P(f_1, f_2, \ldots, f_n)} = \arg\max_{h_j \in H} P(f_1, f_2, \ldots, f_n \mid h_j) P(h_j) \tag{5.5}$$

The most important assumption in naïve Bayes learning is that all attributes are conditionally independent of each other with respect to every hypothesis h_j. Therefore, the joint probability of all attributes is the product of the individual probabilities:

$$h = \arg\max_{h_j \in H} P(h_j) \prod_{i=1}^{n} P(f_i / h_j) \tag{5.6}$$

The Bayesian approach has been widely used in biological problems. Jansen *et al.* (2003) described an approach using Bayesian networks to predict protein–protein interactions. A pair of proteins that interact is defined as 'positive'. Given some positives among the total number of protein pairs, the 'prior' odds of finding one are

$$O_{\text{prior}} = \frac{P(\text{pos})}{P(\text{neg})} = \frac{P(\text{pos})}{1 - P(\text{pos})} \tag{5.7}$$

In contrast, 'posterior' odds are the chance of finding a positive after considering N features with values $f_1 \cdots f_n$:

$$O_{\text{post}} = \frac{P(\text{pos} \mid f_1 \cdots f_n)}{P(\text{neg} \mid f_1 \cdots f_n)} \tag{5.8}$$

(The terms 'prior' and 'posterior' refer to the situation before and after knowing the information in the N features.) The likelihood ratio L is defined as

$$L(f_1 \cdots f_N) = \frac{P(f_1 \cdots f_n \mid \text{pos})}{P(f_1 \cdots f_n \mid \text{neg})} \tag{5.9}$$

It relates prior and posterior odds according to Bayes' rule, $O_{\text{post}} = L(f_1 \cdots f_n)O_{\text{prior}}$. In the special case where the N features are conditionally independent (i.e., they provide uncorrelated evidence), the Bayesian network is a so-called 'naïve' network, and L can be simplified to

$$L(f_1 \cdots f_n) = \prod_{i=1}^{N} L(f_i) = \prod_{i=1}^{N} \frac{P(f_i \mid \text{pos})}{P(f_i \mid \text{neg})} \tag{5.10}$$

L can be computed from contingency tables relating positive and negative examples with the N features (by binning the feature values $f_1 \cdots f_n$ into discrete intervals). Simply put, consider a genomic feature f expressed in binary terms (i.e. 'present' or 'absent'). The likelihood ratio $L(f)$ is then defined as the fraction of gold-standard positives having feature f divided by the fraction of negatives having f. For two features f_1 and f_2 with uncorrelated evidence, the likelihood ratio of the combined evidence is simply the product $L(f_1, f_2) = L(f_1)L(f_2)$. A protein pair is predicted as positive if its combined likelihood ratio exceeds a particular cut-off ($L > L_{\text{cutoff}}$) (negative otherwise). The likelihood ratios are computed for all possible protein pairs in the yeast genome. Based on previous estimates, we think that 30 000 positives is a conservative lower bound for the number of positives (i.e. pairs of proteins that are in the same complex). Given that there are approximately 18 million protein pairs in total, the prior odds would then be about 1 in 600. With $L > L_{\text{cutoff}} = 600$ we would thus achieve $O_{\text{post}} > 1$.

Cross-validation with the reference datasets shows that naïve Bayesian integration of multiple genomic sources leads to an increase in sensitivity over the high-throughput data sets it combined for comparable true positive (TP)/false positive (FP) ratios. ('Sensitivity' measures coverage and is defined as TP over the number of gold-standard positives, P.) This means that the Bayesian approach can predict, at comparable error levels, more complex interactions *de novo* than are present in the high-throughput experimental interaction datasets. The predicted dataset (PIP) was also compared with a voting procedure where each of the four genomic features contributes an additive vote towards positive classification. The results showed that the Bayesian network achieved greater sensitivity for comparable TP/FP ratios (Jansen *et al.*, 2003).

5.4 The Missing Value Problem

The naïve Bayes procedure presented above lends itself naturally to the addition of more features, possibly further improving results. As more sparse data are incorporated, however, the missing value problem becomes severe: the number of protein pairs with complete feature data decreases, and the number of possible missing feature patterns increases exponentially with the number of features.

Some classification methods, such as decision trees, can handle missing values in an automated fashion. Most other classification methods, however, require a full data matrix as input. It is therefore necessary to first fill in missing data with plausible values, a process called imputation. It is important for the imputed values to be consistent with the observed data and preserve the overall statistical characteristics of the feature table, for example the correlations between features. Below we will discuss different mechanisms of missing values, followed by a brief description of several representative methods for missing data imputation.

Mechanisms of missing values

There are two broad categories of missing value mechanisms (Little and Rubin, 1987). The first category is called Missing at Random (MAR). Here, the probability of a feature being missing can be determined entirely by the observed data. A special case is Missing Completely at Random (MCAR), where the patterns of missing data are completely random. Since most missing value analysis methods assume MAR, it is important to assess whether or not this assumption holds for a given set of missing values. In general, missing values are approximately MAR for pair protein features. In one example, synthetic lethal features (Tong *et al.*, 2004) for some protein pairs are missing because the experiments were not performed due to limited resources. In another example, structure-based features, such as multimeric threading scores (Lu, Lu and Skolnick, 2002), can only be computed for proteins with a solved structural homologue, and will become missing otherwise. These missing values can all be approximated as MAR.

In certain cases, however, the probability of a feature being missing is directly related to the missing feature itself and cannot be determined entirely by the observed data. In this case, the missing data are not MAR. For example, consider a situation where a protein pair feature is computed for all protein pairs, and only the best scores (indicative of protein interaction) are kept and the rest of the scores are thrown away and thus become missing. Here the missing data are no longer MAR, and they cannot be treated in the same way as missing due to incomplete coverage. On the other hand, simply recording all the scores, no matter good or bad, will solve this problem. Most methods for missing value analysis assume MAR; we briefly summarize a few representative methods below. Let us suppose that instance x has a missing attribute A.

Mean substitution, *k* nearest neighbours and regression imputation

In mean substitution, the missing attribute A in instance x is replaced with the most common value of attribute A in the whole data matrix, or in the subset of instances that x belongs to. A major disadvantage of this simple method is that the correlations between features are not preserved.

In k nearest neighbours (KNN), the distance between instance x and all instances with complete attributes are calculated, based on the observed attributes in x. The k nearest neighbours are identified, and the missing attribute A in instance x is replaced with the weighted average of attribute A in these k nearest neighbours.

In regression imputation, attribute A is regressed against all other attributes based on all instances with complete attributes. Afterwards, the missing attribute A in instance x is replaced with its predicted value based on the regression model.

SVD imputation, expectation maximization and Bayesian multiple imputation

In SVD imputation, all missing values in the data matrix are filled with initial guesses. Using singular value decomposition (SVD), all instances are then projected to a low-dimensional feature space spanned by the first k principal components. To update the missing attribute A in instance x, instance x is regressed against the first k principal components using observed attributes in x, and the resulting regression model is used to predict the missing attribute A. After all missing values in the data matrix are updated, another round of SVD is performed and the whole process is iterated until convergence. In the case of imputing missing values in DNA microarrays, SVD imputation is found to perform better than mean substitution, but worse than KNN (Troyanskaya *et al.*, 2001).

The expectation maximization (EM) algorithm is a popular method for finding maximum-likelihood estimates for parametric models with missing data. Here, all instances are assumed to be independently and identically distributed based on a parametric model (for example, a normal distribution) with unknown parameters θ. The EM algorithm makes point estimates for missing data and parameters θ in the following way. First, parameters θ are initialized. In the E-step, the missing attribute A in instance x is replaced by its expected value calculated from the estimates for θ

and observed attributes in x. In the M-step, parameters θ are estimated that maximize the complete-data likelihood. This process is iterated till convergence.

The EM algorithm only provides point estimates for missing data and parameters θ. In Bayesian multiple imputation, the posterior probability distributions for the missing values and parameters θ can be directly simulated using Markov chain Monte Carlo methods (MCMC), thereby taking into account the uncertainties associated with the missing values and parameters θ. Details on the EM algorithm and Bayesian multiple imputation can be found in the work of Schafer (1997).

5.5 Network Analysis of Protein Interactions

The recent explosion of genome-scale protein interaction screens has made it possible to construct a map of the interactions within a cell. These interactions form an intricate network. A crucial challenge as these data continue to flood in is how to reduce this complex tangle of interactions into the key components and interconnections that control a biological process. For example, in developing a drug to attack a disease, a molecular target that is a central player in the disease is required. The target must be as specific as possible to reduce unanticipated side-effects.

Below we will introduce two approaches that are particularly important to analyse protein interaction networks: topological analysis of networks and modular analysis of networks.

Topological analysis of networks

In addition to protein networks, complex networks are also found in the structure of a number of wide-ranging systems, such as the internet, power grids, the ecological food web and scientific collaborations. Despite the disparate nature of the various systems, it has already been demonstrated that all these networks share common features in terms of topology (Barabasi and Albert, 1999). In this sense, networks and topological analysis can provide the framework for describing biological systems in a format that is more transferable and accessible to a broader scientific audience.

As mentioned previously, the topological analysis of networks is a means of gaining quantitative insight into their organization at the most basic level. Of the many methods in topological statistics, four are particularly pertinent to the analysis of networks. They are average degree (K), clustering coefficient (C), characteristic path length (L) and diameter (D). Chapter 8 will give formal definitions to these methods.

Earlier analyses of complex networks were based on the theory of classical random networks. The idea was introduced by Erdos and Renyi (1959). The theory assumes that any two given nodes in a network are connected at random, with probability p, and the degrees of the nodes follow a Poisson distribution. This means that there is a strong peak at the average degree, K. Most random networks are of a highly homogenous nature, that is, most nodes have the same number of links, $k(\mathrm{i}) = K$, where $k(i)$ is the ith node. The chance of encountering nodes with k links decreases exponentially for large values of k, i.e., $P(k) = \mathrm{e}^{-k}$. This shows that it is highly unlikely to encounter nodes of a degree that is significantly higher than the average.

Recently, theories other than the classical random network theory were proposed. One such attempt is the 'scale-free' model by Barabasi and Albert (1999) to explain the heterogeneous nature of some complex networks. In their 'scale-free' model, the degree distribution of networks is assumed to follow a power-law relationship ($P(k) = k^{-r}$), rather than the Poisson distribution assumed under earlier classical random network theory. One advantage of having such an assumption is that most of the nodes within such networks are highly connected via hubs, with very few links between them. This attribute makes the model particularly applicable to complex biological networks such as those involving protein–protein interactions. Many aspect of genomic biology have such a scale-free structure (Qian, Luscombe and Gerstein, 2001; Rzhetsky and Gomez, 2001; Koonin, Wolf and Karev, 2002).

In a concurrent effort by Watts and Strogatz (1998), it is found that many networks can be attributed with a 'small-world' property, which means that they are both highly clustered in nature and contain small characteristic path lengths (i.e. large values of C, and small values of L).

Finally, the analysis of complex networks can be further divided into two broad categories: that is, undirected versus directed. In the former, there is a commutative property: the statement 'node A is linked to node B' is the exact equivalent to the statement 'node B is linked to node A'. In contrast, in a directed network, the edges have a defined direction, and thus the clustering coefficient is not applicable for directed networks.

There are many complex networks in biology that can be analysed using graph-topological tools. Recent advances in large-scale experiments have generated a great variety of genome-wide interaction networks, especially for *S. cerevisiae*. Moreover, there exist a number of databases (e.g. MIPS, BIND, DIP) that provide manually curated interactions for the yeast organism. Beyond experimentally derived protein–protein interactions, there are also predicted interactions (Valencia and Pazos, 2002; Lu *et al.*, 2003), literature-derived interactions (Friedman *et al.*, 2001) and regulatory interactions (Lee *et al.*, 2002). All of these networks are amenable to topological analysis.

In order to facilitate the topological analysis of interaction networks, we constructed a web tool, TopNet, to perform automatic comparisons. It is available at: http://topnet.gersteinlab.org/.

TopNet takes an arbitrary undirected network and a group of node classes as an input to create sub-networks. Then it computes all four topological statistics mentioned above and draws a power-law degree distribution for each sub-network. The results of these calculations are plotted in the same format for each statistic to facilitate direct comparison. TopNet also enables the user to explore complex networks by sections. For example, all neighbours of a certain node can be shown on a simple graph. Alternatively, the user can select two nodes and request that all paths not exceeding some specified length be displayed as an independent graph. Figure 5.2 shows a snapshot of TopNet.

Clearly, the great variety and complexity of biological networks present a wealth of interesting problems and challenges for the application of topological analysis, which would lead to better understanding of many aspects of modern biology.

Modelling networks as biological modules

Cellular interaction networks are composed of modules. Biological modules are conserved groups of proteins and other molecules that are responsible for a common structure and function. Experimental study of the signalling network of the budding yeast, *S. cerevisiae*, sparked the conception of modular signalling networks. These well studied pathways are an ideal proving ground for computational study of modularity in biological networks.

Yeast signalling pathways are composed of five distinct mitogen-activated protein kinase (MAP kinase) pathways. The yeast MAP kinase pathways are composed of two modules, an upstream sensing component that is responsible for detecting signals in the environment and a downstream kinase module that amplifies and propagates the signal while maintaining its specificity. MAP kinase pathways and their modules are highly conserved in eukaryotes. These modules are key determinants of the specificity of signalling. The filamentation pathway is one of the least understood of the pathways. Under certain conditions yeast cells undergo a morphological transition from a round form to a filamentous invasive form. This is accompanied by an altered cell cycle and bipolar budding pattern. Little is known about how the signal that mediates the altered cell cycle is transmitted from the MAP kinase module.

Modelling a biological system as a network of modules reduces the complexity of the network and pinpoints the crucial interconnections between modules. These connections can be reprogrammed in evolution, or the laboratory to create new biological responses. For example a drug that targeted the connection between a growth factor detection module and a signal amplification module could stop the progression of cancer by ablating the link responsible for transmitting the aberrant growth signal. Rives and Galitski (2003) show an example of a network clustering method that has been successfully applied to biological networks to model their modular structure. The goal of the clustering method is to define a similarity metric between each possible pair of proteins in the network. This similarity metric is a function that can range from -1 to 1, where a higher score represents a greater prediction that the two proteins are in the same module. Proteins can then be clustered based on their similarity scores.

This network clustering method was applied to the yeast filamentation response network to model its modular structure. Modelling a biological system as a network

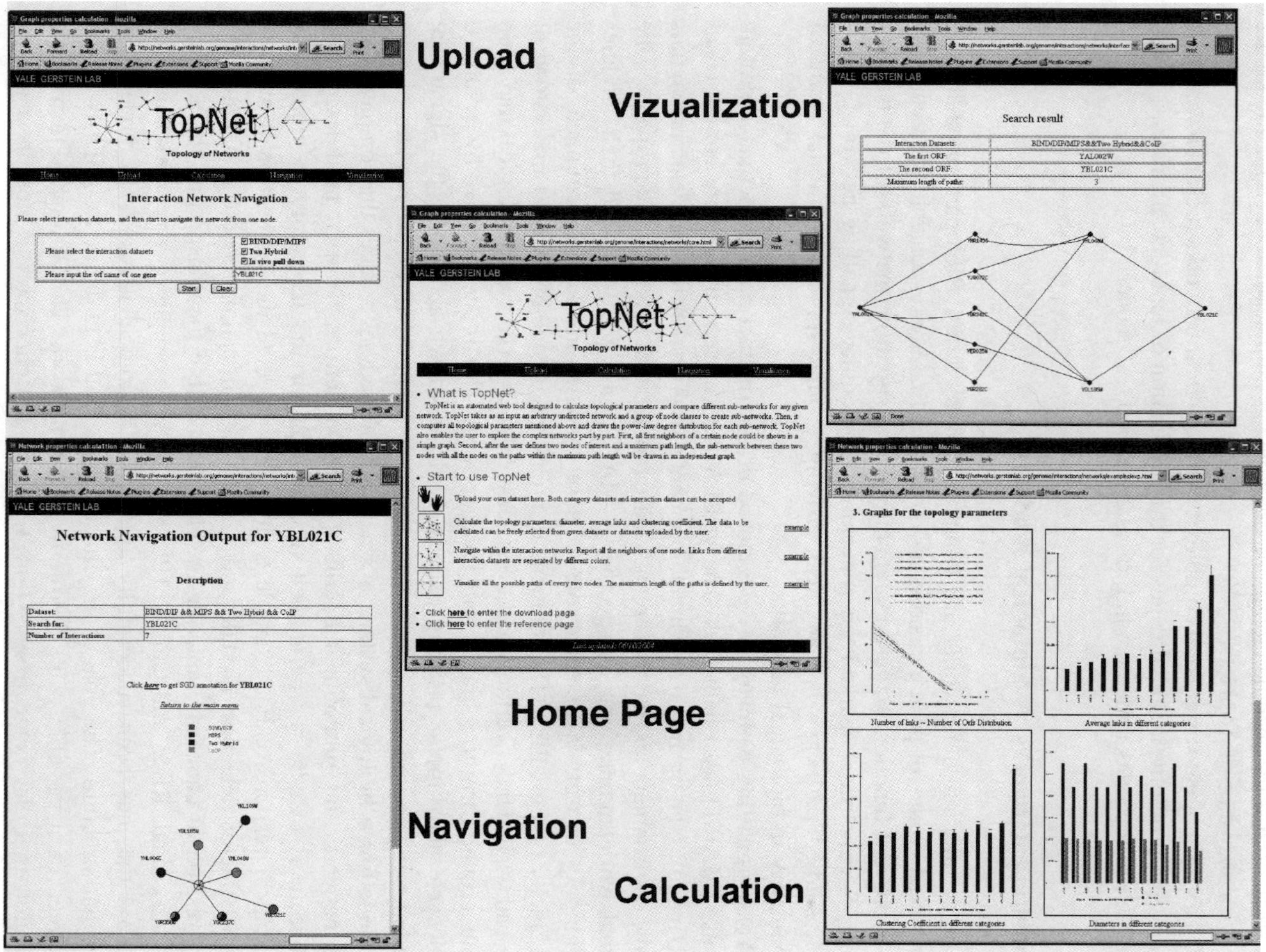

Figure 5.2 A snapshot of TopNet. TopNet consists of four major parts: Upload, Navigation, Calculation and Visualization

of modules identifies key proteins and interconnections. Two important features are hubs and intermodule connections. Modules tend to have one or a few proteins that are highly connected within the module. These hubs are essential to the functions of their modules. While there are many interactions within modules, there is a relative paucity between modules. The biological functions of proteins that appear as connections between modules suggest they are crucial points of information flow constriction and cross-talk.

Protein interaction networks reflect the modular structure of biological systems. They have a high clustering coefficient and a low frequency of direct connections between high-connectivity nodes. Members of biological modules have a high frequency of interactions with other members of their module and a paucity of interactions with members of other modules. Modules form clusters in the interaction network that can be identified using network-clustering methods. Biological modules can be identified in complex protein interaction networks. They can be used to reduce the complexity of a network by moving up in the biological hierarchy. An induced graph is a graph in which some nodes are collapsed together into a single node that shares all of their interactions. It allows a complex interaction network to be reduced to the interactions between emergent modular components. This preserves important information while reducing the complexity. Modular modelling opens many avenues for the investigation of biological systems. As genome-scale interaction data continue to be produced, methods are required which can identify the important interactions and proteins. Computational network clustering approaches can be used to identify these essential proteins and crucial points of cross talk. Furthermore they can be used to generate testable biological hypotheses.

5.6 Discussion

Among the machine-learning approaches that could be applied to predicting interactions described above, Bayesian networks have clear advantages: (1) they allow for combining highly dissimilar types of data (i.e. numerical and categorical), converting them to a common probabilistic framework, without unnecessary simplification; (2) they readily accommodate missing data and (3) they naturally weight each information source according to its reliability. In contrast to 'black-box' predictors, Bayesian networks are readily interpretable, as they represent conditional probability relationships among information sources.

In a naïve Bayesian network, the assumption is that the different sources of evidence (i.e. our datasets with information about protein interactions) are conditionally independent. Conditional independence means that the information in the N datasets is independent given that a protein pair is either positive or negative. From a computational standpoint, the naïve Bayesian network is easier to compute than the fully connected network. As we add more features, we will find more sources of evidence that are strongly correlated. This issue can be addressed in two ways: (1) we will use a fully connected network or subnetwork to handle the correlated features; (2) we will use principal component analysis (PCA), in which the first principal component of the vector of the two correlations will be used as one independent source of evidence for the protein interaction prediction. For example, in analysing expression correlations, we found that two of the main datasets were strongly correlated; however, using the first component of the PCA removed this issue.

Another challenge in extending our naïve Bayesian integration to incorporate additional genomic features is the missing value problem (see Section 5.4).

Determination on protein interactions is the initial step and cornerstone towards mapping molecular interaction networks. Three challenges in the network analysis remain: (1) 3D view of interaction networks in a cell; (2) dynamics and context-dependent nature of interaction networks; (3) quantitative measures of networks. Molecular interaction networks lay the foundation for analysis of the cell in systems biology. With combined experimental, computational and theoretical efforts, a complete mapping of interaction networks, and ultimately a rational understanding of cellular behaviour, will become reality.

References

Ashburner, M., Ball, C. A., Blake, J. A. *et al.* (2000) Gene ontology: tool for the unification of biology. The Gene Ontology Consortium. *Nat Genet*, **25**, 25–29.

Bader, G. D., Donaldson, I., Wolting, C. *et al.* (2001) BIND – the Biomolecular Interaction Network Database. *Nucleic Acids Res*, **29**, 242–245.

Bader, G. D. and Hogue, C. W. (2002) Analyzing yeast protein–protein interaction data obtained from different sources. *Nat Biotechnol*, **20**, 991–997.

Barabasi, A. L. and Albert, R. (1999) Emergence of scaling in random networks. *Science*, **286**, 509–512.

Cho, R. J., Campbell, M. J., Winzeler, E. A. *et al.* (1998) A genome-wide transcriptional analysis of the mitotic cell cycle. *Mol Cell*, **2**, 65–73.

Drawid, A. and Gerstein, M. (2000) A Bayesian system integrating expression data with sequence patterns for localizing proteins: comprehensive application to the yeast genome. *J Mol Biol*, **301**, 1059–1075.

Edwards, A. M., Kus, B., Jansen, R. *et al.* (2002) Bridging structural biology and genomics: assessing protein interaction data with known complexes. *Trends Genet*, **18**, 529–536.

Eisenberg, D., Marcotte, E. M., Xenarios, I. and Yeates, T. O. (2000) Protein function in the post-genomic era. *Nature*, **405**, 823–826.

Erdos, P. and Renyi, A. (1959) On random graphs I. *Publ Math*, **6**, 290–297.

Friedman, C., Kra, P., Yu, H., Krauthammer, M. and Rzhetsky, A. (2001) GENIES: a natural-language processing system for the extraction of molecular pathways from journal articles. *Bioinformatics*, **17** (Suppl. 1), S74–S82.

Ge, H., Liu, Z., Church, G. M. and Vidal, M. (2001) Correlation between transcriptome and interactome mapping data from *Saccharomyces cerevisiae*. *Nat Genet*, **29**, 482–486.

Hartwell, L. H., Hopfield, J. J., Leibler, S. and Murray, A. W. (1999) From molecular to modular cell biology. *Nature*, **402**, C47–C52.

Hughes, T. R., Marton, M. J., Jones, A. R. *et al.* (2000) Functional discovery via a compendium of expression profiles. *Cell*, **102**, 109–126.

Jansen, R., Lan, N., Qian, J. and Gerstein, M. (2002) Integration of genomic datasets to predict protein complexes in yeast. *J Struct Funct Genomics*, **2**, 71–81.

Jansen, R., Yu, H., Greenbaum, D. *et al.* (2003) A Bayesian networks approach for predicting protein–protein interactions from genomic data. *Science*, **302**, 449–453.

Koonin, E. V., Wolf, Y. I. and Karev, G. P. (2002) The structure of the protein universe and genome evolution. *Nature*, **420**, 218–23.

Kumar, A. and Snyder, M. (2002) Protein complexes take the bait. *Nature*, **415**, 123–124.

Lee, T. I., Rinaldi, N. J., Robert, F. *et al.* (2002) Transcriptional regulatory networks in *Sacchoromyces cerevisiae. Science*, **298**, 799–804.

Little, R. J. A. and Rubin, D. B. (1987) *Statistical Analysis with Missing Data.* Wiley, New York.

Lu, L., Arakaki, A. K., Lu, H. and Skolnick, J. (2003) Multimeric threading-based prediction of protein–protein interactions on a genomic scale: application to the *Saccharomyces cerevisiae* proteome. *Genome Res*, **13**, 1146–1154.

Lu, L., Lu, H. and Skolnick, J. (2002) MULTIPROSPECTOR: an algorithm for the prediction of protein–protein interactions by multimeric threading. *Proteins*, **49**, 350–364.

Marcotte, E. M., Pellegrini, M., Thompson, M. J., Yeates, T. O. and Eisenberg, D. (1999) A combined algorithm for genome-wide prediction of protein function. *Nature*, **402**, 83–86.

Mewes, H. W., Frishman, D., Guldener, U. *et al.* (2002) MIPS: a database for genomes and protein sequences. *Nucleic Acids Res*, **30**, 31–34.

Mitchell, T. M. (1997) *Machine Learning*. McGraw-Hill, New York.

Qian, J., Luscombe, N. M. and Gerstein, M. (2001) Protein family and fold occurrence in genomes: power-law behaviour and evolutionary model. *J Mol Biol*, **313**, 673–681.

Quinlan, J. R. (1986) Induction of decision trees. *Machine Learning*, **1**, 81–106.

Rives, A. W. and Galitski, T. (2003) Modular organization of cellular networks. *Proc Natl Acad Sci USA*, **100**, 1128–1133.

Rzhetsky, A. and Gomez, S. M. (2001) Birth of scale-free molecular networks and the number of distinct DNA and protein domains per genome. *Bioinformatics*, **17**, 988–996.

Schafer, J. L. (1997) *Analysis of Incomplete Multivariate Data.* Chapman and Hall, London.

Schwikowski, B., Uetz, P. and Fields, S. (2000) A network of protein–protein interactions in yeast. *Nat Biotechnol*, **18**, 1257–1261.

Tong, A. H., Lesage, G., Bader, G. D. *et al.* (2004) Global mapping of the yeast genetic interaction network. *Science*, **303**, 808–813.

Troyanskaya, O., Cantor, M., Sherlock, G. *et al.* (2001) Missing value estimation methods for DNA microarrays. *Bioinformatics*, **17**, 520–525.

Valencia, A. and Pazos, F. (2002) Computational methods for the prediction of protein interactions. *Curr Opin Struct Biol*, **12**, 368–373.

Vazquez, A., Flammini, A., Maritan, A. and Vespignani, A. (2003) Global protein function prediction from protein–protein interaction networks. *Nat Biotechnol*, **21**, 697–700.

von Mering, C., Krause, R., Snel, B. *et al.* (2002) Comparative assessment of large-scale data sets of protein–protein interactions. *Nature*, **417**, 399–403.

Watts, D. J. and Strogatz, S. H. (1998) Collective dynamics of 'small-world' networks. *Nature*, **393**, 440–442.

Xenarios, I., Salwinski, L., Duan, X. J. *et al.* (2002) DIP, the Database of Interacting Proteins: a research tool for studying cellular networks of protein interactions *Nucleic Acids Res*, **30**, 303–305.

Yu, H., Zhu, X., Greenbaum, D., Karro, J. and Gerstein, M. (2004) TopNet: a tool for comparing biological sub-networks, correlating protein properties with topological statistics. *Nucleic Acids Res*, **32**, 328–337.

6
Integration of Genomic and Phenotypic Data

Amanda Clare

Abstract

Phenotypic data is perhaps the least analysed form of bioinformatics data, but new laboratory techniques are providing more opportunities for genome-scale phenotype observations. The combination of phenotype data together with other sources of bioinformatics data and with whole system and network analysis will open new possibilities for understanding and modelling the emergent and complex properties of the cell.

Keywords

phenotype, genomics, integration, data mining, systems biology

6.1 Phenotype

The phenotype of an organism is its observable characteristics. This can comprise a whole range of properties, from the obviously visible whole-organism characteristics such as flower colour, leaf length and wing size, through drug resistance, to measurements of metabolic flux.

Phenotype data is the oldest form of data available to biologists, preceding the current wave of 'omics' by almost 150 years. As phenotype data is observable data, this form of data has been available for as long as anyone had the insight and patience to observe it. In 1866 Gregor Mendel published his famous work on the analysis of pea plant phenotypes. The mathematician and biologist monk carefully observed seven traits by growing and cross-breeding thousands of pea plants, the traits

Data Analysis and Visualization in Genomics and Proteomics Edited by Francisco Azuaje and Joaquin Dopazo

including whether the peas were smooth or wrinkled, the length of the stem, and the colour of the unripe pods. These observations gave birth to the science of genetics and the realization that different hereditary factors (genes) controlled different aspects of the plant.

The scientific community took 34 years to catch up with and accept this work, which was ahead of its time. Mendel had discovered dominant and recessive traits along with the idea of alleles, and had paved the way for scientists in the early 1900s to experiment by crossing organisms and looking for changes or mutants.

Now, in the computational and genomic world, we integrate phenotype data with a wide range of other omic data in order to better understand biology. In recent times computational biology and bioinformatics has mainly concentrated on the more reductionist omics: on genomics, transcriptomics and proteomics. Now we are beginning to see a shift toward whole-organism understanding, and higher-level omic data, of which phenomics can occupy a variety of positions, from total physical appearance of an organism to specific traits. However, phenotype is still perhaps one of the most underexploited sources of information in bioinformatics. This is due in part to the recent enthusiasm for analysis of the new genomic data, and to the relative lack of large-scale data when compared to the other omics. Its usage is now growing with the introduction of new laboratory techniques for measurement. Phenotypic data generation is currently undergoing the same technology revolution as the rest of whole-genome biology, and is now available on a genome-wide scale for many of biology's model organisms.

The phenotype of an organism is its observable characteristics, but what is observable depends on our current technology. Hellerstein (2004) reviews recent methods for observing metabolic pathway fluxes in order to better define the phenotype of an organism. Genes and proteins give a static picture of an organism, and high-level observable characteristics such as growth measurements of mutants give a blunt picture of the phenotype, but metabolic fluxes can give a much more detailed picture of the properties of the organism. He states

> What we call a phenotype depends completely on the tools that we have for seeing. If we lack tools for measuring biochemical fluxes (i.e. biochemical 'motion detectors'), we cannot see or control phenotype in a complete way.

The phenotype is influenced both by an organism's genotype and by its interaction with the environment. This relationship can be exploited to make models and predictions for gene function and organism behaviour. Classical genetics, or *forward genetics*, is the process of starting with a phenotype, for example 'yellow coloured pea', and then searching for the gene(s) responsible for this phenotype. When phenotypes are quantitative in nature, such as 'flowering time', this is known as QTL analysis (quantitative trait loci). The converse is *reverse genetics*, which begins with knowledge of the gene of interest, and then tries to look at the phenotype of the organism with this gene disrupted.

The following sections give an introduction to the different ways that phenotype data has been used together with other sources of bioinformatics data in computational biology recently. First we consider forward genetics and QTL analysis, then reverse genetics, and then look at how other sources of data are being used to predict phenotype. Section 6.5 describes how phenotype data can be used as part of systems biology, and Section 6.6 describes the current state of integration of phenotype data with bioinformatics databases. Finally, we draw some conclusions.

6.2 Forward Genetics and QTL Analysis

Forward genetics makes use of knowledge of the phenotype in order to find the gene responsible. In the past, in Mendel's era, this was the only knowledge that could be used, but now we have a wide variety of other sources of data to assist.

In the genomic era, we can use knowledge of a phenotype that exists for many organisms, together with detailed knowledge of the genomes of those organisms in order to pinpoint the common genes. An example of this is the work of Jim *et al.* (2004), who integrate phenotypic information gathered from the literature with phylogenetic data in order to identify individual genes associated with that phenotype. Given several organisms exhibiting a phenotype, they look for proteins that are conserved across these organisms, using a BLAST search. They then calculate for each protein the ratio of the fraction of genomes with phenotype f containing that protein compared with the fraction of genomes containing that protein. This gives the propensity that a particular protein is associated with phenotype f. These scores are screened for statistical significance. Phenotype assignments were derived from PubMed abstracts and research Web pages. The phenotypes they investigated were flagella, pili, thermophily and respiratory tract trophism, and many genes were identified as being associated with these phenotypes. For example, several of the genes they found to be associated with thermophily were annotated as being involved in DNA repair or ferredoxin oxidoreductases, both of which are known to be important to thermophiles and their ability to survive at high temperatures.

Identification of genes involved in human disease phenotypes is a very important goal. New large-scale medical–genetic databases, often known as *biobanks*, collect data about human populations and their phenotypes in order to find the genetic factors responsible. The UK Biobank[1] will be a database created for the study of the contributions played by environment and genetics toward human health and disease. From 2005, half a million people will be followed in a study spanning ten years, and records will be made of their diet and lifestyle habits, medical history and biology. The data will enable researchers to compare genetic and environmental influence on health phenotypes. Similar initiatives have been considered or implemented in several other countries, including Iceland, Estonia, Tonga and Singapore (Austin, Harding

[1]See: http://www.ukbiobank.ac.uk

and McElroy, 2003). As with any other project of such importance, these projects are not without their controversy, in terms of privacy, scope, cost, consent, patent issues and exploitation by pharmaceutical industries.

Many phenotypic characteristics of an organism are quantitative in nature. Examples of such characteristics (or 'traits') for the plant *Arabidopsis thaliana* include height of the plant, flowering time, seed weight and phosphate content, and it is hoped that such traits can be linked to specific genes. Phenotypic analysis of gene deletion mutants may not be enough to locate the genes, as there may be many genes involved in creating a particular trait, at several different loci along the genome. These polygenes may also mask the expression of each other (epistasis), or interact with each other.

Molecular markers, spread throughout the genome and used as a genetic map, can be used to map the traits to certain areas of the genome. QTL mapping is the process of relating the quantitative traits with the molecular markers. The use of such markers requires their map positions to be known accurately. The markers are either RFLPs (restriction fragment length polymorphisms), PCR-based markers, AFLPs (amplified fragment length polymorphisms) or SNPs (single-nucleotide polymorphisms). RFLPs are reliable, but take time to analyse; PCR-based markers are much faster to analyse, but can be either unreliable or limited in number; AFLPs are efficient, reliable and effectively unlimited (Alonso-Blanco *et al.*, 1998); but SNPs are the currently favoured type of marker, due to the ability to genotype them in a high-throughput manner, and to their availability (every 500 base pairs in *Arabidopsis* (Borevitz & Nordborg, 2003)).

QTL data comprises data about the markers (their locations in the genome) and phenotypic and genotypic data for each of the recombinant inbred lines (RILs) that have been tested. The phenotypic data gives the measurements for the trait in question (such as leaf area), and the genotypic data gives the genotype of the RIL at each marker position (whether both alleles of the marker were the same, or the marker is heterozygous or missing).

The earliest method of analysis was simply to see which markers were associated with the phenotypes. If, for a particular marker M, all individuals homozygous AA were significantly taller than those that were homozygous aa, then it could be inferred that this marker is associated with this trait. The problems with this were the following.

- The significance level must be corrected for the fact that many markers are being tested.
- Due to linkage among the genes on a chromosome any one QTL may be associated with several markers.
- The actual loci for this trait may not be at a marker, but between markers. The method cannot distinguish between strong association of a marker with a small effect and weak association of a marker with a strong effect.

Since then, other more appropriate techniques for analysis of QTL data have included interval mapping, least-squares regression, marker regression, composite and multiple QTL mapping, Bayesian models, genetic algorithms, Cockerham's model and many more. Surveys of the state of the art include those by Carlborg (2002) and Page *et al.* (2003).

The production of RILs and marker maps has increased (Loudet *et al.*, 2002; Jander *et al.*, 2002), and the prospects for the future of QTL analysis in proving associations between phenotype and the genes responsible look rosy. Glazier, Nadeau and Aitman (2002) give a summary of the stages that must be used in order to prove particular genes are responsible for a complex phenotypic trait using QTL analysis and give a discussion of several such genes in different organisms that have been found so far. Paran and Zamir (2003) look at the future of QTL analysis and conclude that, after more than 500 publications on mapping QTL in the past 10 years, the next stage will be a framework for understanding the statistical outputs of QTL analysis.

6.3 Reverse Genetics

As described above, forward genetics is the process of finding the genes that control a known phenotype, and as such is an important tool for tracking the causes of known characteristics, such as human disease. Reverse genetics provides the opposite: given our detailed knowledge of the human genome and genomes of other organisms, can we use knowledge of the genes to discover knowledge of phenotype?

In phenotypic growth experiments, specific genes are disrupted or removed from an organism to create mutant strains. These single-gene mutants, or 'knockout organisms', grown under different environmental conditions, give us a picture of each gene's individual contribution to the phenotype, and can indicate the function of the missing gene. Current technology, such as disruptive insertion, gene deletion, mutation or RNA interference, provides single-gene disruption mutants on a genome-wide scale for model organisms such as yeast (Oliver, 1996; Giaever *et al.*, 2002), *C. elegans* (Kamath *et al.*, 2003) and *A. thaliana* (Alonso *et al.*, 2003), and now even for mammalian cells (Medema, 2004; Silva *et al.*, 2004). These new and comprehensive methods provide an enormously valuable source of information and complement the lower-level genome analysis in allowing the measurements of effects to which causes are sought.

Organisms can have genetic redundancy (where more than one gene has the same role), or functional composition (where another gene can step in for a missing gene, but would not normally have played this role). This means that single-gene mutants do not always show a phenotype different from that of the wild type. Double- and even triple-gene mutants are now being generated in yeast on a large scale (Tong *et al.*, 2004) in order to identify genes whose deletion effects can be buffered by other genes. Lethal double mutants indicate genes involved in the same pathway, or functionally overlapping with each other. Tong *et al.*, compared pairs of genes shown by double mutants to interact and their Gene Ontology classes and found that over 27 per cent of interactions were between genes with a similar or identical GO annotation.

Identification of gene function is a primary aim of computational biology, and reverse genetics via knockout mutants can be an invaluable tool for this aim. Single-gene deletion phenotypic growth data have been used to make predictions for gene function (Clare and King, 2002). This work used decision trees in order to predict MIPS functional annotation for unannotated yeast genes, from the phenotype data. Phenotype growth data can still be sparse across the genome, due to experimental constraints. The techniques for producing mutants are new and improving, but have in the past been labour intensive in identifying the gene responsible and prone to residual gene activity. The phenotypes of mutants can often show no obvious differences from that of the wild type because of buffering effects from other genes. Furthermore, growth under a wide variety of conditions can be time consuming and expensive. Therefore this research used bootstrap techniques and a multi-label learning method to make better use of the sparse phenotypic data set. The decision trees produced prediction rules that could be easily comprehended. Several rules were analysed and shown to be consistent with known biology, for example that sensitivity or resistance to calcofluor white meant that the knocked-out gene was likely to be involved in cell wall biogenesis.

6.4 Prediction of Phenotype from Other Sources of Data

Both forward and reverse genetics make use of phenotype data to learn about the genetics of an organism. Conversely, phenotype can be predicted from other sources of data. Phenotype prediction is useful when the actual growth and observation experiments would be costly or time consuming, or difficult to do.

King *et al.* (2003) have made predictions of phenotype from functional annotations. Decision trees were used to learn from Gene Ontology (GO) annotations, and results were shown to be generally supported by literature searches and experimental phenotype assays. An example of one of their decision trees is given in Figure 6.1. They note the close relationship between function annotation and phenotype annotation. GO annotations with evidence codes such as IMP (inferred from mutant phenotype) are based on phenotype annotation. When integrating these sources of data, it must be realized that they are not necessarily independent and care must be taken to avoid circular data. In this case, King *et al.* removed genes with GO annotations with IMP, TAS, NAS, IC and NR evidence codes from their data. Understanding of the relationships in the data that is to be integrated is vital to ensure that results are not biased unfairly. If the data is not independent then care must be taken not to use machine learning methods that assume independence of attributes, such as naïve Bayes. If the data is strongly relational, then multi-relational data-mining tools can be applied instead, such as those described in the July 2003 special

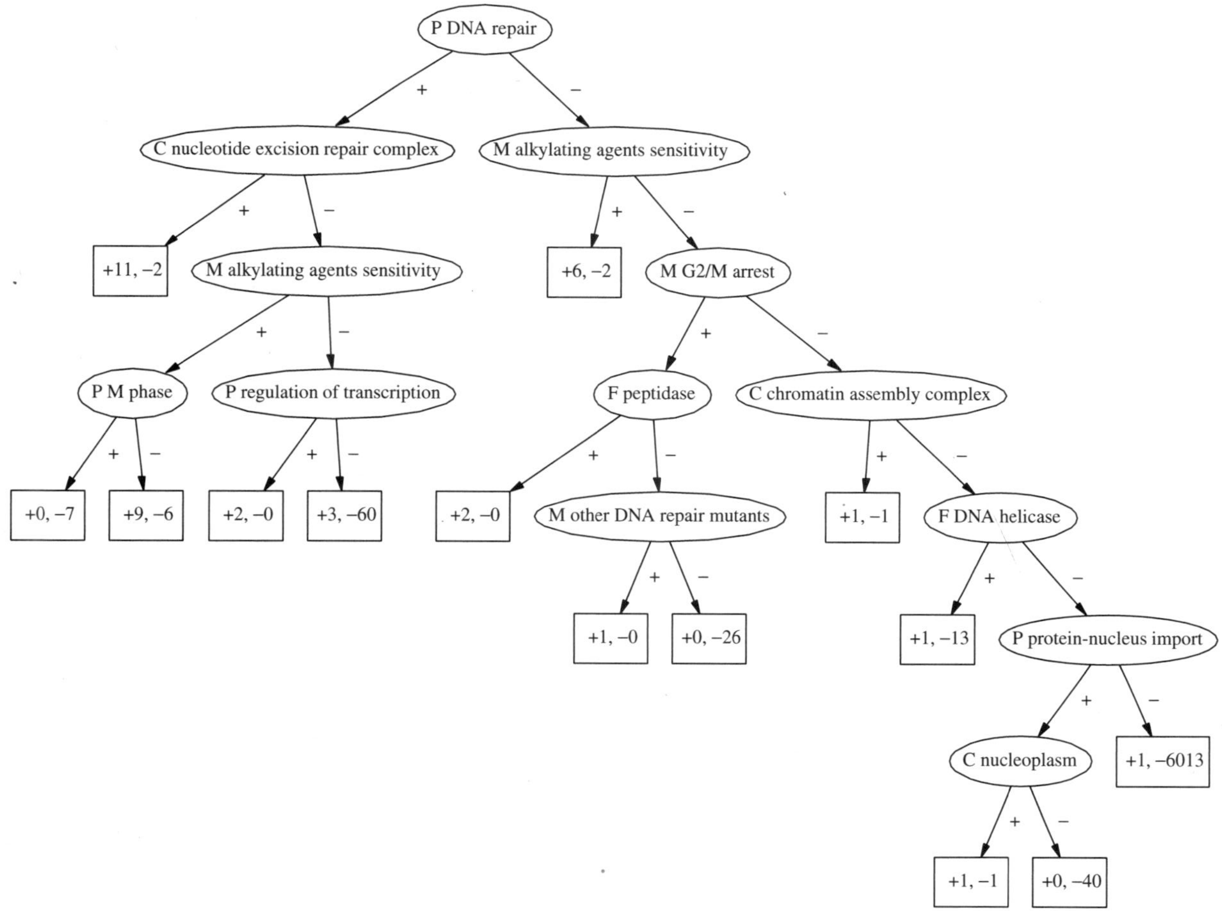

Figure 6.1 A decision tree used to predict whether or not a gene had the MIPS 'UV light sensitivity' phenotype, based on other MIPS phenotype annotations and on GO annotations. The letters P, C, F and M that precede the names of the attributes in the internal nodes indicate that the attributes are from the GO biological process branch, GO cellular component branch, GO molecular function branch, or MIPS yeast phenotype catalogue, respectively (courtesy Oliver King, Fritz Roth, Harvard Medical School)

edition of the ACM SIGKDD Explorations Newsletter. Complex and interdependent data sets are now a common feature in bioinformatics.

Combining a variety of sources of data in order to predict a result is a challenge, and requires a variety of data-mining techniques. Biological data sets present real-world problems for computer scientists specializing in data mining, such as sparse data, noisy data and a range of different data types and formats. In acknowledgement of this, the data-mining community has used gene deletion experiments as part of the Eighth ACM SIGKDD International Conference on Knowledge Discovery and Data Mining (KDD 2002) Cup Challenge. This required participants to integrate and mine data from protein interactions, localization, function annotation and MEDLINE abstracts, to predict the activity of the AHR signalling pathway in yeast single-gene deletion strains. Krogel and Scheffer (2003) describe their experiences in analysing this combination of data, including text mining the scientific abstracts, and propositionalizing the relational data.

Phenotype prediction is particularly useful in cases where mutants are difficult or impossible to obtain, such as in the pharmaceutical fields; e.g., how will a person respond to a drug? Beerenwinkel *et al.* (2003) predict drug resistance phenotype by use of genomic sequence data. Some HIV virus variants are resistant to some of the available drugs. Resistance can be determined through activity in the presence or absence of the drug (phenotype), or by analysing the sequences of the enzymes in the virus that the drug is known to target in order to look for mutations (genotypic testing). Genotyping is faster and cheaper, and hence Beerenwinkel *et al.* show how decision tree and support vector machines can be used to predict the relationship between genotype and phenotype, and to support the interpretation of the genotype.

Parsons *et al.* (2004) use phenotypic experiments to test sensitivity of yeast single-gene mutant strains to different drugs, building a 'chemical-genetic profile' for each drug, indicating which genes interact with the drug and can buffer the drug target. Genes that appear in the profiles for more than one drug can be said to be involved in multi-drug resistance, and they identified 65 genes involved in drug resistance to at least four compounds. They then went on to make double-gene mutants for selected genes, and scored these for fitness, in order to create profiles that could be compared to the chemical-genetic profiles. The chemical-genetic profiles indicate gene-drug interaction, and the double-mutant genetic profiles indicate gene-gene interaction. Similarities between profiles could be used to identify target pathways. For example, 75 genes showed sensitivity to the drug flucanzole. ERG11 is a target of flucanzole, and making double mutants with this gene showed that 13 of the 27 genes that interacted with ERG11 also showed sensitivity to flucanzole.

6.5 Integrating Phenotype Data with Systems Biology

Prediction of phenotype is useful where the growth experiments themselves are expensive to perform. With the increasing capability of laboratory technology, the efficiency and reliability of these experiments improves and this method of data generation becomes more cost effective.

Laboratory robots now provide the means to produce accurate and reproducible phenotype growth experiments, under the control of computer software. This opens new opportunities for automation in science. The Robot Scientist (King *et al.*, 2004) is a project that uses a liquid handling laboratory robot and artificial intelligence software to automatically design and execute phenotype growth experiments in order to analyse the functions of gene products involved in a specific yeast metabolic pathway. In this work the results of phenotypic growth experiments of yeast knock-out mutants are combined with data about the ORFs, enzymes and metabolites involved in the metabolic pathway within a logical model. The model is encoded in the logic-based programming language Prolog. Then a machine learning system creates hypotheses pertaining to the possible roles of the gene products within the pathway. The Robot Scientist then chooses the best phenotype experiments to carry out next, choosing growth media to add to reinstate the pathway from the gene that was knocked out. In this case the 'best' experiment is the most discriminatory between the competing hypotheses of gene-enzyme pairings, while minimizing the cost of growth media. Figure 6.2 shows the architecture of the Robot Scientist. In this way, the Robot Scientist closes the whole scientific loop – constructing hypotheses, devising the experiments, conducting the experiments and using the results to construct the next round of hypotheses. All this is integrated into an automatic system, with no human input in the design of experiments or to interpret results (only to move trays in and out of incubators, and provide supplies of media). In this project, the phenotype experiments and resulting data have been integrated as a part of the whole process, from experiment construction to analysis, and provide a vital part of the evidence.

If phenotype can be predicted from other sources of information, then this information too can be included within a system such as the Robot Scientist, and this will improve the choice of the next round of growth experiments. Testing for protein essentiality, or growth/non-growth of a mutant strain, is a very important phenotype. Despite the availability of systematic gene deletion strains, many genes have still not been tested even for this most basic of phenotypes. Jeong *et al.* (2003) use statistics and network analysis to integrate function annotation, mRNA expression data and protein interaction data in order to predict essentiality in yeast. Dezsö, Oltvai and Barabási (2003) conclude from analysis of the combination of protein interaction data, expression data, cellular localization and function annotation that a gene's deletion essentiality phenotype depends upon the role that it plays within protein complexes. They discover essential and non-essential complexes, and note that the deletion of a gene whose protein is involved in the core of an essential complex will be lethal, whereas if the complex is non-essential the protein will be too. Thus the essentiality phenotype is a property of the complex rather than the individual protein. The proteins in the cores of complexes share common functions, localizations and expression as well as essentiality. Then there are proteins surrounding the cores that

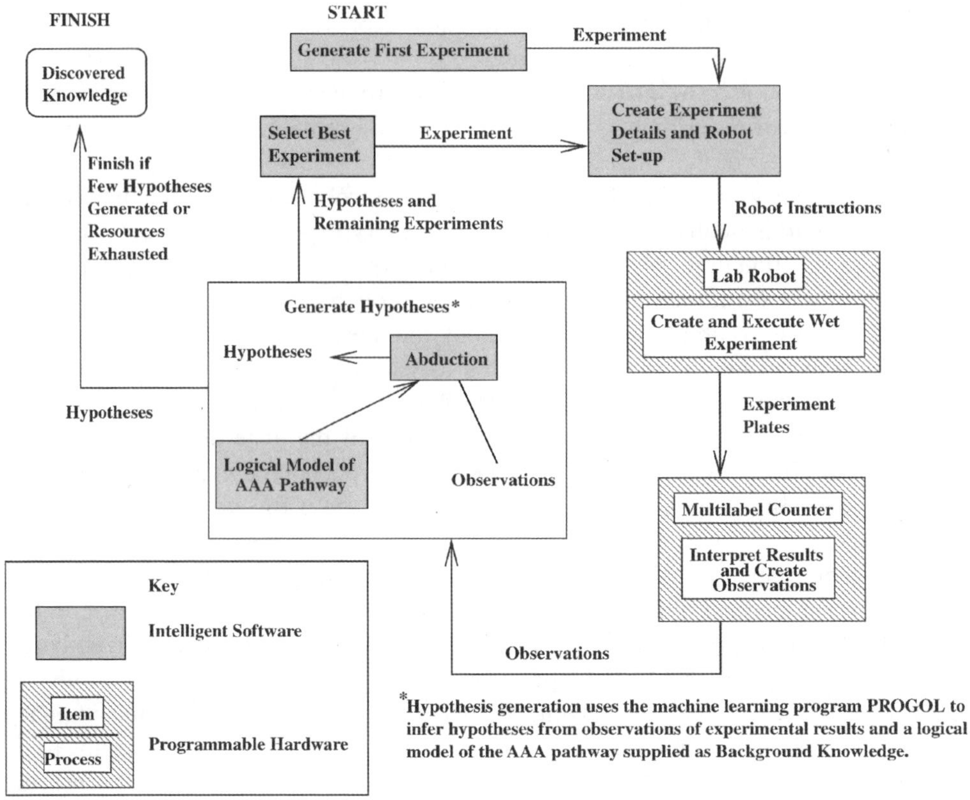

Figure 6.2 The architecture of the Robot Scientist (courtesy Ken Whelan, University of Wales, Aberystwyth)

do not share common essentiality, functions, localizations or expression and these are considered to be just temporary attachments to the complexes.

Network analysis of metabolic networks and systems biology promise to open a whole new way of integrating data and understanding biology at a more comprehensive level. The current vision statement of the BBSRC (the Biotechnology and Biological Sciences Research Council in the UK) expects the realistic modelling of the cell by computer within the next ten years (BBSRC, 2003). Most phenotypic behaviours are a result of multiple interactions and effects caused by multiple components within a cell, and as such we need to use integrated data and models at the level of systems biology if we are to truly understand the cause of phenotypes. The need for integration of data and model in systems biology is also discussed in Chapter 1 of this book.

Famili *et al.* (2003) use a metabolic model of yeast to analyse phenotypic behaviour. A model of the metabolic network for yeast was constructed using genomic, biochemical and physiological information. Then the network was analysed by computer to discover the range of metabolic activity that could be displayed, adding constraints to eliminate invalid behaviours. By placing demands on the network for growth and maintenance, predicted phenotypic behaviour can be calculated, and compared with experimental results. In some cases the model agreed with experimental results. Where the model behaved differently, this led to the researchers adding further constraints in order to improve the model. The metabolic model and the phenotype experiments then become part of an iterative process of improvement of knowledge of metabolism. The model was also used to predict phenotypic behaviour of mutant single-gene deletion strains of yeast, grown on complete media. The predictions agreed with the experimental results from the SGD database in 81.5 per cent of the cases.

Hellerstein (2004), in his analysis of new technology for measuring metabolic fluxes, concludes that recent tools will allow increasingly more detailed measurement and control. These tools will contribute still further to systems biology understanding by allowing measurement of dynamic phenotypes and whole processes, rather than the destructive analysis of parts after an experiment.

6.6 Integration of Phenotype Data in Databases

If phenotypic data sets are to be effectively integrated with other sources of data, they will need to be stored as part of the existing system of databases for biological data.

MIPS[2] provides the Comprehensive Yeast Genome Database (CYGD), which now lists phenotypic information along with the entry for each gene. The results of a search for a yeast gene in the CYGD will describe not only its function, physical features, localization and literature references but also its known disruption phenotypes.

Then there are a wide range of specifically phenotype-oriented databases, which tend to specialize in some way, on either particular organisms or particular phenotypes. For example, the GDPC (Genomic Diversity and Phenotype Connection) database (Casstevens and Buckler, 2003) aims to collect phenotype data from different sources and to integrate it with genomic diversity (e.g. SNPs or molecular markers such as AFLPs or RFLPs). This will be a resource for the plant community, and will allow the data to be made publicly available in a standardized format, using

[2]Munich Information Center for Protein Sequences, see: http://mips.gsf.de

Table 6.1 A small sample of the wide variety of available online phenotype databases. All accessed 18 August 2004

Name	Type	URL
GDPC	Plant	http://www.maizegenetics.net/gdpc/
PharmGKB	Clinical	http://www.pharmgkb.org
Ramedis	Rare metabolic diseases	http://www.ramedis.de
Mouse Phenome Database	Mouse	http://www.jax.org/phenome
Maize Phenotype Database	Maize	http://www.mutransposon.org/project/RescueMu/zmdb/phenotypeDB/
RMAiDB	*C. elegans*	http://nematoda.bio.nyu.edu/
Chinese Gene Variation Database	Human (Chinese)	http://ww.cgvdb.org.tw/
TRIPLES	*S. cerevisiae*	http://ygac.med.yale.edu/triples/triples.htm
Database of Essential Genes	Multiple organisms, essential genes	http://tubic.tju.edu.cn/deg/

XML[3] and SOAP[4] and providing a Java API and a browsing facility. In a similar way, the PharmGKB database (Klein *et al.*, 2001) serves the human community, by storing data about clinical phenotypes, and associating these with genetic information. Data about gene polymorphisms, phenotype variabilities, environmental factors and treatment protocols is stored in the database, and access is provided through a Web front end, a Java API and XML formats. Privacy and confidentiality of the data is assured and no patient-identifying data is stored.

A small sample of the variety of existing databases storing phenotype data can be seen in Table 6.1. Phenotype data sets are currently stored in many disparate databases around the world, each database often describing a single organism, each with different representations, access and levels of detail. On the other hand, genome and proteome data have highly organized and well established databases (Bateman, 2004), microarray experiments have MIAME standards (Brazma *et al.*, 2001) and metabolomics is in the process of developing standards (Hardy *et al.*, 2003). Phenotype data will in the future need to make use of controlled vocabularies and metadata standards in order to describe the experimental conditions and results fully and unambiguously. The use of multiple existing ontologies and vocabularies to describe phenotype are a step in the right direction (see Bard and Rhee, 2004), for a review of the current use of ontologies, particularly in describing phenotype data), but serious thought is needed to ensure that future descriptions will be complete and

[3]XML: eXtensible Markup Language.
[4]SOAP: Simple Object Access Protocol.

useful. Phenotype data will also require well maintained databases to provide standardized access and cross-references to existing bioinformatics data.

6.7 Conclusions

Observations of phenotype are growing both in terms of genome-wide coverage and in terms of level of detail. Phenotype data will have much to contribute to the integrated analysis of systems biology. Integration of phenotypic data sources with other sources of data has already begun for investigation of gene function and prediction of phenotype. Integration of phenotypic experiments together with whole-system and network analysis will open new possibilities for understanding and modelling the emergent and complex properties of the cell. Integration of phenotype data into the standard databases will allow the data to be used as part of automated processes, and introduction of standards for phenotype metadata will enable confidence in the data and results.

As the biological technology for phenotype measurement and high-throughput experiments improves, bioinformatics becomes the ever-closer integration of biology and computer science. 'Wet-lab' experiments become an ongoing part of computational analysis and model construction and the whole scientific loop of hypothesis–experiment–interpretation. Integrated data sources permit a whole-system approach to understanding the cell, and making use of phenotype data is an important part of this process.

References

Alonso, J. M. *et al.* (2003) Genome-wide insertional mutagenesis of *Arabidopsis thaliana. Science*, **301**, 653–657.

Alonso-Blanco, C. *et al.* (1998) Development of an AFLP based linkage map of L*er*, Col and Cvi *Arabidopsis thaliana* ecotypes and construction of a L*er*/Cvi recombinant inbred line population. *Plant J*, **14**(2), 259–271.

Austin, M. A., Harding, S. and McElroy, C. (2003) Genebanks: a comparison of eight proposed international genetic databases. *Community Genet*, **6**(1), 37–45.

Bard, J. B. L. and Rhee, S. Y. (2004) Ontologies in biology: design, applications and future challenges. *Nature Rev Genet*, **5**, 213–222.

Bateman, A. (ed.) (2004) *Nucleic Acids Research: Database Issue*. Oxford University Press, Oxford.

BBSRC (2003) *Bioscience for Society: a Ten-Year Vision*. Available from http://www.bbsrc.ac.uk/about/plans_reports/vision.html [accessed 04/08/04].

Beerenwinkel, N. *et al.* (2003) Geno2pheno: estimating phenotypic drug resistance from HIV-1 genotypes. *Nucleic Acids Res*, **31**(13), 3850–3855.

Borevitz, J. O. and M. Nordborg (2003) The impact of genomics on the study of natural variation in *Arabidopsis. Plant Physiol*, **132**, 718–725.

Brazma, A. *et al.* (2001) Minimum information about a microarray experiment (MIAME)–toward standards for microarray data. *Nature Gene*, **29**, 365–71.

Carlborg, Ö. (2002) *New Methods for Mapping Quantitative Trait Loci*, Ph.D. dissertation, Department of Animal Breeding and Genetics, SLU.

Casstevens, T. M. and E. S. Buckler (2003) *GDPC: The Genomic Diversity and Phenotype Connection: Middleware for Genomic Diversity and Phenotypic Data*, Technical report GDPC Whitepaper, Departments of Statistics and Genetics, North Carolina State University.

Clare, A. and R. D. King (2002) Machine learning of functional class from phenotype data. *Bioinformatics* **18**(1), 160–166.

Dezsö, Z., Z. N. Oltvai, and A-L. Barabási (2003) Bioinformatics analysis of experimentally determined protein complexes in the yeast *Saccharomyces cerevisiae. Genome Res*, **13**, 2450–2454.

Famili, I., J. Förster, J. Nielsen, and B. O. Palsson (2003) *Saccharomyces cerevisiae* phenotypes can be predicted by using constraint-based analysis of a genome-scale reconstructed metabolic network. *Proc Natl Acad Sci*, **100**, 13 134–13 139.

Giaever, G. *et al.* (2002). Functional profiling of the *Saccharomyces cerevisiae* genome. *Nature* **418**, 387–391.

Glazier, A. M., J. H. Nadeau, and T. J. Aitman (2002) Finding genes that underline complex traits. *Science* **298**, 2345–2349.

Hardy, N. *et al.* (2003) Towards a standard representation for metabolomics experiments and their results – ArMet. In *Second International Conference on Plant Metabolomics.*

Hellerstein, M. (2004) New stable isotope-mass spectroscopic techniques for measuring fluxes through intact metabolic pathways in mammalian systems: introduction of moving pictures into functional genomics and biochemical phenotyping. *Metab Eng*, **6**, 85–100.

Jander, G., S. R. Norris, S. D. Rounsley, D. F. Bush, I. M. Levin, and R. L. Last (2002) *Arabidopsis* map-based cloning in the post-genome era. *Plant Physiol*, **129**, 440–450.

Jeong, H., Z. Oltvai, and A-L. Barabási (2003) Prediction of protein essentiality based on genomic data. *ComPlexUs* **1**, 19–28.

Jim, K., K. Parmar, M. Singh, and S. Tavazoie (2004) A cross-genomic approach for systematic mapping of phenotypic traits to genes. *Genome Res*, **14**, 109–115.

Kamath, R. S. *et al.* (2003). Systematic functional analysis of the *Caenorhabdities elegans* genome using RNAi. *Nature* **421**, 231–237.

King, O. D., J. C. Lee, A. M. Dudley, D. M. Janse, G. M. Church, and F. P. Roth (2003) Predicting phenotype from patterns of annotation. *Bioinformatics Suppl*, **1**, I183–I189.

King, R. D. *et al.* (2004). Functional genomic hypothesis generation and experimentation by a robot scientist. *Nature* **427**, 247–252.

Klein, T. E. *et al.* (2001) Integrating genotype and phenotype information: an overview of the PHARMGKB project. *Pharmacogenomics J*, **1**, 167–170.

Krogel, M. and T. Scheffer (2003) Effectiveness of information extraction, multi-relational, and multi-view learning for predicting gene deletion experiments. In *Proceedings of BIOKDD03: 3rd ACM SIGKDD Workshop on Data Mining in Bioinformatics.*

Loudet, O., S. Chaillou, C. Camilleri, D. Bouchez, and F. Daniel-Vedele (2002) Bay-0 × Shahdara recombinant inbred line population: a powerful tool for the genetic dissection of complex traits in *Arabidopsis. Theor Appl Genet* **104**, 1173–1184.

Medema, R. H. (2004) Optimizing RNA interference for application in mammalian cells. *Biochem J*, Immediate Publication 1 April 2004.

Oliver, S. (1996) A network approach to the systematic analysis of yeast gene function. *Trends Genet*, **12**(7), 241–242.

Page, G. F., V. George, R. C. Go, P. Z. Page, and D. B. Allison (2003) 'Are we there yet?': Deciding when one has demonstrated specific genetic causation in complex diseases and quantitative traits. *Am J Human Genet* **73**, 711–719.

Paran, I. and D. Zamir (2003), Quantitative traits in plants: beyond the QTL. *Trends Genet*, **19**, 303–306.

Parsons, A. B. *et al.* (2004) Integration of chemical–genetic and genetic interaction data links bioactive compounds to cellular target pathways. *Nature Biotechnol*, **22**(1), 62–69.

Silva, J. M., H. Mizuno, A. Brady, R. Lucito, and Hannon G. J. (2004) RNA interference microarrays: High-throughput loss-of-function genetics in mammalian cells. *Proc Nat Acad Sci* **101**(17), 6548–6552.

Tong, A. H. Y. *et al.* (2004) Global mapping of the yeast genetic interaction network. *Science* **303**, 808–813.

7

Ontologies and Functional Genomics

Fátima Al-Shahrour and **Joaquín Dopazo**

Abstract

High-throughput methodologies have increased by orders of magnitude the possibility of obtaining data in orders of magnitude. Nevertheless, translating data into useful biological knowledge is not an easy task. We review how bio-ontologies, and in particular gene ontology, can be used to understand the biological roles played by genes that account for the phenotypes studied, which is the ultimate goal of functional genomics. Statistical issues related to high-throughput methodologies, such as the high occurrence of false or spurious associations, are also discussed.

Keywords

gene ontology, multiple testing, annotation, functional genomics

7.1 Information Mining in Genome-Wide Functional Analysis

Molecular biology has addressed functional questions by studying individual genes, either independently or a few at a time. Although it constituted a reductionistic approach, it was extremely successful in assigning functional properties and biological roles to genes and gene products. The recent possibility of obtaining information on thousands of genes or proteins in one sole experiment, thanks to high-throughput methodologies such as gene expression (Holloway *et al.*, 2002) or proteomics (MacBeath, 2002), has opened up new possibilities in querying living systems at the genome level that are beyond the old paradigm 'one gene–one postdoc'. Relevant

Data Analysis and Visualization in Genomics and Proteomics Edited by Francisco Azuaje and Joaquin Dopazo

biological questions regarding gene or gene product interactions or biological processes played by networks of components, etc., can now for the first time be addressed realistically. Nevertheless, genomic technologies are at the same time generating new challenges for data analysis and demand a drastic change in the habits of data management. Dealing with this overabundance of data must be approached cautiously because of the high occurrence of spurious associations if the proper methodologies are not used (see Ge, Walhout and Vidal, 2003, and Chapter 12 for discussions of some related aspects).

Traditional molecular biology approaches tended to mix up the concepts of data and information. This was partially due to the fact that researchers had a great deal of information previously available about the typical data units they used (genes, proteins etc.). Over the last few years the increasing availability of high-throughput methodologies has amplified by orders of magnitude the potential of data production. One direct consequence of this revolution in data production has been to clarify how fictitious the equivalence between data and information actually used to be. System biology approaches emerge then to convert the flood of data into information and knowledge (Bassett, Eisen and Boguski, 1999; Ge, Walhout and Vidal, 2003).

There are, however, several problems related to massive data management. One of them is the lack of accurate functional annotations for a considerable number of genes. Another non-negligible difficulty stems from the fact that, even in the instance of availability of proper functional annotations, processing all the information corresponding to thousands of genes involved in a high-throughput experiment is beyond the human capabilities. Automatic processing of the information therefore becomes indispensable to draw out the biological significance behind the results in this type of experiment. As previously mentioned, the occurrence of false or spurious associations is common when dealing with thousands of elements. Unfortunately, these spurious associations are often considered as evidence of actual functional links, leading to misinterpretation of results. All these features of genomic data must be taken into account for any procedure aiming to properly identify functional roles in groups of genes with a particular experimental behaviour.

7.2 Sources of Information: Free Text Versus Curated Repositories

Any approach using biological information for functional annotation purposes uses two main sources: free text or curated repositories.

The use of techniques of automatic management of biological information to study the coherence of gene groups obtained from different methodologies has been addressed in recent years (Oliveros *et al.*, 2000; Raychaudhuri, Schutze and Altman, 2002; Pavlidis, Lewis and Noble, 2002). Considerable effort has been focused on developing automatic procedures for extracting information from biomedical literature. Information extraction and text mining techniques in particular have been applied to the analysis of gene expression data (Jenssen *et al.*, 2000; Oliveros *et al.*, 2000, Tanabe *et al.*, 1999). It has been claimed that free text processing, essentially using PubMed abstracts as a source of information, has the advantage of providing numerous gene-to-abstract correspondences. Nevertheless, text-mining methodologies still present many drawbacks (Blaschke, Hirschman and Valencia, 2002), such as problems of interpreting terms due to the context of the sentence in which the gene is cited; the lack of a standardized nomenclature of genes in literature that makes it difficult to find all the citations for all the synonyms used for them; a profuse use of acronyms in literature that makes it hard to find accurate citations of genes (for example, for the term STC, corresponding to the gene *secretin*, 158 citations were found in the year 2001, 121 of them corresponded to stem cell transplantation and the rest to other concepts such as spiral computerized tomography, solid cystic tumour etc.; only one of them was a real reference to the gene *secretin*); there are problems surrounding orthography and boundaries for identifying gene names and finally there are irrelevant terms related to non-functional features that appear, nonetheless, to be associated to gene names.

On the other side of the spectra are the repositories with curated functional information, which contain fewer gene-to-term correspondences although these are reliable, consistent and standardized. There are diverse repositories such as pathway databases, among which the KEGG database (Kanehisa *et al.*, 2004) is the paradigm, protein interaction databases (see DIP, Xenarios *et al.*, 2002), protein motif databases (see, for example, InterPro, Mulder *et al.*, 2003) etc.

The most valuable resource is most probably the Gene Ontology (GO) database of curated definitions (Ashburner *et al.*, 2000) and the annotations based on GO.

Diverse genome initiatives and databases are annotating genes according to GO terms (Camon *et al.*, 2003; Xie *et al.*, 2002), constituting a priceless resource for information-mining implementations.

Although some direct applications of free-text mining to data analysis have been proposed (Tanabe *et al.*, 1999; Oliveros *et al.*, 2000; Raychaudhuri *et al.*, 2003), the future of the practical application of these technologies probably resides in its use by information repository curators to help in the annotation process. Methods for predicting GO categories from the analysis of biomedical literature (Raychaudhuri *et al.*, 2002) have therefore been proposed, and there are also similar methods based on the study of different biochemical and physical protein features (Jensen *et al.*, 2003; Schug *et al.*, 2002).

7.3 Bio-Ontologies and the Gene Ontology in Functional Genomics

Applied ontologies are centred around a specific domain of knowledge. These ontologies endeavour to represent a system of categories accounting for a particular vision of a given area, in order to establish rules that describe relationships between these categories, and to instantiate the objects in the categories. In practical terms, these ontologies provide an organizational framework of concepts about biological entities and processes in a hierarchical system in which, associative relations which provide reasoning behind biological knowledge, are included. One of the most powerful features of an ontology is the implementation of a controlled, unambiguous vocabulary. This is extremely useful in an inherently complex and heterogeneous discipline such as biology, where a great deal of sophisticated knowledge, in most cases of a hierarchical nature, needs to be integrated with molecular data (Bard and Rhee, 2004). The most important ontologies in the domain of biology are included under the umbrella of the Open Biological Ontologies (OBO) initiative, which constitutes a *de facto* standard for them (see http://obo.sourceforge.net/). Some known examples are the Unified Medical Language System (UMLS, http://www.nlm.nih.gov/research/umls/), which implements a hierarchy of medical terms used in the indexation of PubMed, or the Microarray Gene Expression Data Society (MGED) Ontology (http://mged.sourceforge.net/), recently popularized by the increasing production of gene expression data with microarrays, etc. Nevertheless, the most relevant ontology in the area of functional genomics is, undoubtedly, the Gene Ontology (GO, http://www.geneontology.org/), which provides a controlled vocabulary for the description of molecular function, biological process and cellular component of gene products (Ashburner *et al.*, 2000). GO terms are used as attributes of gene products by collaborating databases, facilitating uniform queries across them. Because of the existing homologies between proteins among different taxa, GO terms can be thoroughly used across species (Ashburner *et al.*, 2000).

The controlled vocabularies of terms are structured in a hierarchical manner that allows for both attribution (assignment of gene products to particular terms) and querying at different levels of granularity. This hierarchical structure constitutes the representation of the ontology within which each term is a node of a directed acyclic graph (DAG), which is very similar to a tree – the only difference being that in a DAG it is possible for a node to have more than one parent. The deeper a node is in the hierarchy, the more detailed the description of the term. In GO, child to parent relationships can be of two types: 'is a', meaning the child is an instance of the parent (e.g. *chloroplast envelope GO:0009941* is a *membrane GO:0016020*) and 'part of', when the child is a component of the parent (e.g. *inner membrane GO:0019866* and *outer membrane GO:0019867* are part of *membrane GO:0016020*).

The success of an ontology relies largely upon the approval received from the scientific community. The most important achievement of GO is perhaps that the GO consortium has been able to attract a large number of collaborating databases which are actively mapping gene products onto GO terms. These databases with controlled and curated annotations, which can easily be queried by computers, constitute an invaluable though not yet fully exploited resource, for the scientific community. Additional information on the quality of the annotation of gene products is provided by the collaborative databases through the evidence codes (http://www.geneontology.org/GO.evidence.html). The codes represent different types of evidence used in the annotation. Among them, the highest quality codes are for GO–gene correspondences supported by experimental functional assays (IDA, IMP codes) and the lowest quality corresponds to correspondences inferred from electronic annotations (IEA code).

As previously mentioned, the representation of GO resides in its hierarchy. There are different tools available that are useful to browse this hierarchy (see a comprehensive list in: http://www.geneontology.org/GO.tools.html). Such GO browsers allow the viewing of all gene products annotated with a given GO term, or searching for a gene product and view all its associations. In addition, by browsing the ontologies it is also possible to view relationships between terms.

7.4 Using GO to Translate the Results of Functional Genomic Experiments into Biological Knowledge

Functional genomics experiments allow the scaling of the classical functional experiments to a genomic level. Comparison of phenotypes (e.g. patients versus controls, studies of different clinical outcomes etc.) by means of techniques such as DNA microarrays or proteomics provides insight into their molecular basis. Nevertheless, the data obtained in these experiments are measurements of the gene or protein expression levels. To translate this data into information numerical analyses are firstly required to determine which genes (among the thousands analysed) can be considered as significantly related to the phenotypes (see Chapter 12). The second step is to interpret roles played by the targeted genes. The availability of GO annotations for a considerable number of genes helps interpret these results from a biological point of view. The rationale commonly used is as follows: if some genes have been found to be differentially expressed when comparing two different phenotypes (or are correlated to a given continuous phenotypic trait, or to survival etc.) it is because the roles they play at molecular level account (to some extent) for the phenotypes analysed. The GO annotations available for the genes that present the same asymmetrical distribution or correlation serve as a more or less detailed description of these biological roles. For example, if 50 genes from an array of 6500 genes are differentially expressed and 40 of them (80 per cent – a high proportion) are annotated as response to 'external stimulus' (GO:0009605), it is intuitive to conclude that this process must be related to the phenotypes studied. In addition, if the background distribution of this type of gene in the genome is, let us say, of four per cent, one can conclude that most of the genes related to 'external stimulus' have been altered in their expression levels in the experiment.

There are many tools listed on the GO consortium web page that extract lists of GO terms differentially represented when comparing two sets of genes (see http://www.geneontology.org/GO.tools.html) and, in some cases, provide scores or even individual tests for comparisons between two sets of genes. For example, GoMiner (Zeeberg *et al.*, 2003), MAPPFinder (Doniger *et al.*, 2003), GFINDer (http://www.medinfopoli.polimi.it/GFINDer/) or eGON (http://nova2.idi.ntnu.no/egon/), just to cite a few, generate tables that correlate groups of genes to biochemical or molecular functions or GO terms. Some of them are specific for organisms, such as FunSpec (Robinson *et al.*, 2002), which evaluates groups of yeast genes in terms of their annotations in diverse databases, or CLENCH (Shah and Fedoroff, 2004), for *A. thaliana*.

Nevertheless, differences in the distribution of GO terms between groups must, in addition to being spectacular (which is quite a subjective concept), also be significant (which is an objective statistical concept related to the probability of drawing one's observations purely by chance).

7.5 Statistical Approaches to Test Significant Biological Differences

As previously mentioned, much caution should be adopted when dealing with a large set of data because of the high occurrence of spurious associations (Ge, Walhout and Vidal, 2003). Table 7.1 has been constructed using ten datasets obtained by the random sampling of 50 genes from the complete genome of *Saccharomyces cerevisiae*. For each random set, the proportions of all the GO terms (at GO level 4) have been compared between the two partitions (50 genes with respect to the remaining ones), and the GO term showing the most extreme differential distribution was displayed in each case (rows of the table). The first column shows the percentage of genes annotated with the GO term in the random partition of 50 genes, the second column represents the corresponding percentage in the rest of the genome and the

Table 7.1 GO terms found to be differentially distributed when comparing 10 independent random partitions of 50 genes sampled from the complete genome of yeast. See the text for an explanation

% in random set	% in genome	*p*-value	adjusted *p*-value	GO term
8.33	1.86	0.0752	1	ion homeostasis (GO:0050801)
10.00	31.34	0.0096	0.6735	nucleobase, nucleoside, nucleotide and nucleic acid metabolism (GO:0006139)
3.33	0.24	0.075	1	one-carbon compound metabolism (GO:0006730)
4.04	8.00	0.0177	0.6599	energy pathways (GO:0006091)
3.45	0.22	0.0669	1	metabolic compound salvage (GO:0043094)
5.88	0.67	0.024	1	vesicle fusion (GO:0006906)
6.45	1.60	0.09	1	negative regulation of gene expression, epigenetic (GO:0045814)
13.79	3.97	0.028	1	response to external stimulus (GO:0009605)
16.13	4.23	0.0097	1	response to endogenous stimulus (GO:0009719)
2.70	0.13	0.054	1	host–pathogen interaction (GO:0030383)

third column shows the p-value obtained upon the application of a Fisher exact test for 2×2 contingency tables. For many people it still seems staggering that most of the random partitions present asymmetrical distributions of GO terms with significant individual p-values (column 3). This apparent paradox stems from the fact that we are not conducting a single test in each partition, but as many tests as GO terms are being checked (several hundreds). Nevertheless, in this situation the researcher tends to forget about the many hypotheses rejected and only focus on the term for which an apparent asymmetrical distribution was found. In some cases this situation is caused by the way in which some of the above mentioned programs work. To some extent the fact that many tests are really being conducted is hidden to the user and the result is presented as if it were the case of a unique test. If we conduct several hundreds of tests simultaneously, the probability of finding an apparently asymmetrical distribution for a given GO term increases enormously. A very simple example can be used here to illustrate this concept: let us imagine you flip a coin 10 times and you get 10 heads. You would certainly suspect that something was wrong with the coin. If the same operation were repeated with 10 000 different coins one or even several occurrences of 10 heads would not be considered surprising. We intuitively accept this because of the probability of having an unexpected result just by chance is high. If we were interested in checking whether an observation is significantly different from what we could expect simply by chance in a multiple testing situation then the proper correction must be applied. The fourth column of Table 7.1 shows an adjusted p-value using one of the most popular multiple-testing corrections, the false discovery rate (FDR; Benjamini and Yekutieli, 2001), and it is obvious that none of the situations depicted in columns 1 and 2 can be attributed to anything other than random occurrence.

Table 7.1 shows how random partitions, for which no functional enrichment should be expected, yield apparent enrichments in GO terms because the most asymmetrically distributed GO terms among several hundreds are chosen *a posteriori*. These values occur simply by chance and cannot be considered as either biologically authentic or statistically significant. This clearly shows, beyond any doubt, that multiple testing adjustment must be used if several hypotheses are simultaneously tested.

Multiple testing has been addressed in different ways depending on particular cases and the number of simultaneous hypotheses tested. Thus, corrections such as Bonferroni or Sidak are of very simple application but are too conservative if the number of simultaneous tests is high (Westfall and Young, 1993). Another family of methods that allow less conservative adjustments is the family wise error rate (FWER), that controls the probability that one or more of the rejected hypotheses (GO terms whose differences cannot be attributed to chance) is true (that is, a false positive). The minP step-down method (Westfall and Young, 1993), a permutation-based algorithm, provides a strong control (i.e. under any mix of false and true null hypothesis) of the FWER. Approaches that control the FWER can be used in this context although they are dependent on the number of hypotheses tested and tend to be too conservative for a high number of simultaneous tests. Aside from a few cases in which FWER control could be necessary, the multiplicity problem in prospective functional assignation does not require protection against even a single false positive. In this case, the drastic loss of power involved in such protection is not justified. It would be more appropriate to control the proportion of errors among the identified GO terms whose differences among groups of genes cannot be attributed to chance instead. The expectation of this proportion is the false discovery rate (FDR). Different procedures offer strong control of the FDR under independence and some specific types of positive dependence of the test statistics (Benjamini and Hochberg, 1995), or under arbitrary dependency of test statistics (Benjamini and Yekutieli, 2001).

We have shown how important multiplicity issues are in finding functional associations to clusters of genes. Any procedure that does not take this into account is as a consequence considering a high number of spurious relationships as reliable.

7.6 Using FatiGO to Find Significant Functional Associations in Clusters of Genes

The FatiGO (Fast Assignment and Transference of Information using GO, available at http://fatigo.org) tool was the first application for finding significant differences in the distribution of GO terms between groups of genes taking the multiple testing nature of the contrast into account (Al-Shahrour *et al.*, 2004). FatiGO takes two lists of genes (ideally a group of interest and the rest of the genome, although any two groups formed in any way can be tested against each other) and convert them into two lists of GO terms using the corresponding gene–GO association table. Since distinct genes are annotated with more or less detail at the different levels of the hierarchy, it is meaningless to test for different terms that are really descriptions in different detail of the same functional property (e.g., why test apoptosis versus regulation of apoptosis?). To deal with this, FatiGO implements the 'inclusive analysis', in which a level in the DAG hierarchy is chosen for the analysis. Genes annotated with terms that are descendant of the parent term corresponding to the level chosen therefore take the annotation from the parent. Figure 7.1 illustrates this procedure. If apoptosis node level is chosen for the analysis, eight genes, annotated in descendant nodes, will be assigned to the term apoptosis. If inclusive analysis is not used, then four terms: apoptosis (with two genes), regulation of apoptosis (three), negative regulation of apoptosis (one) and induction of apoptosis (two) are taken into account, with the obvious decrease in the power of the test.

A Fisher exact test for 2×2 contingency tables is used. For each GO term the data are represented as a 2×2 contingency table with the rows being presence/absence of the GO term, and each column representing each of the two clusters (so that the numbers in each cell would be the number of genes of the first cluster where the GO term is present, the number of genes in the first cluster where the GO term is absent, and so on).

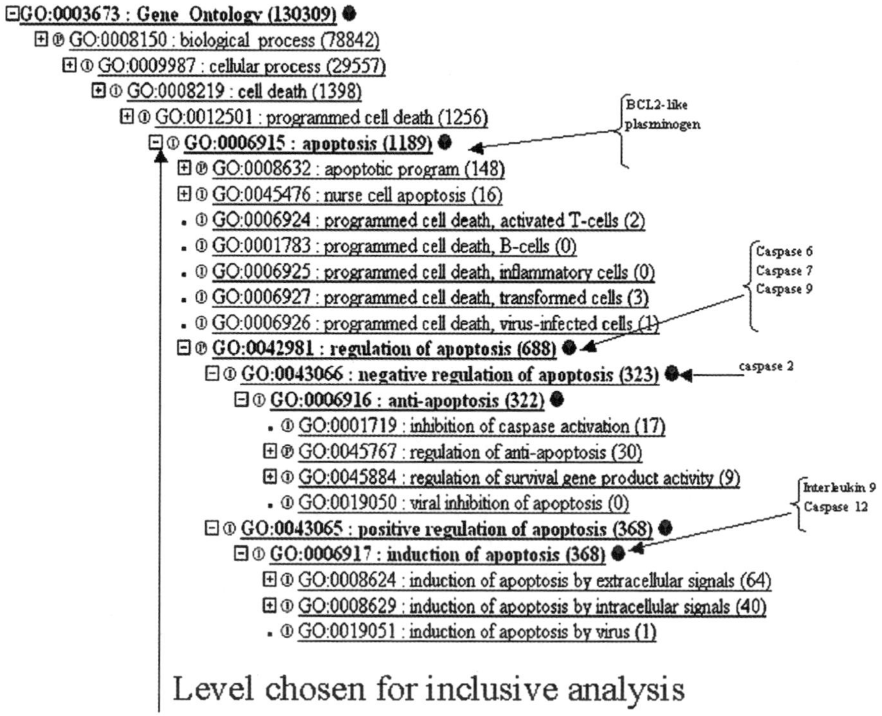

Figure 7.1 Representation of the inclusive analysis concept

In addition to the unadjusted p-values (which are given just because they are obtained as part of the process, but should not be considered as evidence of significant differential distribution of GO terms between clusters), FatiGO returns adjusted p-values based on three different ways of accounting for multiple testing: FDR under independence (Benjamini and Hochberg, 1995), or under arbitrary dependency of test statistics (Benjamini and Yekutieli, 2001) as well as FWER control by the minP step-down method (Westfall and Young, 1993). Results are arranged by p-value to facilitate the identification of GO terms with a significant asymmetrical distribution between the groups of genes studied.

7.7 Other Tools

Recently, other tools have included some multiple-testing possibilities. For example, the latest versions of Onto-Express (Khatri *et al.*, 2002) include Bonferroni and Sidak corrections as well as a permutation test, or GeneMerge (Castillo-Davis and Hartl, 2003), which implements Bonferroni correction. New tools such as FunAssociate (http://llama.med.harvard.edu/cgi/func/funcassociate) include unspecified permutation tests, although others include more established multiple testing controls such as FDR, which is the case of GoSurfer (http://biosun1.harvard.edu/complab/gosurfer/) or GOStat (Beissbarth and Speed, 2004), which has exactly the same functionalities as FatiGO.

7.8 Examples of Functional Analysis of Clusters of Genes

As previously mentioned, a research scientist is continuously interested in understanding the molecular roles played by potentially relevant genes in a given experiment. One of the most popular hypotheses in microarray data analysis is that coexpression of genes across a given series of experiments is most probably explained through some common functional role (Eisen *et al.*, 1998). Actually, this causal relationship has been used to predict gene function from patterns of coexpression (van Noort, Snel and Huynen, 2003; Mateos *et al.*, 2002).

Here is an example using the data from DeRisi, Iyer and Brown (1997), in which they analyse the complete genome of *Saccharomyces cerevisiae* to carry out a comprehensive study of the temporal programme of gene expression accompanying the metabolic shift from fermentation to respiration. With the aim of finding groups of genes that coexpress across the seven time points measured, gene expression patterns were clustered using the SOTA algorithm (Dopazo and Carazo, 1997; Herrero, Valencia and Dopazo, 2001; see also Chapter 10) as implemented in the GEPAS (http://gepas.bioinfo.cnio.es) suite of web tools (Herrero *et al.*, 2003). Figure 7.2 shows the clusters of genes obtained. The parameters used were coefficient of correlation as distance measure and the growth was stopped at 95 per cent of variability (see Herrero, Valencia and Dopazo, 2001, for details of the procedure). The cluster with 21 genes that are initially active and suffer a late repression was analysed with FatiGO (Al-Shahrour, Díaz-Uriarte, and Dopazo, 2004). Seventy-five per cent of these genes were annotated as biosynthesis, and the differences in proportion with respect to the background (30 per cent) were clearly significant. It can be claimed that genes with the described temporal behaviour are involved in the biosynthesis biological process. In the event of not performing p-value adjustment, another three processes (sexual reproduction, conjugation and aromatic compound metabolism) would have been considered as important despite the differences in the proportions between the cluster and the rest of genes that can occur simply by chance (the adjusted p-values are too high).

Genes showing significant differential expression when comparing two or more phenotypes, or genes significantly correlated to a trait (e.g. the level of a metabolite) or to survival, can be analysed in the same way. Comparison of distributions of GO terms helps to understand what makes these genes different from the rest.

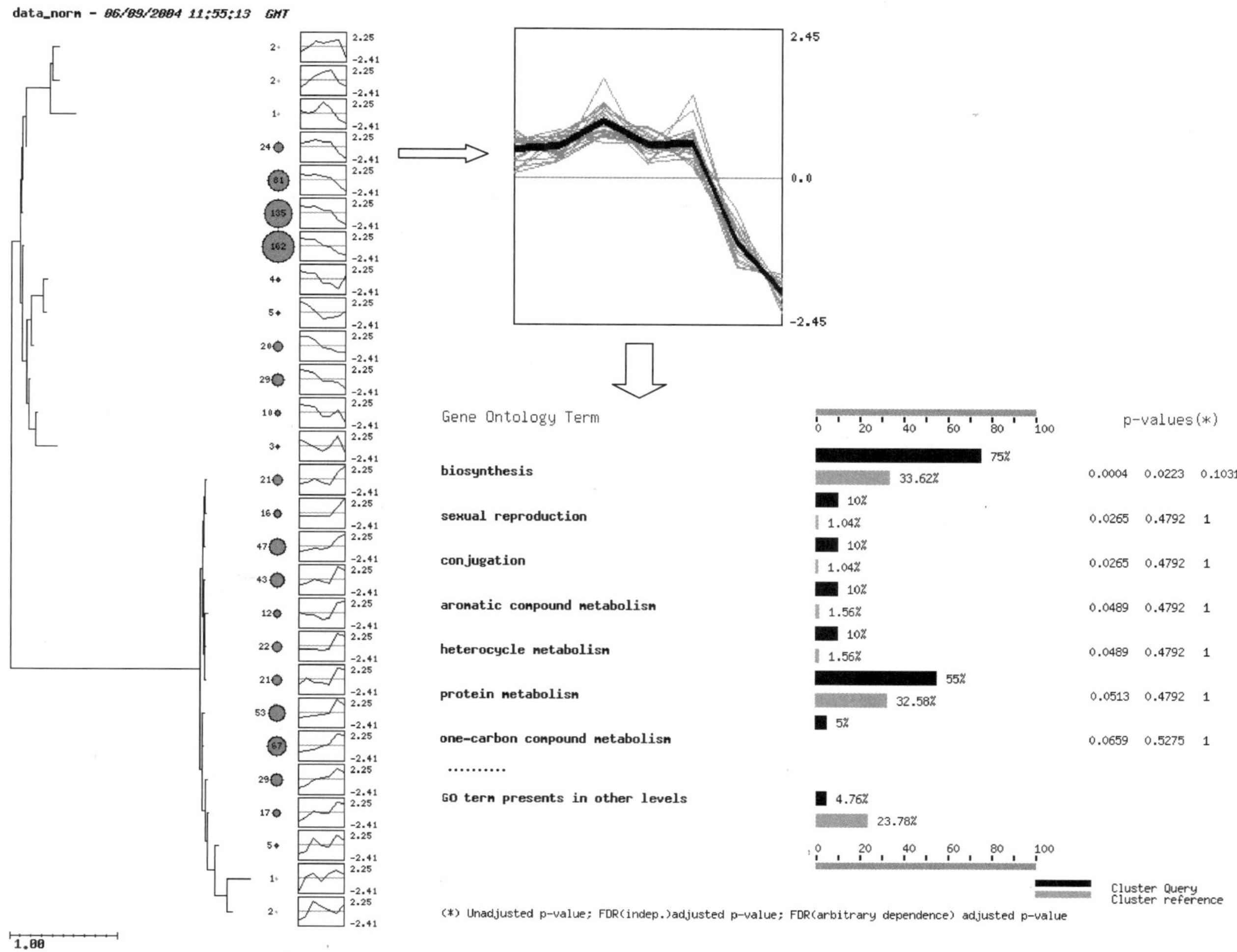

Figure 7.2 Clustering of gene expression patterns from the experiment of diauxic shift of yeast cells (DeRisi, Iyer and Brown, 1997) obtained using the SOTA algorithm (Herrero, Valencia and Dopazo, 2001)

7.9 Future Prospects

The importance of using biological information as an instrument to understand the biological roles played by genes targeted in functional genomics experiments has been highlighted in this chapter. There are situations in which the existence of noise and/or the weakness of the signal hamper the detection of real inductions or repressions of genes. Improvements in methodologies of data analysis, dealing exclusively with expression values, can to some extent help (see Chapter 12). Recently, the idea of using biological knowledge as part of the analysis process is gaining in support and popularity. The rationale is similar to the justification of using biological information to understand the biological roles of differentially expressed genes. What differs here is that genes are no longer the units of interest, but groups of genes with a common function. Let us consider a list of genes arranged according their degree of differential expression between two conditions (e.g. patients versus controls). If a given biological process is accounting for the observed phenotypic differences we should then expect to find most genes involved in this process overexpressed in one of the conditions against the other. In contrast if the process has nothing to do with the phenotypes, the genes will be randomly distributed amongst both classes (for example if genes account for physiological functions unrelated to the disease studied, they will be active or inactive in both patients and controls). Díaz-Uriarte, Al-Shahrour and Dopazo (2003) proposed the use of a sliding window across the list of genes to compare the distribution of GO terms corresponding to genes within the window against genes outside the window. If terms (but not necessarily individual genes) were found differentially represented in the extremes of the list, one could conclude that these biological processes are significantly related to the phenotypes. Al-Shahrour *et al.* (2003) generalized this approach to other types of arrangement based on other types of experiment. Recently, Mootha *et al.* (2003) proposed a different statistic with the same goal. This is part of a more general question, which would be the study of differences on prespecified groups of genes, which is discussed in Chapter 12.

Different creative uses of information in the gene selection process as well as the availability of more detailed annotations will enhance our capability of translating experimental results into biological knowledge.

References

Al-Shahrour, F., Díaz-Uriarte, R. and Dopazo, J. (2004) FatiGO: a web tool for finding significant associations of Gene Ontology terms with groups of genes. *Bioinformatics*, **20**, 578–580.

Al-Shahrour, F., Herrero, J., Mateos, Á., Santoyo, J., Díaz-Uriarte, R. and Dopazo, J. (2003) Using Gene Ontology on genome-scale studies to find significant associations of biologically relevant terms to group of genes. *Neural Networks for Signal Processing XIII*. IEEE Press, New York, 43–52 (see technical report in http://bioinfo.cnio.es/docus/papers/techreports.html#FatiGO-NNSP).

Ashburner, M., Ball, C. A., Blake, J. A. *et al.* (2000) Gene Ontology: tool for the unification of biology. *Nat Genet*, **25**, 25–29.

Bard, J. B. and Rhee, S. Y. (2004) Ontologies in biology: design, applications and future challenges. *Nat Rev Genet*, **5**, 213–322.

Bassett, D. E., Eisen, M. B. and Boguski, M. S. (1999) Gene expression informatics – it's all in your mine. *Nat Genet*, **21**, 51–55.

Beissbarth T., Speed T. (2004) GOstat: find statistically overrepresented Gene Ontologies within a group of genes. *Bioinformatics*, in press.

Benjamini, Y. and Hochberg, Y. (1995) Controlling the false discovery rate: a practical and powerful approach to multiple testing. *J R Stat Soc*, **B 57**, 289–300.

Benjamini, Y. and Yekutieli, D. (2001) The control of the false discovery rate in multiple testing under dependency. *Ann Stat*, **29**,1165–1188.

Blaschke, C., Hirschman, L., Valencia, A. (2002) Information extraction in molecular biology. *Briefings Bioinformatics*, **3**, 154–165.

Camon, E., Magrane, M., Barrell, D. *et al.* (2003) The Gene Ontology Annotation (GOA) project: implementation of GO in SWISS-PROT, TrEMBL, and InterPro. *Genome Res*, **3**, 662–672.

Castillo-Davis, C. I. and Hartl, D. L. (2003) GeneMerge – post-genomic analysis, data mining and hypothesis testing. *Bioinformatics*, **19**, 891–892.

DeRisi, J. L., Iyer, V. R. and Brown, P. O. (1997) Exploring the metabolic and genetic control of gene expression on a genomic scale. *Science*, **278**, 680–686.

Díaz-Uriarte, R., Al-Shahrour, F. and Dopazo, J. (2003) Use of GO terms to understand the biological significance of microarray differential gene expression data. In *Microarray Data Analysis III*, eds. K. F. Johnson and S. M. Lin. Kluwer, Dodrecht, 233–247.

Doniger, S. W., Salomonis, N., Dahlquist, K. D., Vranizan K., Lawlor, S. C. and Conklin, B. R. (2003) MAPPFinder: using Gene Ontology and GenMAPP to create a global gene-expression profile from microarray data. *Genome Biology*, **4**, R7.

Dopazo, J. and Carazo, J. M. (1997). Phylogenetic reconstruction using a growing neural network that adopts the topology of a phylogenetic tree. *J. Mol. Evol*, **44**, 226–233.

Eisen, M., Spellman, P. L., Brown, P. O. and Botsein, D. (1998) Cluster analysis and display of genome-wide expression patterns. *Proc. Natl. Acad. Sci. USA*, **95**, 14 863–14 868.

Ge, H., Walhout, A. J., Vidal, M. (2003) Integrating 'omic' information: a bridge between genomics and systems biology. *Trends Genet*, **19**, 551–560.

Herrero, J., Al-Shahrour, F., Díaz-Uriarte, R., Mateos, A., Vaquerizas, J. M., Santoyo, J. and Dopazo, J. (2003) GEPAS, a web-based resource for microarray gene expression data analysis. *Nucleic Acids Res*, **31**, 3461–3467.

Herrero, J., Valencia, A. and Dopazo, J. (2001) A hierarchical unsupervised growing neural network for clustering gene expression patterns. *Bioinformatics*, **17**, 126–136.

Holloway, A. J., van Laar, R. K., Tothill, R. W., Bowtell, D. D. (2002) Options available – from start to finish – for obtaining data from DNA microarrays II. *Nat Genet*. **32** (Suppl.), 481–489.

Jensen, L. J., Gupta, R., Staerfeldt, H. H., Brunak, S. (2003) Prediction of human protein function according to Gene Ontology categories. *Bioinformatics*, **19**, 635–642.

Jenssen, T.-K., Laegreid, A., Komorowski, J. Hovig, E. (2000) A literature network of human genes for high-throughput analysis of gene expression. *Nature Genet*, **28**, 21–28.

Kanehisa, M., Goto, S., Kawashima, S., Okuno, Y., Hattori, M. (2004) The KEGG resource for deciphering the genome. *Nucleic Acids Res*, **32** (database issue), D277–D280.

Khatri, P., Draghici, S., Ostermeier, G. C. and Krawetz, S. A. (2002). Profiling gene expression using onto-express. *Genomics*, **79**, 1–5.

MacBeath, G. (2002) Protein microarrays and proteomics. *Nat Genet*. **32**, (Suppl). 526–532.

Mateos, Á., Dopazo, J., Jansen, R., Tu, Y., Gerstein, M. and Stolovitzky, G. (2002) Systematic learning of gene functional classes from DNA array expression data by using multilayer perceptrons. *Genome Res*, **12**, 1703–1715.

Mootha, V. K., Lindgren, C. M., Eriksson, K. F., *et al.* (2003) PGC-1alpha-responsive genes involved in oxidative phosphorylation are coordinately downregulated in human diabetes. *Nat Genet*, **34**, 267–273.

Mulder, N. J., Apweiler, R., Attwood, T. K., *et al.* (2003) The InterPro Database, 2003 brings increased coverage and new features. *Nucleic Acids Res*, **31**, 315–318.

Oliveros, J. C., Blaschke, C., Herrero, J., Dopazo, J., Valencia, A. (2000) Expression profiles and biological function. *Genome Informatics*, **10**, 106–117.

Pavlidis, P., Lewis, D. P., and Noble, W. S. (2002) Exploring gene expression data with class scores. *Pacific Symp. Biocomput*, **7**, 474–485.

Raychaudhuri, S, Chang, J. T., Imam, F., Altman, R. B. (2003) The computational analysis of scientific literature to define and recognize gene expression clusters. *Nucleic Acids Res*, **31**, 4553–4560.

Raychaudhuri, S., Chang, J. T., Sutphin, P. D., Altman, R. B. (2002) Associating genes with gene ontology codes using a maximum entropy analysis of biomedical literature. *Genome Res*, **12**, 203–214.

Raychaudhuri, S., Schutze, H., Altman, R. B. (2002) Using text analysis to identify functionally coherent gene groups. *Genome Res*, **12**, 1582–1590.

Robinson, M. D., Grigull, J., Mohammad, N. and Hughes, T. R. (2002). FunSpect: a web-based cluster interpreter for yeast. *BMC Bioinformatics*, **3**, 1–5.

Schug, J., Diskin, S., Mazzarelli, J., Brunk, B. P., Stoeckert, C. J. Jr. (2002) Predicting gene ontology functions from ProDom and CDD protein domains. *Genome Res*, **12**, 648–655.

Shah, N. H., Fedoroff, N. V. (2004) CLENCH: a program for calculating Cluster ENriCHment using the Gene Ontology. *Bioinformatics*, **20**, 1196–1197.

Tanabe, L., Smith, L. H., Lee, J. K., Scherf, U., Hunter, L., Weinstein, J. N. (1999) MedMiner: an internet tool for filtering and organizing bio-medical information, with application to gene expression profiling. *BioTechniques*, **27**, 1210–1217.

van Noort, V., Snel, B. and Huynen M. A. (2003) Predicting gene function by conserved co-expression. *Trends Genet* **19**, 238–242.

Westfall, P. H. and Young, S. S. (1993) *Resampling-Based Multiple Testing*. Wiley, New York.

Xenarios, I., Salwinski, L., Duan, X. J., Higney, P., Kim, S., Eisenberg D. 2002 DIP: the Database of Interacting Proteins. A research tool for studying cellular networks of protein interactions. *Nucl. Acids Res*, **30**, 303–305.

Xie, H., Wasserman, A., Levine, Z., Novik, A., Grebinskiy, V., Shoshan, A., Mintz, L. (2002) Large-scale protein annotation through gene ontology. *Genome Res*, **2**, 785–794.

Zeeberg, B. R., Feng, W., Wang, G., *et al.* (2003) GoMiner: a resource for biological interpretation of genomic and proteomic data. *Genome Biology*, **4** (4), R28.

8

The *C. elegans* Interactome: its Generation and Visualization

Alban Chesneau and **Claude Sardet**

Abstract

This chapter focuses on the generation and utilization of a *C. elegans* interactome network. The first part describes the high-throughput ORF cloning and high-throughput two-hybrid technique in order to determine the protein–protein interactions and to generate such a protein–protein interaction map. In the second part, elements of the topological structures of interactomes in general as well as the *C. elegans* interactome more specifically are presented. The biological utility of such an approach is discussed. The last part focuses on the integration of protein–protein interactions with other post-genomics data in order to filter these datasets as well as to give a dynamical view of this interactome.

Keywords

C. elegans, two hybrid, protein–protein interaction networks, post-genomics data integration

8.1 Introduction

In 1966, Sydney Brenner, whose initial interest was to study the development of the nervous system, chose the nematode *C. elegans* as a model for its simplicity and experimental tractability. Almost 40 years later, thanks to the availability of many mutants, *C. elegans* has allowed us to achieve the larger goal of delineating the exact course of development at an unanticipated level. The work performed with this model can now quickly consolidate and extend the knowledge obtained from studying

Data Analysis and Visualization in Genomics and Proteomics Edited by Francisco Azuaje and Joaquin Dopazo

several non-model organisms. The extensive use of molecular techniques as well as genetic approaches rendered *C. elegans* an organism of choice for sequencing the whole genome of a metazoan.

In 1983, John Sulston and Alan Coulson began to construct a complete physical map of the genome of *C. elegans* and initiated the so-called '*C. elegans* Genome Project'. Fifteen years after its foundation, this project has achieved the ambitious goal of providing to the scientific community with the entire genomic sequence of this model organism (*C. elegans* Sequencing Consortium, 1998). This sequencing revealed that 83 per cent of the 17 000 genes of *C. elegans* have human homologues, strongly supporting the notion that studies performed in this metazoan model organism could provide the basis to test new hypotheses relevant to human biology (Lai *et al.*, 2000). Although the availability of these sequences has prompted many new outside investigators to study *C. elegans* gene functions, only seven per cent of the genes in the *C. elegans* genome have already been associated with a specific biological function based on classical forward genetics or biochemical analyses.

To provide a universal tool to accelerate this search, Marc Vidal's laboratory (Dana Farber Cancer Institute, Boston) developed a collection of 12 000 ORFs (open reading frames) or the ORFeome (Reboul *et al.*, 2003) of *C. elegans.* The availability of this resource, combined with the recent development of high-throughput two-hybrid screening, provides the opportunity to establish large-scale protein–protein interaction maps. These maps (the interactome) open novel perspectives to explore a protein's functions and to formulate global biological hypotheses based on the existence of unexpected networks of protein interactions. Historically, the *E. coli* bacteriophage T7 was the first organism for which a small-scale protein–protein interaction map was generated in 1996 (Bartel *et al.*, 1996), an interactome that included only 25 protein interactions identified by a two-hybrid approach in yeast.

A couple of years later, the yeast *S. cerevisiae* was chosen as a model to establish the first interactome of an eukaryotic organism. Various large-scale proteome-wide approaches have been used to reach this goal, including (1) high-throughput two-hybrid screenings (Uetz *et al.*, 2000; Ito *et al.*, 2001), (2) tandem affinity purifications of multiprotein complexes identified by mass spectrometry analysis (Gavin *et al.*, 2002; Ho *et al.*, 2002) and (3) *in silico* computational predictions (Dandekar *et al.*, 1998; Marcotte *et al.*, 1999; Pellegrini *et al.*, 1999). Altogether, these experiments have already generated more than 28 000 protein interactions, 2500 of which were identified by at least two different methods.

In multicellular organisms, the possibility of using the large-scale two-hybrid technique to examine protein interaction networks was first successfully applied to a set of *C. elegans* ORFs corresponding to proteins involved in vulval development (Walhout *et al.*, 2000). After this initial success, several other studies dealing with various integrated biological processes such as DNA damage repair (Boulton *et al.*, 2002), complexes such as the proteasome (Davy *et al.*, 2001) or specific biological states (Walhout *et al.*, 2002) were performed in Marc Vidal's laboratory. Finally, the first version of a global genome-wide interactome of *C. elegans* was recently

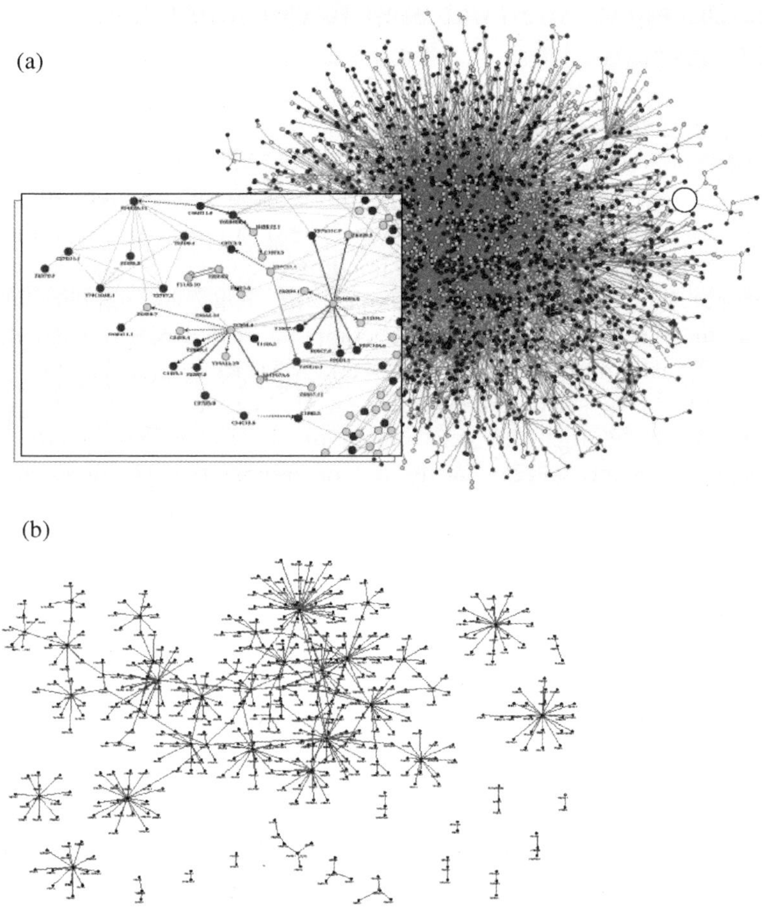

Figure 8.1 The *C. elegans* interactome. (a) The initial version of the *C. elegans* interactome is composed of more than 4000 protein iterations classified into different categories depending on the level of confidence (Li *et al.*, 2004). (b) The first version of the chromatin remodelling network is a part of the *C. elegans* interactome. Like the entire network, the chromatin remodelling interactome exhibits small-world properties: a few proteins (hubs) are highly connected whereas the majority is poorly connected

published (Li *et al.*, 2004) (Figure 8.1). Interestingly, the *D. melanogaster* interaction map generated simultaneously also exhibits the same network properties (scale-freeness, connectivity) (Giot *et al.*, 2003).

As participants in the international effort co-ordinated by the Vidal laboratory that led to the publication of the first version of the *C. elegans* interactome, we now return to the technical developments that helped to generate this map and discuss the scientific conclusions, caveats and future directions that follow from this publication.

8.2 The ORFeome: the First Step Toward the Interactome of *C. elegans*

The sequencing of the entire *C. elegans* genome (error rate of 1 nucleotide/30 kb) revealed that it contains more than 17 000 predicted genes separated by intergenic regions that, on average, are shorter than those from other multicellular model organisms (*C. elegans* Sequencing Consortium, 1998). A large fraction of these genes appeared to be organized in operons corresponding to clusters of co-expressed genes. The efforts put into the sequencing of eukaryotic genomes are parallel to that dedicated to the development of reliable computational predictions to determine the number and position of exons, introns, 5′ and 3′ UTRs (untranslated regions) and alternatively spliced sites (for an extensive, well documented review, see Zhang, 2002). Like other model organisms, this annotation of the *C. elegans* genome is under continuous improvement, notably by the use of comparative genomics approaches with related genomes, e.g. with *C. briggsae* in the case of *C. elegans* (Stein *et al.*, 2003).

Although far from perfect, the first annotated version of the *C. elegans* genome was used in the late 1990s by M. Vidal's laboratory to initiate the first large-scale ORFs amplification project (ORFeome) (Figure 8.2). The goal of this project was to provide experimenters with an ordered collection of individualized and unique DNA fragments representing the full-length coding sequences of all the genes annotated in the *C. elegans* genome. Practically, these ORFs were isolated by PCR from a cDNA library representative of several developmental stages of the nematode (i.e. larval L1, L2, L3, L4, egg, dauer, male and hermaphrodite worms) using specific pairs of primers designed to fit the 3′ and 5′ end boundaries of each annotated gene. Each amplification product, OST (ORF sequence tag), was then checked for its size and sequenced. Notably, a significant proportion of PCR (polymerase chain reaction) products did not fit the size predicted by the initial bioinformatics predictions, although the sequenced tag indicated that it was the expected target gene. Among other possible explanations, these variations reflect the presence of splice variants and of mispredicted annotated ORFs.

Thus, beside its usefulness for post-genomic studies, this OST approach turned out to be useful as an alternative strategy to verify *in silico* the predictions made by the Genescan (Burge and Karlin, 1997) or Genefinder (unpublished work of Colin Wilson, LaDeana Hilyer and Phil Green) interfaces concerning the existence of a gene and of the position of the intron/exon junctions (mispredicted for more than 50 per cent of the genes). A first version of the ORFeome (1.1) was completed in 2003 and corresponds to 12 000 full-length ORFs out of 17 000 predicted sequences (Reboul *et al.*, 2003).

To facilitate the use of this resource in various functional genomics screenings, the 3′ and 5′ ends of these ORFs were tailed to be compatible with the Gateway-based recombinational cloning procedure (Invitrogen). This cloning design allow a rapid transfer of the complete ORFs collection from an entry

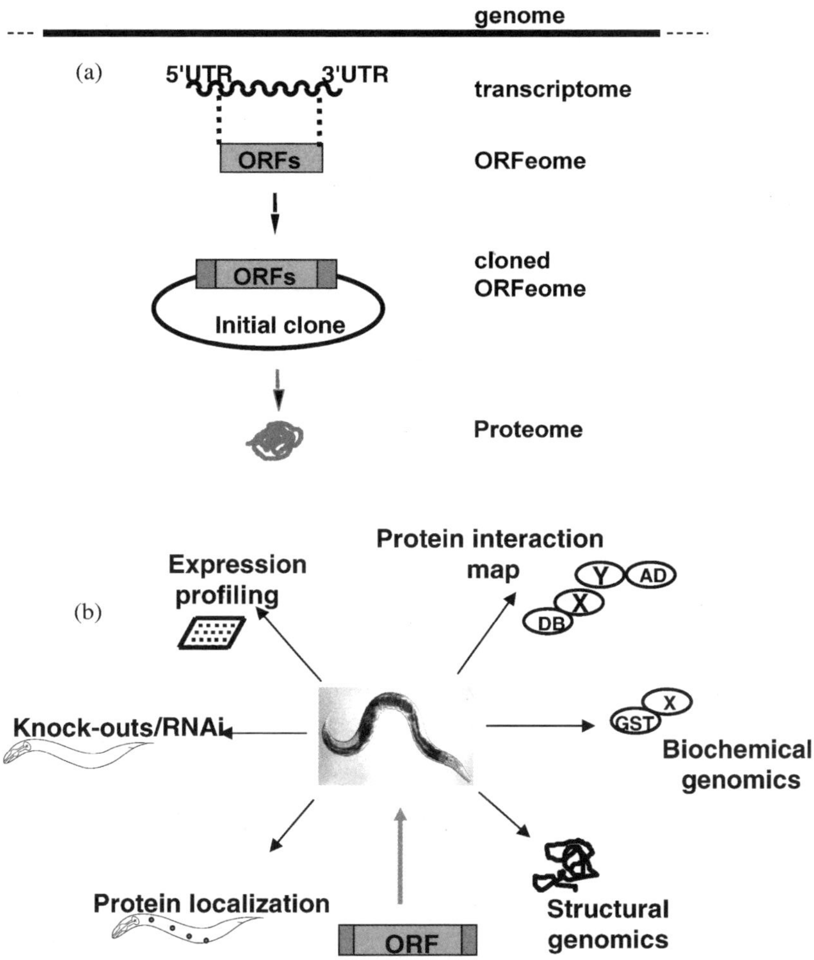

Figure 8.2 The *C. elegans* ORFeome and the Gateway transfer. (A) The first version of the *C. elegans* ORFeome was obtained by cloning a set of 12 000 ORFs predicted by different algorithms (e.g. Genescan and Genefinder) from the complete genome of *C. elegans*. This collection of ORFs is extremely helpful in order to express a protein of interest and so to envisage proteomic as well as post-genomics approaches. (B) The Gateway recombinational cloning allows a convenient transfer of the ORFs of interest from a donor vector to a destination vector (e.g. GST fusion vectors for biochemical genomics approaches or GFP fusion vectors for determining the protein's localization)

plasmid vector into many specific, existing or future gateway compatible destination vectors. This makes the ORFeome a fantastic tool to develop high-throughput functional genomic and proteomic strategies at a genome-wide scale. The following chapter describes the use of this resource to realize large-scale two-hybrid screens in yeast.

8.3 Large-Scale High-Throughput Yeast Two-Hybrid Screens to Map the *C. elegans* Protein–Protein Interaction (Interactome) Network: Technical Aspects

The yeast two-hybrid approach (Y2H) to identify interactions between two proteins X and Y, was initially described by S. Fields and co-workers (Fields and Song, 1989). It is based on a yeast genetic system in which expression of Gal4-responsive reporter genes depends on the reconstitution of Gal4 activity via the interaction of X with Y. This is accomplished by the co-expression in the same yeast cell of the two interacting proteins X and Y fused to the Gal4 DNA binding (DB) and Gal4 transcriptional activation (TA) domains, respectively.

A large-scale yeast two-hybrid screen aiming to map part of the *C. elegans* interactome was coordinated by Marc Vidal's laboratory and performed as follows. In order to be as complementary as possible to the already available interactome of yeast, it was chosen to select more than 3000 predicted bait proteins of *C. elegans* ($\sim$15 per cent of the proteome). The function of these selected baits might be either specific for metazoans (embryogenesis, differentiation, immunity, pharynx, sexual reproduction etc.) or involved in evolutionary conserved biological processes whose outcome and context are different between yeast and multicellular organisms (meiosis, mitosis, chromatin remodelling). 860 proteins were chosen according to previous genetics or literature-based data supporting the view that they fit these criteria. In addition, 2200 others were selected on the criteria that they have a clear homologue in the human but not in the yeast proteome.

For proteins involved in evolutionary conserved biological processes that were used in this screen ($\sim$400 proteins), such as many factors involved in chromatin remodelling, the criterion of choice was either (1) the existence of a clear orthologue in *S. cerevisiae* and human (orthology: blast value $<10^{-10}$) involved in this process according to genetics or literature-based data, or (2) the presence of evolutionary conserved specific domains identified *in silico* (e.g. chromodomains, bromodomains or SET domains that are clearly involved in the regulation of transcription at the chromatin level).

Approximately 2000 of these selected proteins were present in the Gateway-cloned ORFs. They were retrieved from the ORFeome library and transferred into the yeast two-hybrid vector ppC97-dest (Gal4DB, bait vector) using a Gateway LR recombinational reaction. Each Gal4DB-ORF bait plasmid was then transferred into the Y2H yeast strain MaV203, which carries several integrated Gal4-responsive reporter genes (GAL1::HIS3, GAL1::LACZ, SPAL10::URA3) suitable to monitor Y2H interactions. To eliminate Gal4DB-ORFs that were either toxic for the yeast cells or had the characteristics of autoactivators (i.e. activate transcription on their own without the need of interacting proteins), all these baits were first tested for their capacity to activate the reporter genes in absence of any AD-containing vector. Finally, $\sim$1900 baits were used in high-throughput Y2H screenings (Figure 8.3). (A) The two hybrid is based on the transcriptional activity of the Gal4p protein. Using proteins (X and Y)

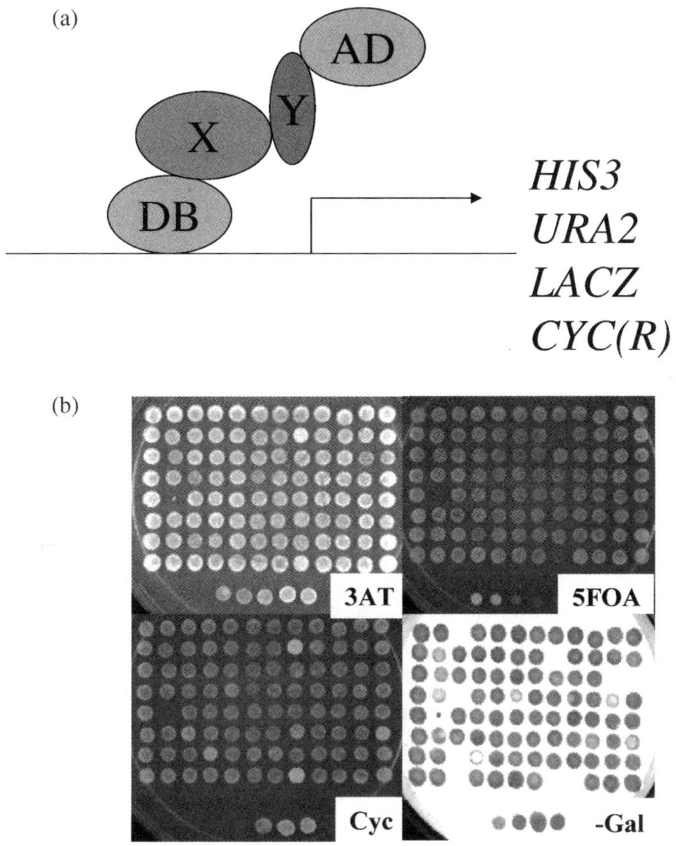

Figure 8.3 Generalities about the two-hybrid and phenotypic tests

fused to the DNA binding domain (DB) or the activation domain of Gal4p (AD), a functional Gal4p protein can be reconstituted if X and Y interact. Several gene reporters can be used to test this interaction: *HIS3, URA2, LACZ* and the resistance marker to cycloheximide. (B) Shows the result of the phenotypic tests. 3-AT (3-aminotriazole) is used for a positive selection of the *HIS3* gene activity, 5-FOA (5-fluorouracil) for a negative selection of the *URA2* gene, the β-Gal assay for testing the activity of the *LACZ* gene and the cycloheximide resistance marker for evaluating the self-activation of the preys.

In order to improve both the coverage and accuracy of these screens, they were performed against two different, yet complementary, Gal4-AD-cDNA libraries. The first one, AD-wormcDNA, is a classic Y2H cDNA library prepared from *C. elegans* mRNAs of several developmental stages (Walhout *et al.*, 2000b). The usual limitation intrinsic to this type of library is an uneven representation of the transcripts, which can lower the number of recorded true positives (interactions that really occur). To tackle this problem, the second library of Y2H preys, AD-ORFeome 1.0, was built from a pool of all the ORFs present in the previously described ORFeome. Although better in terms of normalization, this second library also presents drawbacks, especially because the ORFeome does not cover all the ORFs. Moreover, many interactions can be missed because numerous Y2H interactions may be masked when strictly using full-length proteins (Uetz *et al.*, 2000; Legrain and Selig, 2000).

In order to increase the reliability of the determination of true positives based on phenotypic tests, three reporter genes were monitored (GAL1::HIS3, GAL1::LACZ and SPAL10::URA3) (Vidal, 1997). Bait–prey pairs were considered as positives if they activated at least two out of these three reporters (Vidal *et al.*, 1996) and if they were able to reproduce the same phenotype upon their re-expression in fresh yeast cells. The corresponding AD–Y interactor sequence was then amplified by PCR (polymerase chain reaction) and sequenced in order to obtain an IST (interaction sequence tag). The IST was retained only if (1) the sequence quality was sufficient (the Phred score, which is a computer program for automated base calling, should be at least 20 over at least 15 per cent of its length; Erwin and Green, 1998; Erwin *et al.*, 1998), (2) it showed a blast E value $<10^{-10}$ against the WormPep WS100 and (3) its AD–prey reading frame was correct. In total, 16 000 ISTs were obtained.

Interactions were classified into confidence categories (core, non-core and scaffold datasets) corresponding on one side to the core data set of high-confidence interactions and on the other side to all other Y2H identified interactions. The former high-confidence interactions are represented by in-frame bait–prey found either at least three times independently or less than three times but passing the retest into fresh yeast cells. Altogether, more than 4000 distinct interactions were generated by these screens with half of them fitting the criteria of high confidence. Surprisingly, only six per cent of these interactions were identified with both the AD-wrmcDNA and AD-ORFeome 1.0 libraries, emphasizing the importance of using both types of library for such large-scale screenings.

These Y2H interaction data was pooled with that previously published by the Vidal laboratory about various biological processes in *C. elegans* (namely the proteasome, vulval development, DNA damage responses and the germline formation). Finally, *in silico* searches were performed to identify conserved interactions (termed 'interologues'), whose orthologous pairs have been shown to interact in other species in at least two of the following datasets: yeast two hybrid (Uetz *et al.*, 2000; Ito *et al.*, 2001), large-scale pull-down MS (Gavin *et al.*, 2002; Ho *et al.*, 2002), computational methods (Dandekar *et al.*, 1998; Marcotte *et al.*, 1999; Pellegrini *et al.*, 1999) and the MIPS complex list (http://mips.gsf.de/genre/proj/yeast/index.jsp). Worm orthologues were identified by reciprocal best-hit BLASTP searches, with E-value $\leq 10-6$.

Altogether, interaction data concerning $\sim$1000 *C. elegans* 'interologues' were combined with the Y2H experimental data to constitute the Worm Interactome version 5 (WI5). In total, this version of the *C. elegans* interactome connects 15 per cent of the predicted proteome of *C. elegans* through a network of $\sim$3000 nodes connected to each other by more than 5400 interaction edges.

8.4 Visualization and Topology of Protein–Protein Interaction Networks

Visualization of the *C. elegans* protein–protein interaction network

All the information about the experimental features of each interaction (i.e. phenotypic screening, yeast plates, retesting conditions, degree of confidence), as well as the corresponding IST sequence information (chromatogram, blast values, Phred scores), were stored in a database termed I-view developed by M. Vidal's laboratory (http://vidal.dfci.harvard.edu/interactomedb/i-View.pl). This tool is flexible and many other libraries or projects can be added to the existing information. This data visualization tool is also linked to several genome-wide post-genomics datasets available about *C. elegans* genes and proteins, including those concerning mRNA localization by *in situ* hybridization, the phenotypic consequences of gene inactivation by RNAi (RNA interference) and their profile of expression (cDNA microarrays). Finally, I-view provides a graphic representation of the interactome itself as a network of nodes and links generated using a specific software derived from Leda Graphwin. Leda Graphwin is an interface that forms a bridge between the graph data types, the graph algorithms and the graphics interface of LEDA: http://www.algorithmic-solutions.com/enleda.htm

Other graphic interfaces could be used to visualize these biomolecular interaction networks, such as Osprey (http://biodata.mshri.on.ca/ osprey/servlet/Index) or Interviewer (http://wilab.inha.ac.kr/interviewer) which both offer a 3D representation of the networks, and more interestingly Cytoscape (http://www.cytoscape.org), which also allows an integration of this interaction data with gene expression profiles and other state data. They all provide basic functionality that could lay out and query this network.

Network topology: an overall view

Several types of interaction network, such as transcriptional, metabolic and protein–protein interaction networks, are emerging from high-throughput post-genomic studies. Several tools are currently developed to quantitatively analyse and dissect these networks, allowing them to be compared. This will help us to understand their topologies, which should reflect their functions (see Barabasi and Oltvai, 2004, for a review) (Figure 8.4). (A) Shows a graphical representation of different types of network. (1) Scale-free network. (2) Modular network made of four interlinked modules. Its organization is not scale free. (3) Hierarchical network with a scale-free topology. This network is typical of the metabolic organization. (Adapted from Ravasz *et al.*, 2002.) (B) Different types of motif were observed for several networks (transcriptional networks, neuron connectivity, food webs and electronical circuits),

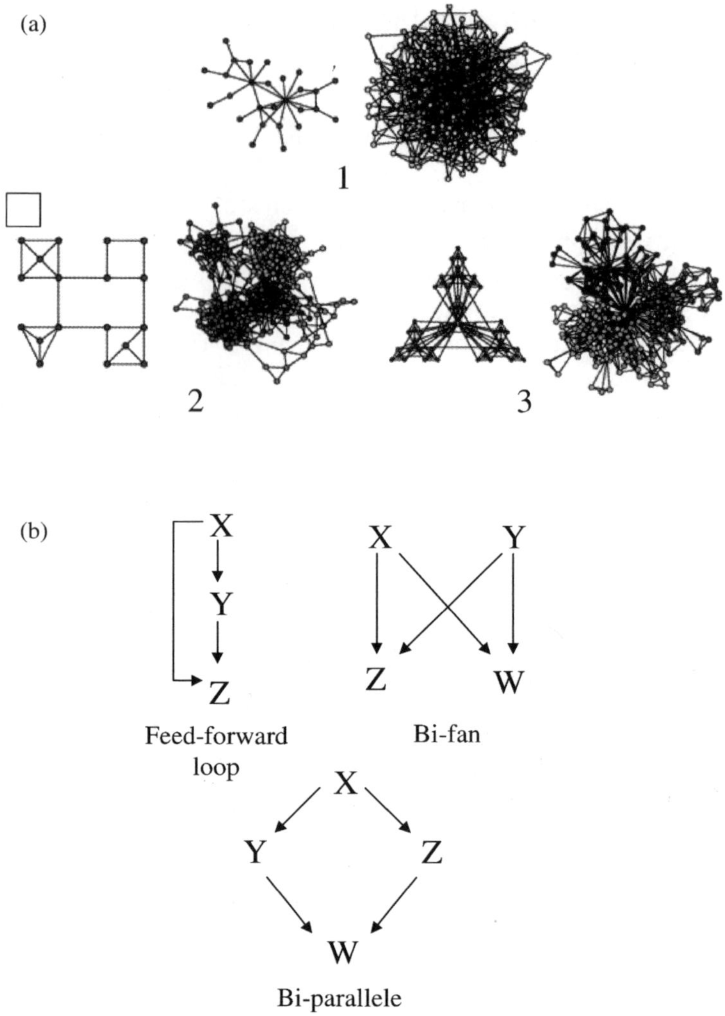

Figure 8.4 Different types of networks

underlying the fact that many networks share the same structural specificity (adapted from Milo *et al.*, 2002).

The parameters that are used to compare and characterize these networks are the following. (1) The degree k (also called average connectivity), which indicates how many links a given node has with the others. (2) The distribution degree $P(k)$, which gives the probability for a selected node to have exactly k links. Most biological networks are scale free and are thus governed by a power-law distribution that predicts that a few nodes (hubs) are highly interconnected, whereas a few links connect a wide proportion of nodes. (3) The degree exponent of the typical power-law distribution of biological networks corresponds to the proportionality existing between the number of nodes and the number of links. It reflects the number of hubs that support the global organization of the network. (4) The shortest, largest and mean path lengths indicate respectively the smallest, largest and average number of links that we have to pass to travel between two nodes. (5) The clustering coefficient is a measure of the connectivity of a node. It corresponds to the fraction of the existing links compared to all possible links of a node. In a genomic context network, the clustering coefficient is much higher (0.6) than that of a random network with the same number of nodes and connections (0.005) (Snel, Bork and Huynen, 2002).

Biological network properties

At present, the application of these tools to set the parameter of the global organization of the yeast and *C. elegans* protein–protein networks identified by Y2H reveals that they fit with the structure of a small-world network. Moreover, these networks seem to adopt a scale-free degree distribution (Li *et al.*, 2004; Yook, Oltvai and Barabasi, 2004) compared with randomly generated protein networks (Goldberg and Roth, 2003; Li *et al*, 2004). This topology is characterized by densely connected neighbourhoods and short characteristic path lengths. In other words, it appears clearly that a few proteins are highly connected (hubs), whereas the others are poorly connected. Notably, most of these hubs do not interact directly. It is still unclear whether this last observation is biologically relevant or whether, at least in part, it reflects a systematic bias in the experimental procedure used to detect the links. Indeed, in Y2H screens, it has been observed that proteins corresponding to hubs are preferentially bait rather than prey (Aloy and Russell, 2002). Another surprising observation about these highly connected proteins is the quasi-absence of membrane proteins and enzymes among them, a feature that, again, might reflect the limited capacity of Y2H to identify these interactions, or alternatively the real biological situation of these proteins. Interestingly, these small-world and scale-free properties also seem to apply to other biological networks, such as gene regulatory and metabolic networks. Thus, it appears a that few transcription factors regulate an important number of genes whereas, conversely, many transcription factors regulate a few genes. Similarly, in the case of metabolic networks, a few molecules such as ATP, GTP, SAM or NAD are involved in many biological reactions (Ravasz *et al.*, 2002).

8.5 Cross-Talk between the *C. elegans* Interactome and other Large-Scale Genomics and Post-Genomics Data Sets

Interactome and gene expression data

Comparison of results obtained from various large-scale genomics and post-genomics data sets, such as gene expression data arising from transcriptome analyses, physical protein interactions found by two-hybrid, mass-spectrometry analyses of protein complexes or genetic interactions, should help to improve the accuracy of predicted biological networks. Obviously, the comparison of the WI5 Y2H data set with other genome-wide protein–protein interaction data sets should be considered first. However, these large-scale data sets are not yet available for the metazoan *C. elegans*.

Several attempts are currently being tested at a small scale (Boulton *et al.*, 2002) to create the molecular tools and protocols to performed proteomic studies at the scale of the ORFeome. These approaches use the gateway-based transfer of the ORFeome in proteomic-dedicated expression vectors that allow, for example, mass spectrometry analyses of protein complexes immunoprecipitated or purified from transgenic worms expressing tagged proteins and pull-down experiments using GST fused ORF products (Li *et al*, 2004).

At present, the integration of the WI5 interactome data set with other large-scale data sets has only been done with the *C. elegans* transcriptome and phenome data sets (Walhout *et al.*, 2002) (Figure 8.5). An accurate, well adapted statistical analysis, as well as a thorough organization of the results of gene expression studies (see Leung and Cavalieri, 2003, for a review) might be a bottleneck. Nevertheless, several attempts have already been made to integrate protein interaction datasets with gene expression profiles of various physiological conditions or different developmental stages, obtained by DNA microarray analyses (for examples, refer to Hill *et al.*, 2000; Reinke *et al*, 2000; Blumenthal *et al.*, 2002; Gaudet and Mango, 2002; Roy *et al.*, 2002, and Zhang, 2002). This type of analysis was first applied to the *S. cerevisiae* datasets using statistical tools including Pearson's correlation coefficient (Gregoriev, 2001), Cosine's coefficient (Kemmeren *et al.*, 2002), normalized difference (Jansen, Greenbaum and Gerstein, 2001), clustering and protein interaction density (Ge *et al.*, 2001). The last analysis was performed at a genome-wide scale by Ge *et al.* by organizing clusters derived from a set of related transcriptional profiling experiments and calculating the protein interaction density for each square of this cluster. This analysis showed that co-regulated genes have a higher tendency to interact together. Similar conclusions were obtained for *C. elegans* (Walhout *et al.*, 2002; Li *et al.*, 2004). This conclusion was obtained by overlapping part of the WI5 interactome map (core, non-core, scaffold and literature datasets) with the *C. elegans* transcriptome microarray data contained in the *C. elegans* topomap (Kim *et al.*, 2001). The topomap provides expression profiling data organized into 'mountains' of genes clustered according to the degree of similarity of their expression. However, numerous genes (up to more than 1000) are gathered into each 'mountain' (Kim *et al.*, 2001) and data sets are imperfectly overlapping (Kemmeren *et al.*, 2002). Thus, this conclusion should be considered as a preliminary result that only gives the 'flavour' of the correlation that might exist between transcriptome and interactome in *C. elegans*.

In a first approximation, WI5 interactions corresponding to *C. elegans* gene pairs that are highly correlated in expression profile tend to be of higher confidence. Interestingly, several experimental examples found in the literature indicate that this type of correlation is more obvious for proteins belonging to permanent complexes

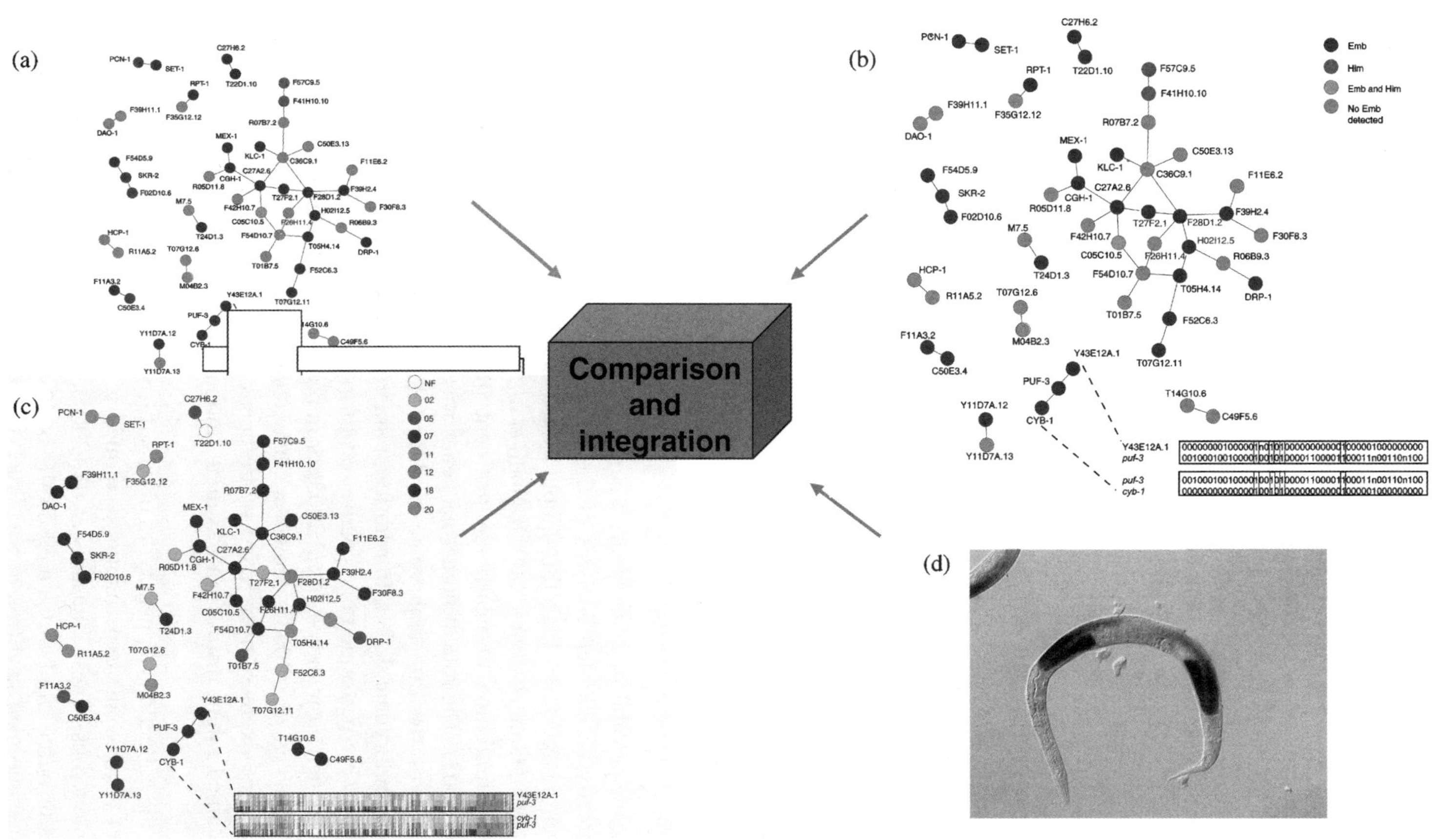

Figure 8.5 Data comparison and integration, example of the germline-specific proteins. (A) The germline-specific gene interactome from Walhout *et al.* (2002). (B) Comparison between phenome information (each observation on the phenotype was annotated by a Boolean and then the succession of Booleans was compared between the interacting proteins. (C) Comparison between gene expression data for interacting proteins. Each protein was affiliated to a mountain of expression (Kim *et al.*, 2001). It appears that several proteins in this interactome tends to belong to the same mountain. (D) The data of *in situ* hybridization can also be compared for the proteins belonging to this interactome, since they are involved in the germline process, which is restricted to specific tissues

than for those participating in transient complexes. Conversely, the proteins encoded by genes displaying similar expression patterns tend to form stable complexes. This was demonstrated for yeast proteins (Jansen *et al.*, 2002) and also proven to be true for *C. elegans* proteins (Li *et al*, 2004). In conclusion, although poorly documented at present, there is no doubt that these types of overlapping analysis will soon help to estimate the accuracy of each data set and will provide hypotheses of functional links between proteins of known and unknown functions. Examples of such predictions are illustrated in the *C. elegans* WI5 protein–protein interaction publication (Li *et al.*, 2004).

Protein–protein interaction and genetics analysis

Overlapping the *C. elegans* protein–protein dataset with phenotype datasets present in the WormBase was performed at both a small (Walhout *et al.*, 2002; Boulton *et al.*, 2002) and a more global scale (Li *et al.*, 2004) by the Vidal laboratory. *C. elegans* geneticists have traditionally used forward or reverse genetics as standard approaches to obtain information about gene function. Several genome-wide tools were developed by the *C. elegans* community to perform these analyses: (1) a collection of ds (double-strand) RNA-expressing plasmids that block protein expression by mRNA interference; (2) a set of more than 700 knock-out strains generated by the *C. elegans* Knock-Out Consortium; (3) the ORFeome that can be transferred into various expression vectors.

Historically, the possibility to inactivate a gene function by RNA interference in *C. elegans* has greatly accelerated analysis of the phenotypic consequences of loss of function. Several large-scale high-throughput RNAi studies have been performed (Fraser *et al.*, 2000; Gonczy *et al.*, 2000; Kamath *et al.*, 2003), the largest one scoring for the possible inhibition of nearly 86 per cent of predicted *C. elegans* genes (Kamath *et al.*, 2003). The role of numerous genes was uncovered by scoring specific phenotypes such as sterility, embryonic or larval lethality, slow post-embryonic growth or a post-embryonic defect. In addition, numerous other smaller-scale studies of the same type were performed on specific sets of genes, such as those involved in transposon silencing (Vastenhouw *et al.*, 2003) or in protection of the genome against mutations (Pothof *et al.*, 2003). However, one should keep in mind that the inactivation of gene expression by RNAi is, in most cases, incomplete, and that the phenotype obtained might result from residual amounts of message of the targeted gene.

Altogether, the collection of phenotypes obtained from these small- and large-scale RNAi and KO (knock-out) experiments is termed a 'phenome' and archived in a database for numerous genes (Gunsalus *et al.*, 2004). Interestingly, the information present in this resource can be clustered on the basis of phenotypic data to identify groups of genes enriched in different functions of classes. Moreover, this database provide a phenotypic blast tool, making it possible to generate phenotypic maps and to identify which genes display the most similar phenotypic signatures relative to a reference gene, simply by looking at their phenotypic descriptors.

Therefore, by overlapping this database with the WI5 map, it becomes possible to speculate on whether interacting proteins belong to the same phenotypic classes. One of the most successful analyses of this type was performed for the proteins involved in the DNA damage repair (DDR) machinery (Boulton *et al.*, 2002). In this study, phenotypes were recorded after a short exposure to radiation of animals for which DDR genes were silenced by RNAi. Several proteins that are interacting in the DDR network appear to exhibit the same phenotypes. The same conclusions applied to the germline-specific genes (Walhout *et al.,* 2002) and, more recently, to several protein networks identified in the WI5 map, such as the proteins involved in the meiosis process (unpublished results). Overall, these reports show that interacting protein pairs in the high-confidence dataset are several times more likely than non-interactors to share an annotated RNAi phenotype.

Notably, a similar relationship between interactome and phenome was observed in yeast (Jeong *et al.*, 2001). A comparison of the number of connections per node with the data on the lethality of mutations indicates that the larger the number of physical interactions, the higher is the probability that the mutated gene is essential for survival (Jeong *et al.*, 2001). Thus, it appears that essential genes in this organism are often associated with hubs in the interactome.

Several other attempts are now underway in *C. elegans* to integrate the phenome data with protein interaction data. New phenotypic assays need to be developed in order to evaluate more precisely the phenotypic effect of thousands of genes tested simultaneously, e.g. intracellular localization, life span, general morphology and growth rate. One solution might be the development of new high-throughput formats, such as living cell microarrays, which are now proposed and which could be applied to *C. elegans* biology.

Other integrative approaches

The WI5 interactome dataset should also soon be challenged by other genome-wide biological datasets. Thus, since the co-localization of two proteins is a prerequisite for their interaction in the animal context, one must consider all information about the localization of the corresponding proteins or, although less informative, their mRNA. Indeed, comparisons between protein interactions and mRNA localization should already be possible, since an *in situ* hybridization experimental dataset covering approximately 8000 transcripts of *C. elegans* is already available at the Genome Biology Laboratory of the National Institute of Genetics of Mishima, Japan (http://nematode.lab.nig.ac.jp). At the protein level, only a very small fraction of the proteome has been precisely localized in the cell. However, this situation could rapidly improve since Gateway GFP (green fluorescent protein) fused destination vectors, compatible with the *C. elegans* ORFeome, are now available. This makes possible the creation of a large collection of GFP fusion proteins that will help to collect information about protein localization in worms.

Additionally, the rapid development of structural genomics approaches may soon provide very useful datasets to overlap with the WI5 map. The structural analyses of known molecular complexes using NMR spectroscopy or X-ray crystallography will provide us with a defined and very accurate set of protein interactions that will strongly challenge the relative accuracy of the various interactome datasets generated by genome-wide scale approaches (Edwards *et al.*, 2002). The results of pilot experiments of this type were recently presented for several protein complexes of *S. cerevisiae* (Jansen *et al.*, 2002) and similar studies are under way for *C. elegans* proteins in the Luo laboratory (University of Alabama at Birmingham).

It would also be interesting to integrate genomic context information and protein interaction datasets. Indeed, it appears that functional modules identified using a genomic-context method, i.e. genomic neighbourhood, fusion of genes or co-occurrence, often contain genes involved in the same biological process (Matthews *et al.*, 2001). In addition, the stronger the genomic environment conservation, the higher is the probability that proteins encoded by these genes functionally interact. Since the *C. elegans* genome is organized into operons, one can propose that at least some of these operons encode interacting proteins. Supporting this hypothesis, it has been observed that similar RNAi phenotypes are shared by genes present in the same operons.

Finally, the conservation of interactions between two proteins whose sequences are conserved is observed for a significant number of genes between *S. cerevisiae* and *C. elegans*. It would now be interesting to extend such an observation to other model organisms and to human datasets. The data obtained from these 'orthologous pairs' would be useful to confirm scaffold protein networks and undoubtedly would shed new light on the mechanism of evolution that led to the construction of these essential protein networks.

Very few attempts have been made to integrate more than two types of data. This mainly results from the format used to transcribe the data, which limits the comparison of information arising from completely different methods of observation. Another limitation is that the datasets often cover sub-proteomes. Nonetheless, it is likely that this type of integration should greatly enhance our degree of confidence with respect to the compiled data. While several examples of integration of protein–protein interaction data with more than one other dataset are available (Jansen, Greenbaum and Gerstein, 2001; Ge *et al.*, 2001 for examples), the only attempt for a metazoan was performed on the *C. elegans* germline specific genes by Walhout *et al.* (2002). The interactome, transcriptome and phenome datasets concerning 65 genes were overlapped and multiple correlations were established for about 20. However, the format used to classify each gene into a specific set of phenotypes or clusters of expression was not accurate enough to unambiguously conclude that a strong biological link exists between these genes.

8.6 Conclusion: from Interactions to Therapies

Biomolecules are physical and chemical objects that interact with one another and with their environment. These interactions set the basis for the dynamics of life. Many pathological states are related to a loss of interactions within a particular pathway. To globally visualize metazoan organism functions, we needed to screen, at a genome-wide scale, all potential protein interactions, using high-throughput approaches. The generation of a first version of the *C. elegans* interactome by high-throughput two-hybrid screenings uncovered several novel features.

First, the global structure of a metazoan network is scale free, as previously observed in the yeast *S. cerevisiae*. This is reminiscent of the structure of the *D. melanogaster* protein–protein interaction network.

Second, the acquisition of such information is fundamental for gaining biological insights into *C. elegans* biology. Knowledge about human protein interactions can also be improved by analysing conserved proteins shown to interact in other metazoan organisms. Moreover, these datasets can be combined with other post-genomic information, such as the phenotypes obtained from genome-wide RNAi studies or transcriptome analysis using microarrays. The correlation and the integration of such data are major steps in the modelling of *C. elegans*. It appears that integration is also necessary to improve the accuracy of high-throughput methods that, taken individually, generate high levels of both false positives and negatives.

The last, but possibly not least, point is the combined efforts and the scientific collaborations needed to reach each milestone, i.e. the genome sequence, the ORFeome and now the interactome as well as the integration of post-genomic data. This illustrates the tremendous amount of work, as well as the diversity of techniques, that is required to generate a human interactome map and to improve existing maps. Finally, looking at conserved interactions or interologues (Matthews *et al.*, 2001) between *C. elegans* and human can be a way to complete the annotation of the human genes and proteins. For example, several candidates remain to be validated for their potential roles as tumour suppressors or oncogenes.

We are currently investigating the involvement of several conserved proteins between *C. elegans* and human for their role in aberrant gene expression due to their chromatin remodelling activity. We chose several families of proteins in order to perform this study. Several interactors were found and they can be grouped in a sub-network containing transcription factors, histone modifying enzymes and transcriptional coregulators. Interestingly, their orthologues in human seem to be implicated in many pathologies, such as leukaemia (Cho, Elizondo and Boerkoel, 2004). We tested their biological activities and the preliminary results suggest a modulator role in gene expression through their enzymatic activities at the chromatin level, possibly due to their potential partners. Hence, beyond the improvement of the genome annotation by the ORFeome, it is possible to go further by using interactome data to highlight the biological mechanisms disrupted in certain pathologies. This would be the first step to propose therapies derived from high-throughput protein–protein interaction data.

References

Aloy, P. Russell, R. B. (2002) Potential artefacts in protein-interaction networks. *FEBS Lett*, **530** (1–3), 253–254.

Barabasi, A. L. and Oltvai, Z. N. (2004) Network biology: understanding the cell's functional organization. *Nat Rev Genet*, **5** (2), 101–113.

Bartel, P. L., Roecklein, J. A., SenGupta, D. and Fields, S. (1996) A protein linkage map of *Escherichia coli* bacteriophage T7. *Nat Genet*, **12** (1), 72–77.

Blumenthal, T., Evans, D., Link, C. D., Guffanti, A., Lawson, D., Thierry-Mieg, J., Thierry-Mieg, D., Chiu, W. L., Duke, K., Kiraly, M. and Kim, S. K. (2002) A global analysis of *Caenorhabditis elegans* operons. *Nature*, **417** (6891), 851–854.

Boulton, S. J., Gartner, A., Reboul, J., Vaglio, P., Dyson, N., Hill, D. E. and Vidal, M. (2002) Combined functional genomic maps of the *C. elegans* DNA damage response. *Science*, **295** (5552), 127–131.

Burge, C. and Karlin, S. (1997) Prediction of complete gene structures in human genomic DNA. *J Mol Biol*, **268**, 78–94.

C. elegans Sequencing Consortium. (1998) Genome sequence of the nematode *C. elegans*: a platform for investigating biology. *Science*, **282** (5396), 2012–2018 [review].

Cho, K. S., Elizondo, Li and Boerkoel, C. F. (2004) Advances in chromatin remodelling and human disease. *Curr Op Gen Dev*, **14**, 308–315.

Dandekar, T., Snel, B., Huynen, M. and Bork, P. (1998) Conservation of gene order: a fingerprint of proteins that physically interact. *Trends Biochem Sci*, **23** (9), 324–328.

Davy, A., Bello, P., Thierry-Mieg, N., Vaglio, P., Hitti, J., Doucette-Stamm, L., Thierry-Mieg, D., Reboul, J., Boulton, S., Walhout, A. J., Coux, O. and Vidal, M. (2001) A protein-protein interaction map of the *Caenorhabditis elegans* 26S proteasome. *EMBO Rep*, **2** (9), 821–828.

Edwards, A. M., Kus, B., Jansen, R., Greenbaum, D., Greenblatt, J., Gerstein, M. (2002) Bridging structural biology and genomics: assessing protein interaction data with known complexes. *Trends Genet*, **18** (10), 529–536.

Erwin, B. and Green, P. (1998) Base-calling of automated sequencer traces using phred 2 – Error probabilities. *Genome Res*, **8** (3), 186–194.

Erwin, B., Hillier L. *et al.* (1998) Base-calling of automated sequencer traces using phred 1 – Accuracy assessment. *Genome Res*, **8** (3), 175–185.

Fields, S. and Song, O. (1989) A novel genetic system to detect protein – protein interactions. *Nature*, **340** (6230), 245–246.

Fraser, A. G., Kamath, R. S., Zipperlen, P., Martinez-Campos, M., Sohrmann, M. and Ahringer, J. (2000) Functional genomic analysis of *C. elegans* chromosome I by systematic RNA interference. *Nature*, **408** (6810), 325–330.

Gaudet, J. and Mango, S. E. (2002) Regulation of organogenesis by the *C. elegans* FoxA protein PHA-4. *Science*, **295** (5556), 821–825.

Gavin, A. C., Bosche, M., Krause, R., Grandi, P., Marzioch, M., Bauer, A., Schultz, J., Rick, J. M., Michon, A. M., Cruciat, C. M., Remor, M., Hofert, C., Schelder, M., Brajenovic, M., Ruffner, H., Merino, A., Klein, K., Hudak, M., Dickson, D., Rudi, T., Gnau, V., Bauch, A., Bastuck, S., Huhse, B., Leutwein, C., Heurtier, M. A., Copley, R. R., Edelmann, A., Querfurth, E., Rybin, V., Drewes, G., Raida, M., Bouwmeester, T., Bork, P., Seraphin, B., Kuster, B., Neubauer, G. and Superti-Furga, G. (2002) Functional organization of the yeast proteome by systematic analysis of protein complexes. *Nature*, **415** (6868), 141–147.

Ge, H., Liu, Z., Church, G. M. and Vidal, M. (2001) Correlation between transcriptome and interactome mapping data from *Saccharomyces cerevisiae*. *Nat Genet*, **29** (4), 482–486.

Giot, L., Bader, J. S., Brouwer, C., Chaudhuri, A., Kuang, B., Li, Y., Hao, Y. L., Ooi, C. E., Godwin, B., Vitols, E., Vijayadamodar, G., Pochart, P., Machineni, H., Welsh, M., Kong, Y., Zerhusen, B.,

Malcolm, R., Varrone, Z., Collis, A., Minto, M., Burgess, S., McDaniel, L., Stimpson, E., Spriggs, F., Williams, J., Neurath, K., Ioime, N., Agee, M., Voss, E., Furtak, K., Renzulli, R., Aanensen, N., Carrolla, S., Bickelhaupt, E., Lazovatsky, Y., DaSilva, A., Zhong, J., Stanyon, C. A., Finley RL, Jr., White, K. P., Braverman, M., Jarvie, T., Gold, S., Leach, M., Knight, J., Shimkets, R. A., McKenna, M. P., Chant, J. and Rothberg, J. M. (2003) A protein interaction map of *Drosophila melanogaster. Science*, **302** (5651), 1727–1736.

Goldberg, D. S. and Roth, F. P. (2003) Assessing experimentally derived interactions in a small world. *Proc Natl Acad Sci USA*, **100** (8), 4372–4376.

Gonczy, P., Echeverri, C., Oegema, K., Coulson, A., Jones, S. J., Copley, R. R., Duperon, J., Oegema, J., Brehm, M., Cassin, E., Hannak, E., Kirkham, M., Pichler, S., Flohrs, K., Goessen, A., Leidel, S., Alleaume, A. M., Martin, C., Ozlu, N., Bork, P. and Hyman, A. A. (2000) Functional genomic analysis of cell division in *C. elegans* using RNAi of genes on chromosome III. *Nature*, **408** (6810), 331–356.

Gregoriev, A. (2001) A relationship between gene expression and protein interactions on the proteome scale: analysis of the bacteriophage T7 and the yeast *Saccharomyces cerevisiae. Nucleic Acids Res*, **29**, 3513–3519.

Gunsalus, K. C., Yueh, W. C., MacMenamin, P. and Piano, F. (2004) RNAiDB and PhenoBlast: web tools for genome-wide phenotypic mapping projects. *Nucleic Acids Res*, **32** (database issue), D406–D410.

Hill, A. A., Hunter, C. P., Tsung, B. T., Tucker-Kellogg, G., Brown, E. L. (2000) Genomic analysis of gene expression in *C. elegans. Science*, **290** (5492), 809–812.

Ho, Y., Gruhler, A., Heilbut, A., Bader, G. D., Moore, L., Adams, S. L., Millar, A., Taylor, P., Bennett, K., Boutilier, K., Yang, L., Wolting, C., Donaldson, I., Schandorff, S., Shewnarane, J., Vo, M., Taggart, J., Goudreault, M., Muskat, B., Alfarano, C., Dewar, D., Lin, Z., Michalickova, K., Willems, A. R., Sassi, H., Nielsen, P. A., Rasmussen, K. J., Andersen, J. R., Johansen, L. E., Hansen, L. H., Jespersen, H., Podtelejnikov, A., Nielsen, E., Crawford, J., Poulsen, V., Sorensen, B. D., Matthiesen, J., Hendrickson, R. C., Gleeson, F., Pawson, T., Moran, M. F., Durocher, D., Mann, M., Hogue, C. W., Figeys, D. and Tyers, M. (2002) Systematic identification of protein complexes in *Saccharomyces cerevisiae* by mass spectrometry. *Nature*, **415** (6868), 180–183.

Ito, T., Chiba, T., Ozawa, R., Yoshida, M., Hattori, M. and Sakaki, Y. (2001) A comprehensive two-hybrid analysis to explore the yeast protein interactome. *Proc Natl Acad Sci USA*. **98** (8), 4569–4574.

Jansen, R., Greenbaum, D. and Gerstein, M. (2001) Relating whole-genome expression data with protein–protein interactions. *Science*, **293** (5537), 2087–2092.

Jansen, R., Lan, N., Qian, J. and Gerstein, M. (2002) Integration of genomic datasets to predict protein complexes in yeast. *J Struct Funct Genomics*, **2** (2), 71–81.

Jeong, H., Mason, S. P., Barabasi, A. L. and Oltvai, Z. N. (2001) Lethality and centrality in protein networks. *Nature*, **411** (6833), 41–42.

Kamath, R. S., Fraser, A. G., Dong, Y., Poulin, G., Durbin, R., Gotta, M., Kanapin, A., Le Bot, N., Moreno, S., Sohrmann, M., Welchman, D. P., Zipperlen, P. and Ahringer, J. (2003) Systematic functional analysis of the *Caenorhabditis elegans* genome using RNAi. *Nature*, **421** (6920), 231–237.

Kemmeren, P., van Berkum, N. L., Vilo, J., Bijma, T., Donders, R., Brazma, A. and Holstege, F. C. (2002) Protein interaction verification and functional annotation by integrated analysis of genome-scale data. *Mol Cell*, **9** (5), 1133–1143.

Kim, S. K., Lund, J., Kiraly, M., Duke, K., Jiang, M., Stuart, J. M., Eizinger, A., Wylie, B. N. and Davidson, G. S. (2001) A gene expression map for *Caenorhabditis elegans. Science*, **293** (5537), 2087–2092.

Lai, C. H., Chou, C. Y., Ch'ang, L. Y., Liu, C. S. and Lin, W. (2000) Identification of novel human genes evolutionarily conserved in *Caenorhabditis elegans* by comparative proteomics. *Genome Res*, **10** (5), 703–713.

Legrain, P. and Selig, L. (2000) Genome-wide protein interaction maps using two-hybrid systems. *FEBS Lett*, **480** (1), 32–36.

Leung, Y. F. and Cavalieri, D. (2003) Fundamentals of cDNA microarray data analysis. *Trends Genet*, **19** (11), 649–659.

Li, S., Armstrong, C. M., Bertin, N., Ge, H., Milstein, S., Boxem, M., Vidalain, P. O., Han, J. D., Chesneau, A., Hao, T., Goldberg, D. S., Li, N., Martinez, M., Rual, J. F., Lamesch, P., Xu, L., Tewari, M., Wong, S. L., Zhang, L. V., Berriz, G. F., Jacotot, L., Vaglio, P., Reboul, J., Hirozane-Kishikawa, T., Li, Q., Gabel, H. W., Elewa, A., Baumgartner, B., Rose, D. J., Yu, H., Bosak, S., Sequerra, R., Fraser, A., Mango, S. E., Saxton, W. M., Strome, S., Van Den Heuvel, S., Piano, F., Vandenhaute, J., Sardet, C., Gerstein, M., Doucette-Stamm, L., Gunsalus, K. C., Harper, J. W., Cusick, M. E., Roth, F. P., Hill, D. E. and Vidal, M. (2004) A map of the interactome network of the metazoan *C. elegans*. *Science*, **303** (5657), 540–543.

Marcotte, E. M., Pelligrini, M. *et al.* (1999) Detecting protein function and protein–protein interactions from genome sequences. *Science*, **285** (5428), 751–753.

Matthews, L. R., Vaglio, P., Reboul, J., Ge, H., Davis, B. P., Garrels, J., Vincent, S. and Vidal, M. (2001) Identification of potential interaction networks using sequence-based searches for conserved protein – protein interactions or 'interologs'. *Genome Res*, **11** (12), 2120–2126.

Milo, R., Shen-Orr, S., Itzkovitz, S., Kashtan, N., Chklovskii, D. and Alon, U. (2002) Network motifs: simple building blocks of complex networks. *Science*, **298** (5594), 824–827.

Pellegrini, M., Marcotte, E. M., Thompson, M. J., Eisenberg, D. and Yeates, T. O. (1999) Assigning protein functions by comparative genome analysis: protein phylogenetic profiles. *Proc Natl Acad Sci USA*, **96** (8), 4285–4288.

Pothof, J., van Haaften, G., Thijssen, K., Kamath, R. S., Fraser, A. G., Ahringer, J., Plasterk, R. H. and Tijsterman, M. (2003) Identification of genes that protect the *C. elegans* genome against mutations by genome-wide RNAi. *Genes Dev*, **17** (4), 443–448.

Ravasz, E., Somera, A. L., Mongru, D. A., Oltvai, Z. N. and Barabasi, A. L. (2002) Hierarchical organization of modularity in metabolic networks. *Science*, **297** (5586), 1551.

Reboul, J., Vaglio, P., Rual, J. F., Lamesch, P., Martinez, M., Armstrong, C. M., Li, S., Jacotot, L., Bertin, N., Janky, R., Moore, T., Hudson J. R., Jr, Hartley, J. L., Brasch, M. A., Vandenhaute, J., Boulton, S., Endress, G. A., Jenna, S., Chevet, E., Papasotiropoulos, V., Tolias, P. P., Ptacek, J., Snyder, M., Huang, R., Chance, M. R., Lee, H., Doucette-Stamm, L., Hill, D. E. and Vidal, M. (2003) *C. elegans* ORFeome version 1.1: experimental verification of the genome annotation and resource for proteome-scale protein expression. *Nat Genet*, **34** (1), 35–41.

Reinke, V., Smith, H. E., Nance, J., Wang, J., Van Doren, C., Begley, R., Jones, S. J., Davis, E. B., Scherer, S., Ward, S. and Kim, S. K. (2000) A global profile of germline gene expression in *C. elegans*. *Mol Cell*, **6** (3), 605–616.

Roy, P. J., Stuart, J. M., Lund, J. and Kim, S. K. (2002) Chromosomal clustering of muscle-expressed genes in *C. elegans*. *Nature*, **418**, 975–979.

Snel, B., Bork, P. and Huynen, M. A. (2002) The identification of functional modules from the genomic association of genes. *Proc Natl Acad Sci USA*, **99** (9), 5890–5895.

Stein, L. D., Bao, Z., Blasiar, D., Blumenthal, T., Brent, M. R., Chen, N., Chinwalla, A., Clarke, L., Clee, C., Coghlan, A., Coulson, A., D'Eustachio, P., Fitch, D. H., Fulton, L. A., Fulton, R. E., Griffiths-Jones, S., Harris, T. W., Hillier, L. W., Kamath, R., Kuwabara, P. E., Mardis, E. R., Marra, M. A., Miner, T. L., Minx, P., Mullikin, J. C., Plumb, R. W., Rogers, J., Schein, J. E., Sohrmann, M., Spieth, J., Stajich, J. E., Wei, C., Willey, D., Wilson, R. K., Durbin, R. and Waterston, R. H. (2003) The genome sequence of *Caenorhabditis briggsae*: a platform for comparative genomics. *PLoS Biol*, **1** (2), E45.

Uetz, P., Giot, L., Cagney, G., Mansfield, T. A., Judson, R. S., Knight, J. R., Lockshon, D., Narayan, V., Srinivasan, M., Pochart, P., Qureshi-Emili, A., Li, Y., Godwin, B., Conover, D., Kalbfleisch, T., Vijayadamodar, G., Yang, M., Johnston, M., Fields, S. and Rothberg, J. M. (2000) A comprehensive analysis of protein–protein interactions in *Saccharomyces cerevisiae*. *Nature*, **403** (6770), 623–627.

Vastenhouw, N. L., Fischer, S. E., Robert, V. J., Thijssen, K. L., Fraser, A. G., Kamath, R. S., Ahringer, J. and Plasterk, R. H. (2003) A genome-wide screen identifies 27 genes involved in transposon silencing in *C. elegans*. *Curr Biol*, **13** (15), 1311–1316.

Vidal M. (1997). The reverse two-hybrid system. (2000) Protein interaction mapping in *C. elegans* using proteins involved in vulval development. *Science*, **287** (5450), 116–122.

Vidal, M., Brachmann, R. K., Fattaey, A., Harlow, E. and Boeke, J. D. (1996) Reverse two-hybrid and one-hybrid systems to detect dissociation of protein–protein and DNA–protein interactions. *Proc Natl Acad Sci USA*, **93** (19), 10 315–10 320.

Walhout, A. J., Reboul, J., Shtanko, O., Bertin, N., Vaglio, P., Ge, H., Lee, H., Doucette-Stamm, L., Gunsalus, K. C., Schetter, A. J., Morton, D. G., Kemphues, K. J., Reinke, V., Kim, S. K., Piano, F. and Vidal, M. (2002) Integrating, phenome, and transcriptome mapping data for the *C. elegans* germline. *Curr Biol*, **12** (22), 1952–1958.

Walhout, A. J., Sordella, R., Lu, X., Hartley, J. L., Temple, G. F., Brasch, M. A., Thierry-Mieg, N. and Vidal, M. (2000a) Protein interaction mapping in *C. elegans* using proteins involved in vulval development. *Science*, **287** (5450), 116–122.

Walhout, A. J., Temple, G. F., Brasch, M. A., Hartley, J. L., Lorson, M. A., van den Heuvel, S. and Vidal, M. (2000b) GATEWAY recombinational cloning: application to the cloning of large numbers of open reading frames or ORFeomes. *Methods Enzymol*, **328** 575–592.

Yook, S. H., Oltvai, Z. N. and Barabasi, A. L. (2004) Functional and topological characterization of protein interaction networks. *Proteomics*, **4** (4), 928–942.

Zhang, M. Q. (2002) Computational prediction of eukaryotic protein-coding genes. (2002) *Nat Rev Genet*, **3** (9), 698–709.

III Integrative Data Mining and Visualization – Emphasis on Combination of Multiple Prediction Models and Methods

9

Integrated Approaches for Bioinformatic Data Analysis and Visualization – Challenges, Opportunities and New Solutions

Steve R. Pettifer, James R. Sinnott and **Teresa K. Attwood**

Abstract

Bioinformaticians are routinely required to analyse vast amounts of information held both in large remote databases and in flat data-files hosted on local machines. The contemporary toolkit available for this task consists of an *ad hoc* collection of data-manipulation tools, scripting languages and visualization systems, each with its own mode of operation and conceptual models. Much of the user's cognitive capacity is therefore focused on controlling the tool rather than on performing the research. In this chapter, we review some of these issues and introduce UTOPIA, a project in which reusable software components are being built and integrated closely with the familiar desktop environment to make easy-to-use visualization tools for the field of bioinformatics.

Keywords

visualization, human–computer interaction, data analysis

9.1 Introduction

In recent years, computer technology has become commonplace and has completely transformed the nature of biological research. Tasks, such as simulating protein folding for example, that not so long ago were a challenge for the world's super-computer

Data Analysis and Visualization in Genomics and Proteomics Edited by Francisco Azuaje and Joaquin Dopazo

facilities can now be distributed worldwide to thousands of computers on the Internet, to be performed during their idle-cycles while simultaneously providing novelty screen-savers. Driven by the games market, recent advances in consumer-grade graphics hardware have meant that visualizing the structure of a molecule – a job originally only possible by painstaking manual assembly of plastic ball-and-stick models and then 'revolutionized' by monster graphics packages running on specialized hardware, such as Evans and Sutherland machines and more recently Silicon Graphics workstations – can now be achieved through real-time 3D rendering on even the most modest PC.

All manner of biological resources are now available to the masses via the Web. Routine applications of high-throughput technologies are pouring increasing quantities of data into the public domain at unprecedented rates – a new nucleotide sequence is deposited every 10 seconds. Not only is the volume of information growing, but so are the types of data being collected; consequently, the tools required to manipulate, store and analyse them are also proliferating. The challenge facing bioinformaticians today is to harness the new technologies to help rationalize this growing mass of information, to derive more efficient means of data storage and to design more incisive and reliable analysis tools. The imperative that drives these developments is to convert raw data into biochemical and biophysical knowledge, and to use that knowledge to provide new insights into, and understanding of, dynamic biological systems, from the level of individual genes to the levels of whole genomes and whole organisms. It is no longer enough just to know what a genome is: we must understand what its components mean, how they function and how they relate to the whole, and how to repair the system when parts of it fail.

Biologists have been eager to adopt technology to help generate and analyse their data – witness the huge amount of electronically generated biological information that already exists. However, using the data to understand biological complexity requires co-operation and interaction between scientific communities that have, in the past, largely kept themselves to themselves: e.g., to be able to answer these higher-level questions, researchers in the fields of nucleotide sequence analysis, protein sequence and protein structure analysis now need tools that can access one another's data repositories in coherent and consistent ways. Finding appropriate resources, and learning how to use and combine them, is a major obstacle to a biologist wishing to make the best use of the specialist resources that are now available. This is difficult enough, but the real challenge lies in the semantic complexity of biological data. Many scientific disciplines are axiom based, using fundamental laws and equations as a basis for communication between communities, but biology tends to be more qualitative. Though there are petabytes of well formatted biological data generated from techniques such as DNA microarrays, crucial information is often attached in the form of free-text descriptions, easily readable by humans who know where to look, but hard to access or process automatically. To complicate matters, results are often published in a preliminary or speculative form, and here again, additional free-text annotations give scope for including suitable caveats and disclaimers that would be difficult to capture in any machine-readable way.

When communities worked in isolation, this style of archiving and publishing data was manageable; now that information needs to cross community boundaries, the situation has become increasingly chaotic. Database curators infer or copy data from other evolving databases, and complex interdependencies have emerged. Results derived from such labile information are subject to uncertainty, yet often enter the scientific arena stripped of the necessary 'health warnings' in the journey from community to community. Consequently, tools to help manage the migration of data, to track its provenance and to ensure its integrity are becoming fundamental to progress. Moreover, if we are to make sense of biological data, we also need computers to be able both to understand what the data means, and to handle multiple data types, and multiple prediction models and methods; but how can we create computer programs with sufficient sophistication to be able to model organic complexity meaningfully? How can we provide cross-community tools that are actually usable and that will interact with the data in ways that users can comprehend?

To illustrate the challenges of building easy-to-use scientific software, we will concentrate on the field of sequence analysis. This was the progenitor of bioinformatics, having grown out of Margaret Dayhoff's pioneering evolutionary studies in the 1960s (Dayhoff *et al.*, 1965), for which she manually collected and compared hundreds of protein sequences; almost 20 years later, this collection spawned the first sequence database (Dayhoff *et al.*, 1980). Today, comparison and alignment of protein sequences are still fundamental to evolutionary studies, but so too is the comparison and alignment of protein structures. Interestingly, however, the field of protein structure analysis grew up from different roots, and it is only relatively recently (largely prompted by the advent of fold classification databases) that it has converged with sequence analysis. Now, sequence and structure analysis are almost inseparable, and their interweaving provides good examples of the problems that can be thrown up by the requirements of cross-domain interaction.

In a Utopian world, a user would expect a sequence analysis package to include all possible ways of analysing or visualizing data, to access all possible databases, to read all possible inputs, to provide standard, publication-quality outputs and to be 'easy to use'. To reach this Utopian state, tools and data need to be seamlessly integrated. In pursuing this goal, two main approaches have traditionally been used: in one, integration is at the level of databases, which are then presented to the world through 'portal' mechanisms, usually by exploiting the ubiquity of the Web; in the other, integration is at the level of tools. Looking back, it is clear that there are problems with both solutions. Here, we analyse these two views, discussing both the challenges they present and the opportunities they afford to find better solutions.

9.2 Sequence Analysis Methods and Databases

Before looking in detail at portal- and tool-based approaches, we will consider for a moment the key role played in many of the underpinning analysis methods by multiple sequence alignments, to illustrate how these approaches became necessary.

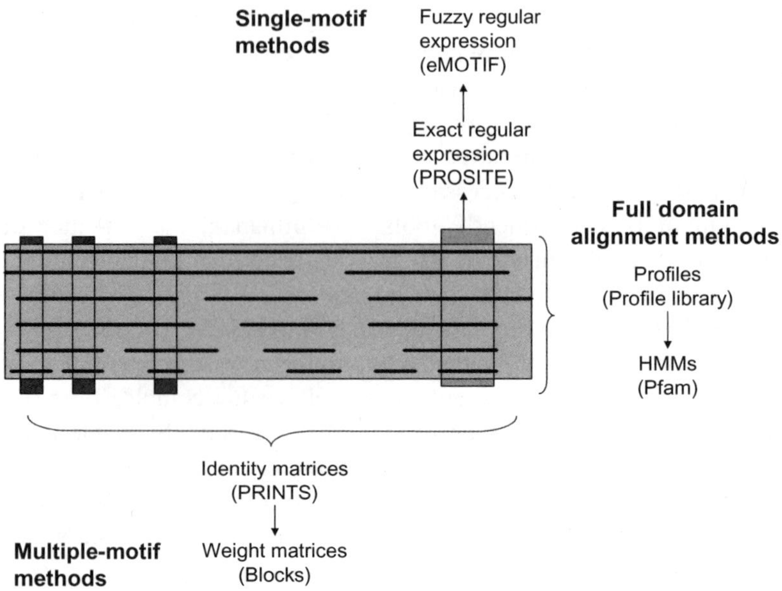

Figure 9.1 Approaches for classifying protein sequences into families (using single- or multiple-motif- or domain-based methods) and the databases (parentheses) to which they give rise

Alignments are important as they reveal the evolutionary relationships between members of protein families in terms of the similarities (or patterns of conservation) they share. When building an alignment, as more distantly related sequences are included, insertions are often required to bring equivalent parts of adjacent sequences into vertical register, as illustrated schematically in Figure 9.1. As a result of this gap-insertion process, islands of conservation emerge from a backdrop of mutational change. These regions (or motifs) tend to correspond to the core structural or functional elements of the protein.

The conserved nature of motifs effectively provides a familial blueprint, and different pattern-recognition techniques have evolved to exploit this fact. As shown in Figure 9.1, the methods fall broadly into three categories, depending on whether they use single motifs, multiple motifs or full domain alignments. All of these methods involve the derivation of some kind of discriminatory representation of the conserved features of the alignment, providing a characteristic signature for the family that can be used to diagnose future query sequences (Attwood, 2000).

The different methods of encoding protein families have given rise to different databases (parentheses in Figure 9.1), to store both the family signatures and annotation relating to their structural and functional significance. The problem is that to analyse a new sequence requires accessing each of these disparate resources, gathering the different outputs, reading the different annotations and arriving at some sort of consensus view. For many users, the effort required to perform all of these searches and to rationalize the outputs is simply too great – ultimately, it is much easier to perform a quick-and-dirty BLAST (Altshchul *et al.*, 1990) search and to hope that this will give the right, if superficial, sort of answer.

9.3 A View Through a Portal

To make sequence analysis more straightforward, the curators of the family databases eventually worked together to create a unified database of protein families, termed InterPro (Mulder *et al.*, 2003). Overall, the aim was to provide a single, central annotated resource for protein family diagnosis, with pointers to its satellite databases, accessible via the Web: in other words, a centralized 'portal' providing a one-stop shop for protein family analysis. Following the integration of the initial partner resources and the first successful applications of InterPro to the analysis and annotation of the fly and human genomes, several other databases were integrated.

The goal of InterPro was twofold: it aimed not only to make sequence analysis easier for the user, but also to make database maintenance easier for the partners – curators could hand over the messy business of annotation to InterPro, and users would have only one database to search (see Figure 9.2). To some extent, this is now

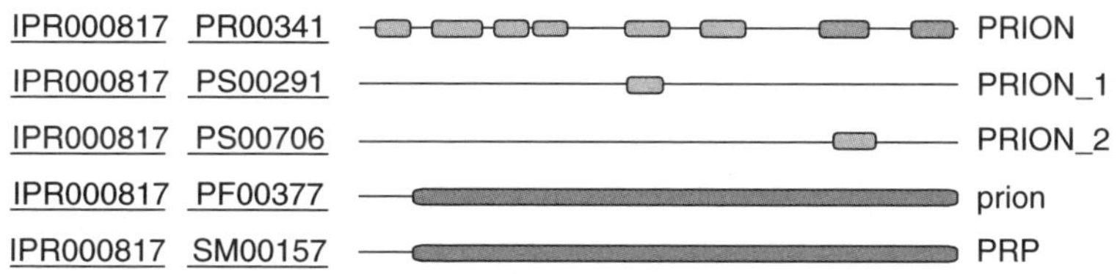

Figure 9.2 Unified graphical output from InterPro, providing an at-a-glance diagnosis of a query sequence, with motif-based matches to the PRINTS (top line) and PROSITE (second line) prion protein signatures, and domain-based matches to the Pfam (third line) and SMART (bottom line) signatures

true, but unfortunately the source curators still have to annotate their own family entries, and InterPro curators have to rationalize and merge the different source views, often with different levels of success – is the family of sequences identified by a single motif the same as that identified by several motifs, and are they the same as the family of sequences that share a particular domain? If the results from a motif-based family analysis are different from those of a domain-based analysis, then why do they differ, and what does this difference mean (see Figure 9.3)? If the answers are not the same, then what exactly is a family? And what is a domain? These are surprisingly difficult biological and philosophical questions to answer, challenging not only our habit of trying to neatly pigeonhole biology, but also our ability to formalize the concepts we use to describe biology. Consequently, InterPro can be

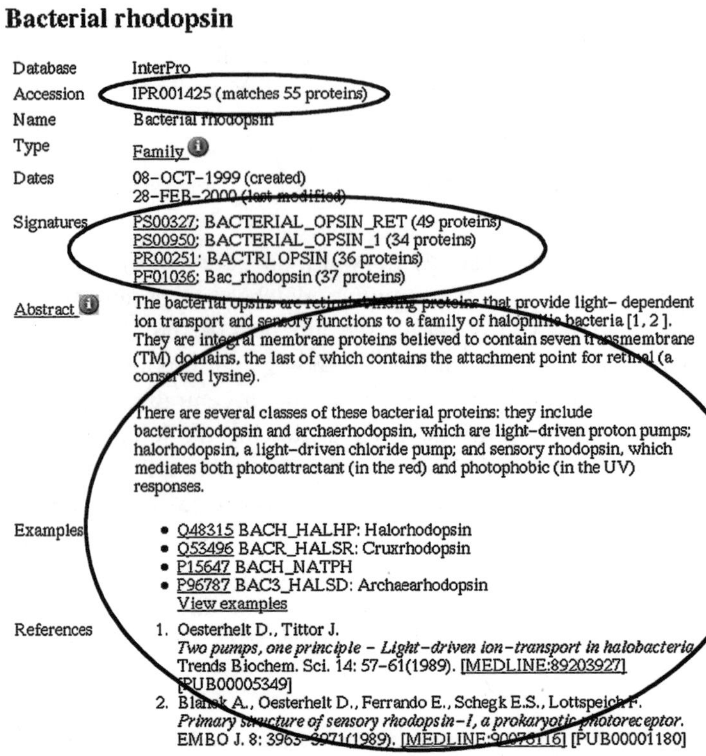

Bacterial rhodopsin

Database	InterPro
Accession	IPR001425 (matches 55 proteins)
Name	Bacterial rhodopsin
Type	Family
Dates	08-OCT-1999 (created) 28-FEB-2000 (last modified)
Signatures	PS00327; BACTERIAL_OPSIN_RET (49 proteins) PS00950; BACTERIAL_OPSIN_1 (34 proteins) PR00251; BACTRLOPSIN (36 proteins) PF01036; Bac_rhodopsin (37 proteins)
Abstract	The bacterial opsins are retinal-binding proteins that provide light- dependent ion transport and sensory functions to a family of halophilic bacteria [1, 2]. They are integral membrane proteins believed to contain seven transmembrane (TM) domains, the last of which contains the attachment point for retinal (a conserved lysine). There are several classes of these bacterial proteins: they include bacteriorhodopsin and archaerhodopsin, which are light-driven proton pumps; halorhodopsin, a light-driven chloride pump; and sensory rhodopsin, which mediates both photoattractant (in the red) and photophobic (in the UV) responses.
Examples	• Q48315 BACH_HALHP: Halorhodopsin • Q53496 BACR_HALSR: Cruxrhodopsin • P15647 BACH_NATPH • P96787 BAC3_HALSD: Archaearhodopsin View examples
References	1. Oesterhelt D., Tittor J. *Two pumps, one principle - Light-driven ion-transport in halobacteria.* Trends Biochem. Sci. 14: 57-61(1989). [MEDLINE:89203927] [PUB00005349] 2. Blanck A., Oesterhelt D., Ferrando E., Schegk E.S., Lottspeich F. *Primary structure of sensory rhodopsin-I, a prokaryotic photoreceptor.* EMBO J. 8: 3963-3971(1989). [MEDLINE:90078116] [PUB00001180]

Figure 9.3 InterPro entry illustrating different 'family' viewpoints from the source databases. The merged entry indicates a family of 55 proteins (first circle), while the separate family signature databases indicate 49, 34, 36 or 37 family members (second circle), depending on the underlying analysis method used. Beneath the results of the constituent databases, a merged annotation abstract describes the unified 'family', with examples of typical members and supporting literature (third circle). The reasons for the disparities in the numbers of family members identified by each of the source databases and by InterPro itself are not obvious to the casual user

quite demanding both of its curators and of its users, especially if the relationships between a merged InterPro entry and its individual source database entries are not clear.

9.4 Problems with Monolithic Approaches: One Size Does Not Fit All

There are many technological difficulties associated with maintaining centralized resources such as InterPro, not least the problems of keeping them up to date and consistent with their contributing databases (InterPro inevitably lags behind its sources). These are, however, tractable problems with solutions that are already understood in principle, even if they are hard to put into practice. More fundamental issues, however, concern the way in which users and curators are forced to interact with these monolithic structures. The difficulties are not so much to do with the technology of the portal, which for the most part works well, but rather with its interface metaphor. By exploiting the ubiquity of the Web browser, portal developers simplify the provision of graphical user interface distribution (the portal's server decides what interface components should be displayed, and the user's local browser decides how to display these in a style that suits the current desktop environment). This allows portal designers to expose powerful functionality to the user in a controlled way. However, by choosing what to expose, the developer is equally choosing what to hide – thus, while empowered in some ways, the user is, at the same time, being constrained in others. These constraints are felt most severely when users want to develop their own interfaces to the portal's data, or want to use the information in a way not predicted by its provider (the desire to use data in novel ways is common, innovation being a fundamental part of the scientific process); for these users, the metaphor of accessing data 'through a portal' hinders seamless integration both conceptually and technologically.

Another issue to consider is the balance between 'expert' and 'novice' use for a particular task. The functionality 'behind the scenes' is likely to be complex: expose all of it in one go, and the novice will struggle to get started; hide too much of it, however, making the interface too simplistic, and the power user will soon discard it in frustration. An interesting every-day analogy here is to that of a car. Cars have a 'turn left, turn right' control, a 'go faster' and a 'go slower' control, all of which are clearly related to, and easily explained in terms of, the purpose of the vehicle. But for manual-transmission cars, we are also faced with an 'allow the change of ratio between engine power and torque' pedal and a 'select one of five ratios of engine power to torque' lever, neither of which are initially very interesting to someone who just wants to make the car go forward, but both of which allow an expert driver more control over their ride than afforded by automatic transmission.

In bioinformatics, as in so many things, one size does not necessarily fit all.

9.5 A Toolkit View

An alternative to the portal-based approach, which very much keeps operations 'out there' and 'on the other side of the portal', is a tool-based approach, where the metaphor is of working with a collection of tools and utilities. Instead of having pre-packaged tasks that can be invoked through a portal, the user is presented with a collection of individual tools that can be used together in any combination to solve a particular problem.

Let us once more take as an example a familiar scenario from the field of protein sequence analysis. When trying to characterize an unknown query sequence, programs such as FastA (Lipman & Pearson, 1985) and BLAST (Altschul *et al.*, 1990) are used to generate pair-wise alignments between the query and target sequences in a database. The user examines the alignments of the best-scoring matches to determine the biological significance of the hits; if a group of related sequences is identified, he or she then creates a multiple alignment, in order to be able to visualize the most conserved regions (motifs) of the family, which may be indicative of particular structural or functional features. The user then performs more sensitive database searches using just these conserved motifs, allowing more distant family members to be retrieved and analysed. If a three-dimensional structure of a member of the family is known, the next step might involve alignment of the query with the sequence of known structure, and subsequently pinpointing conserved residues within the protein fold: this might give clues as to the whereabouts, say, of molecular interaction or binding sites, so shedding light on possible aspects of the unknown protein's functionality. Another step in his or her analysis might also involve the construction of phylogenetic trees from multiply-aligned family members, thereby helping to elucidate their evolutionary relationships and, again, potentially facilitating functional characterization of the unknown protein.

In principle, such tasks are relatively straightforward to perform. In practice, however, they usually involve the use of diverse tools and databases, of which some are stored locally, while others must be accessed remotely. For example,

- some require interaction with Web forms, and subsequent retrieval of information from poorly structured HTML pages (e.g., BLAST),
- some involve interaction with applets, and retrieval of results via email (e.g., CINEMA; Parry-Smith *et al.*, 1998),
- others require use of applications available on the user's PC, or on local Unix/ Linux-based servers (e.g., ClustalW (Thompson *et al.*, 1994), PHYLIP (Felsenstein, 1989) and EMBOSS (Rice, Longden and Bleasby, 2000)) and
- still others might require remote login to national facilities, and subsequent file transfer between remote and local machines (e.g., GCG; Devereux, Haeberlie and Smithies, 1984).

There are several reasons why this diversity of approaches has arisen: many users still do not know (and do not want to have to find out) how much they can do via the Internet (they are comfortable with a self-contained, desktop package, supplemented with an occasional BLAST search); some users are not allowed to make extensive use of the Internet (e.g., industrialists who live behind robust firewalls), so must have resources and tools available on platforms in house; many bioinformatic tools have only been written as Unix/Linux-based applications and have not been ported to the Windows or MacOs worlds inhabited by most biologists; other tools have only been written as applets, ostensibly to obviate portability problems; most packages come bundled with tools and databases that date quickly, making access to the latest algorithms and data via the Internet still essential; and some packages and databases have prohibitively expensive or restrictive licensing arrangements, and are therefore only feasibly accessible at remote multi-user national or international resource centres.

Turning our attention away from the working environment itself to focus on the types of alignment program available, again we find a bewildering variety, ranging from stand-alone automatic multiple alignment tools, accessible as command-line driven applications or via Web pages (e.g., ClustalW), components of large (often commercial) integrated packages (e.g., pileup in GCG), command-line driven manual alignment editors with X-windows interfaces (e.g., XALIGN; Perkins and Attwood, 1995) and manual editors written in Java as applets (e.g., CINEMA and JalView (Clamp, Cuff and Barton, 1998)) to X-windows Java-based alignment viewers (e.g., BelVu; Sonnhammer, 1999). Each of these has its own idiomatic style, interface and behaviour, and works with its own bespoke file format. To give a trivial example, virtually all of these programs use different input/output formats (e.g., NBRF-PIR, FastA, Clustal, GDE, PHYLIP, MSF, to name but a few) in spite of the fact that they really 'just represent a protein' or 'set of proteins'. Thus, to import an alignment created by an automatic package into a manual editor, it is necessary first to use a program to convert between formats. Similarly, to integrate an automatic alignment tool into an existing manual editor (or vice versa), an appropriate format-exchange program must be written or bundled into the system.

It is tempting to think that the problems of inconsistency and interchange between tools can be solved by integrating them into a form of 'super-tool', but this runs the risk of revisiting the problems associated with the more monolithic portal approaches, and really relies on being able to predict every possible way in which the global community may want to use such a tool. In the real world, this is rarely the approach we actually use: if we want to do a job properly, we use a specific tool for that purpose – we do not expect a single tool to do everything, or, if we do, we do not expect it to do all things equally well. The classic example here is the Swiss Army Knife, which neatly illustrates the contrast between bespoke engineering versus the Jack-of-all-trades compromise – excellent back-up or fix-it tool though it is, its individual components are seldom as good as the real thing (especially the corkscrew!).

In summary, the current state of bioinformatics tools is far from ideal. It is not that the data or algorithms (representing 'the real science') are flawed, but rather that the means by which users are forced to interact with these resources is often cumbersome and confusing: portals and integrated packages are coherent and consistent, but restrict the user; toolkits composed of individual tools are potentially powerful but suffer from inconsistent interfaces and an over-abundance of file formats.

9.6 Challenges and Opportunities

In light of these issues, we felt that a new perspective was needed on the problem of providing bioinformatics tools and databases in a user-friendly environment. We need systems in which the abilities of the user are supported rather than confounded by computational tools, where the user does not feel intimidated by unwieldy or inappropriate interfaces and does not have to worry about underlying file-types or operating systems, but can use whatever tools are needed within a clear, visually supportive and intuitive framework. Within such a system, we need, for example, to be able to (a) align sequences (manually and/or automatically), whether protein, DNA or RNA, (b) search databases, whether sequence, motif, structure, mutation, literature based, etc. and (c) visualize, and interact with, 2D and 3D representations, whether of molecular structures, protein–protein interactions, phylogenetic trees, dot-plots/ surfaces, gene-expression data, etc. The environment needs to offer different views of the user's work-space, for example via a resource browser that indicates the locations, types, sizes and ages not only of databases, but also of input and output files (sequences, structures, alignments, search results or whatever). The system needs to be customizable, so that databases and tools can be updated automatically via appropriate agent software, either without troubling the user or by notifying him that new versions of various resources are now available for installation.

In addressing these issues, there are two fundamental challenges. The first concerns interaction between the user and the tool set – a human–computer interaction problem; the second concerns interaction between the tools themselves – a software and knowledge-engineering problem. There is clearly an opportunity to tackle these problems together, in order to produce a coherent solution; but how do we avoid the 'one size fits all' trap, while at the same time providing continuity between our tools?

We are exploring these problems in a project that aims to build UTOPIA (User-friendly Tools for OPerating Informatics Applications). Our approach is to make the user's computing environment the integrating factor – to extend the familiar features of the computer's desktop environment to address the needs of today's bioinformatician, i.e., to turn the desktop environment into a bioinformatics workbench. The WIMP (windows, icons, menus and pointing device) paradigm desktop interface is the most familiar example of an interface metaphor that 'just works' for most purposes. Assuming the user understands that a computer is a machine capable of storing and of processing data, metaphors such as 'filing system', 'folder', 'document' and 'trash can' lead him quite intuitively through the functionality of the device. 'I can put my stuff in a document, file it in a folder, and dispose of it in the trash can' is a much more straightforward explanation than could ever be achieved by exposing details of the inodes, link structures and workings of the machine's hard drive. There are many well documented and detailed design principles for making a user interface 'easy to use', such as penalty-free exploration of its functionality (e.g., using drop-down menus to browse possible features without the danger of executing one that may do something unwanted) or the provision of complete and consistent 'cancel' and 'undo' functions. There can be little doubt that diligent application of these principles to current tools would improve the situation to some extent. However, these detailed issues are secondary to determining a suitable metaphor to guide the overall structure of the interface, and to understanding the nature of the data that the interface is to present. Identifying concepts that are common to, and interchangeable between, many applications (such as 'protein sequence', 'molecular structure', 'nucleotide sequence') and making these first-class citizens of the interface (regardless of the complexities of their underlying representation) is a vital part of making the tool usable. It is important that the tools are able to avoid revealing the technicalities of underlying file formats and data structures, and are able to communicate with one another in terms of higher-level, user-centric concepts. Understanding which concepts are first-class citizens of the interface, however, is a notoriously difficult task, and can only be reliably achieved by rigorous study of the bioinformatician at work (using techniques such as user-centred design or ethnographic study), and thorough user trials of any software that is produced.

9.7 Extending the Desktop Metaphor

The traditional desktop metaphor neatly encompasses the majority of tasks that a user wishes to perform using his or her computer: 'files' are things that are stored locally (and only change when the user causes them to change); 'applications' are tools that are at the user's disposal on his or her local machine, and other resources are generally accessed remotely via a Web browser. As we have seen, the bioinformatician has often to call upon resources that are remote/networked in order to be able to perform analyses, and doing this solely via the Web is cumbersome. Here, then, the current desktop implementations are limited, and opportunities for innovation lie in turning the local desktop into a bioinformatician's workbench able to deal with this more dynamic environment.

The UTOPIA system (Pettifer, Sinnott and Attwood, 2004) aims both to exploit the familiarity of the desktop environment and to extend its functionality in a way that allows seamless access to remote and rapidly changing resources. The system is built using cross-platform components: Trolltech's Qt (http://www.trolltech.com) widget set for the user interface, OpenGL for 3D rendering, WebDAV (Whitehead and Goland, 1999) for inter-process communication and ANSI C++ as the main programming language. The system (see Figure 9.4) is currently based around sequence analysis packages, though tools for other forms of analysis are planned. At the heart of UTOPIA lies a virtual filing system (the UTOPIA Filing System, or UFS), which provides a bridge between the large number of diverse resources that are 'out there' and the work the user is doing 'on his or her computer'. The UFS integrates with the host machine's existing filing system, tracking and monitoring the manipulations of resources under its control. For example, a protein sequence is downloaded from Swiss-Prot (Boeckmann *et al.*, 2003), and stored in the UFS: although the file from the source database may be in a particular format, the UFS records the important information from this file in its own internal extensible RDF (resource description framework) database, together with information regarding who downloaded the file, where it came from, when it was modified and so on. It is then

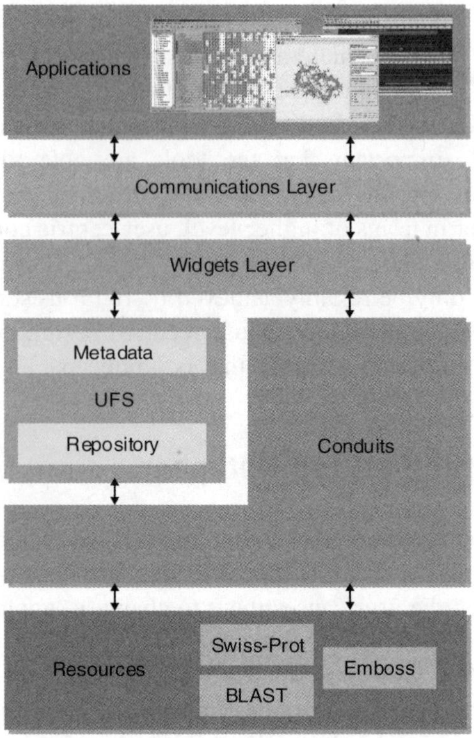

Figure 9.4 Architecture of the UTOPIA system

able to classify the resource as, say, 'a protein' (rather than 'a text file in FastA format'), and extensible rules embedded in the filestore then allow the file to be presented to legacy applications in any suitable format, without the need for an explicit file-conversion program. As the UFS tracks all modifications to these files, and accumulates provenance information behind the scenes, the user can query the system to discover relationships between objects he or she has manipulated. For example, a protein sequence is updated in one of the source databases, but has since been used in an alignment, which in turn has generated motifs and fingerprints that are now published in a new database. As UTOPIA knows which tools have loaded which files to generate which outputs, it is able to identify the fingerprints that may now be invalid because of the source change, and to notify the user. For remote services offering change notification, the process of warning the user may be invoked automatically.

As well as providing an 'intelligent' environment for managing data, UTOPIA provides tools and applications, such as a sequence alignment editor (CINEMA 5) and 3D structure viewer (Ambrosia, 'a molecule browsia') that can interoperate via the UFS and desktop environment – e.g., allowing a protein to be seen and manipulated simultaneously as a 3D molecular structure and as a 2D residue sequence (see Figure 9.5). Although, at first sight, this may not seem very 'clever', achieving this

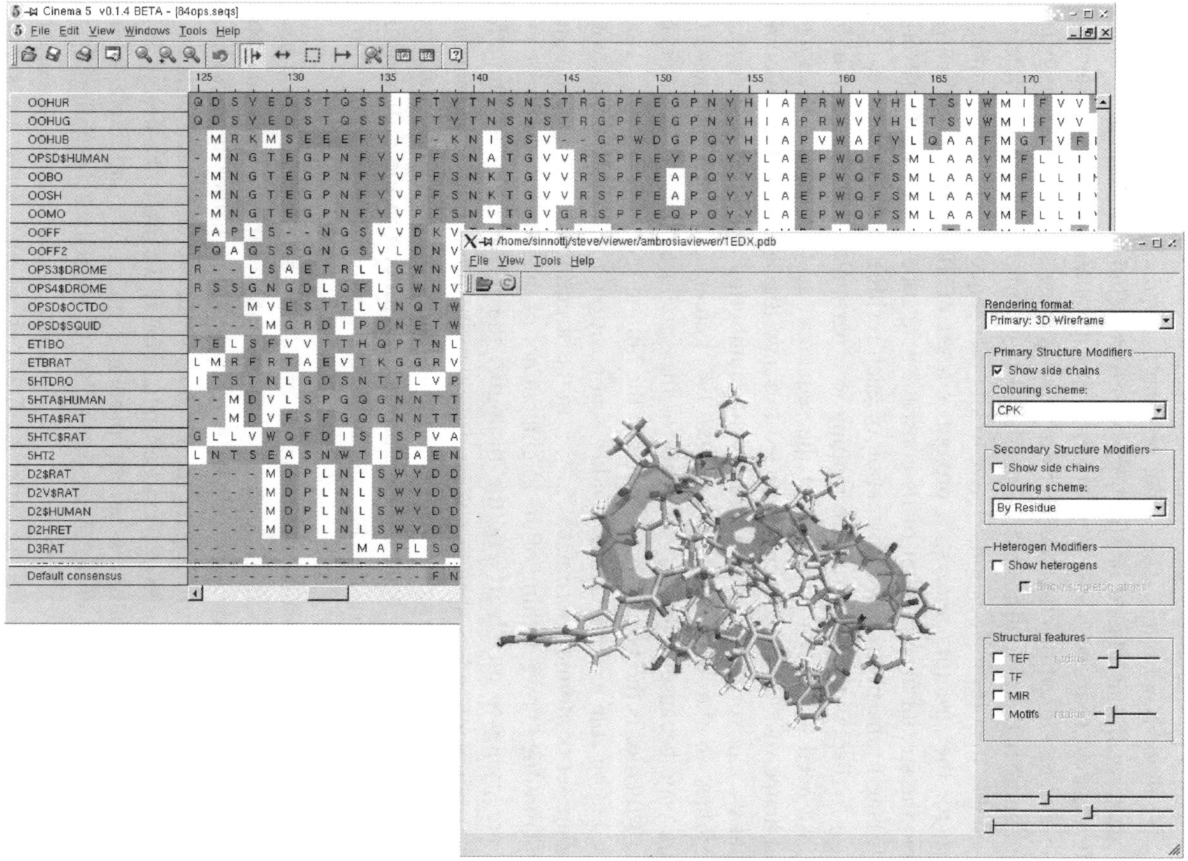

Figure 9.5 Interoperability in UTOPIA: the Ambrosia 3D structure viewer (front) showing residues 1–40 of bovine rhodopsin, and the CINEMA 5 sequence alignment editor (back) showing an alignment containing the bovine rhodopsin sequence. Conserved residues, highlighted in the alignment using CINEMA 5, may be pinpointed within the protein structure in Ambrosia. Similarly, structurally important residues, highlighted using Ambrosia, can be shown in the alignment within CINEMA 5

type of interoperation is in fact non-trivial. As we saw earlier, the sequence and structure communities have evolved along independent paths: both deal with proteins, but each 'sees' them rather differently. For a sequence analyst, a protein is a string of letters, each with a particular biochemical meaning, and each of which is numbered from the so-called N-terminus (i.e., the left-most residue). When comparing sequences, each residue is placed or slid horizontally in relation to similar residues in vertically adjacent sequences and its position in the alignment becomes a function of the number of gap insertions that are necessary to bring the sequences into register – its residue number then identifies which residue it is within its sequence, while its positional number denotes its location relative to all other sequences in the alignment. By contrast, for the structure analyst, a protein is a set of 3D atomic coordinates, which define the spatial arrangement of its constituent residues – obviously, several atoms contribute to the positional information of each residue in the sequence.

If we wish to display the 3D structure relating to a particular sequence within an alignment, we need to know the source of the sequence, so that we can verify its content and hence residue numbering; we need to know the source of the structure, again so that we can verify its content and residue numbering; we need to be aware that the residues in the sequence may not be 100 percent identical to those in the structure (say, because either the sequence or the structure contains an error, or because the structure is the bovine homologue of the human sequence, or whatever); we then need to be able to equivalence residue X at position Y in the alignment with the appropriate set of atomic coordinates, x, y, z, in the structure file, bearing in mind that the structure file knows nothing of the alignment nor of the particular sequence to which we are trying to attach it! For the biologist, the relationship between the sequence, its alignment and its structure is intuitive, but the file formats and contents used to encapsulate these different concepts are completely different; the challenge then is to make the computer aware of the equivalences hidden in the different file formats, and to render the information back to the user in a way that is easy to use and hides all of this ugliness.

To address some of these issues, the tools we are developing are based on user studies that have helped identify the important concepts and metaphors that should be exposed via their interfaces, and conversely to hide anything that is merely an artefact of the implementation. Considering again the issue of sequence alignment, it is easy to forget that even this idea employs a kind of metaphor; if we are to be brutal, it is biologically meaningless to 'align a set of protein sequences', and there is no underlying biological significance to 'inserting a gap' or indeed to the notion of 'a gap'. These concepts are merely a convenient way of engaging the human perceptual system's pattern-matching abilities in order to find out whether there are similar regions in a set of sequences, as answers to this question have genuine biological importance. The functionality of 'inserting a gap', for example, which is a feature provided by the majority of sequence alignment packages, is almost certainly poor interface design. The biologist is not *really* interested in the business of inserting gaps; rather, the task at hand is actually of determining whether regions of a set of sequences are similar, and it happens to be that a process of sliding them around relative to each other in an attempt to spot and to align similar regions is a good way of doing this – gaps just appear as a side-effect of the sliding process (the whole business of 'inserting gaps' in any case is likely to be the underlying implementation of the editor accidentally being exposed through the interface, as 'inserting a gap into a string of characters' is the natural and most likely way in which the sequences will be represented within the program's data structures). Gaps, then, are at best merely a by-product of the attempt to align similar regions, and at worst are the exposed innards of the editing program. The interface should really be built around features that allow the user to 'slide this back and forth', or 'align this bit of the sequence with that bit of that sequence' rather than 'insert 139 gaps here'; this is the approach taken in the CINEMA 5 alignment editor.

9.8 Conclusions

Despite enormous progress in computer technology, the work of the jobbing bioinformatician today is still often confounded by the disparate nature of the tools at his or her disposal, most of which have different interfaces, use their own file formats, do not communicate readily with each other, can only perform some of the necessary tasks and cannot easily be customized. Traditionally, the methods used to tackle these issues have tended to be either monolithic portals or toolkits assembled from *ad hoc* sources. However, although portals provide consistent user interfaces, they are difficult to scale and make it hard for users to innovate (they constrain users to the functionality offered by the interface); toolkits, by virtue of the different tools they integrate, are difficult to keep uniform.

Integration at the level of portals versus integration at the level of tools represent different approaches to similar problems: each has different challenges and opportunities. In UTOPIA, our approach is not a panacea, but a response to a challenge – the difficulty of using dynamic data from a host of different databases with different analysis tools on different machines in different locations. It is also an opportunity to make sequence analysis easier for biologists in future. By starting with some of the simplest tasks performed by bioinformaticians – such as aligning protein sequences – it may look old hat, something that has been done numerous times before, but we believe that by making the desktop and the filing system the source of uniformity, UTOPIA will provide the best of both worlds and will bring something unique to the desktop.

Acknowledgements

We are grateful to the ESNW and EMBnet for providing support for James Sinnott. Many thanks also go to our colleagues in the Advanced Interfaces Group and the School of Biological Sciences.

References

Altschul, S. F., Gish, W., Miller, W., Myers, E. W. and Lipman, D. J. (1990) Basic local alignment search tool. *J Mol Biol*, **215** (3), 403–410.

Attwood, T. K. (2000) The role of pattern databases in sequence analysis. *Briefings Bioinformatics*, **1**, 45–59.

Boeckmann, B., Bairoch, A., Apweiler, R., Blatter, M.-C., Estreicher, A., Gasteiger, E., Martin, M. J., Michoud, K., O'Donovan, C., Phan, I., Pilbout, S. and Schneider, M. (2003) The Swiss-Prot protein knowledgebase and its supplement TrEMBL. *Nucleic Acids Res*, **31** (1), 365–370.

Clamp, M., Cuff, J. and Barton, G. (1998) JalView – analysis and manipulation of multiple sequence alignments. *EMBnet News*, **5** (4).

Dayhoff, M. O., Eck, R. V., Chang, M. A. and Sochard, M. R. (1965) *Atlas of Protein Sequence and Structure* Vol. 1. National Biomedical Research Foundation, Silver Spring, MD.

Dayhoff, M. O., Schwartz, R. M., Chen, H. R., Hunt, L. T., Barker, W. C. and Orcutt, B. C. (1980) Nucleic acid sequence bank. *Science*, **209**, 1182.

Devereux, J., Haeberli, P. and Smithies, O. (1984). A comprehensive set of sequence analysis programs for the VAX. *Nucleic Acids Res*, **12** (1 Pt 1), 387–395.

Felsenstein, J. (1989) PHYLIP – Phylogeny Inference Package. *Cladistics*, **5**, 164–166.

Lipman, D. J. and Pearson, W. R. (1985) Rapid and sensitive protein similarity searches. *Science*, **227**, 1435–1441.

Mulder, N. J., Apweiler, R., Attwood, T. K., Bairoch, A., Bateman, A., Binns, D., Biswas, M., Bradley, P., Bork, P., Bucher, P., Copley, R. R., Courcelle, E., Das, U., Durbin, R., Falquet, L., Fleischmann, W., Griffith-Jones, S., Haft, D., Harte, N., Hermjakob, H., Hulo, N., Kahn, D., Kanapin, A., Krestyaninova, M., Lopez, R., Letunic, I., Lonsdale, D., Silventoinen, V., Orchard, S., Pagni, M., Peyruc, D., Ponting, C. P., Servant, F., Sigrist, C. J. A., Vaughan, R. and Zdobnov, E. (2003) The InterPro Database – 2003 brings increased coverage and new features. *Nucleic Acids Res*, **31** (1), 315–318.

Parry-Smith, D. J., Payne, A. W. R., Michie, A. D. and Attwood, T. K. (1998) CINEMA – a novel Colour INteractive Editor for Multiple Alignments. *Gene*, **221**, GC57–GC63.

Perkins, D. N. and Attwood, T. K. (1995) VISTAS – a package for visualising structures and sequences of proteins. *J Mol Graph*, **13**, 73–75.

Pettifer, S. R., Sinnott, J. R. and Attwood, T. K. (2004). UTOPIA – User-friendly Tools for OPerating Informatics Applications. *Comparative and Functional Genomics*, **5** (1), 56–60.

Rice, P. Longden, I. and Bleasby, A. (2000) EMBOSS: The European Molecular Biology Open Software Suite. *Trends in Genetics*, **16** (6), 276–277

Sonnhammer, E. (1999) *BelVu*. Karolinska Institutet, Stockholm. Available from http://www.cgr.ki.se/cgr/groups/sonnhammer/Belvu.html.

Thompson, J. D., Higgins, D. G. and Gibson, T. J. (1994) CLUSTALW: improving the sensitivity of progressive multiple sequence alignment through sequence weighting, position-specific gap penalties and weight matrix choice. *Nucleic Acids Res*, **22** (22), 4673–4680.

Whitehead, E. J. Jr., Goland, Y. Y. (1999) WebDAV A network protocol for remote collaborative authoring on the Web *Proceedings of the Sixth European Conference on Computer Supported Cooperative Work (ECSCW'99)*, 291–310.

10

Advances in Cluster Analysis of Microarray Data

Qizheng Sheng, Yves Moreau, Frank De Smet, Kathleen Marchal and **Bart De Moor**

Abstract

Clustering genes into biological meaningful groups according to their pattern of expression is a main technique of microarray data analysis, based on the assumption that similarity in gene expression implies some form of regulatory or functional similarity. We give an overview of various clustering techniques, including conventional clustering methods (such as hierarchical clustering, *k*-means clustering and self-organizing maps), as well as several clustering methods specifically developed for gene expression analysis.

Keywords

microarray, clustering, biclustering

10.1 Introduction

The first question in microarray data analysis is to identify genes whose expression levels are significantly changed under different experimental conditions. Basic statistical techniques can solve this problem efficiently (Baldi and Brunak, 2001). However, such an analysis treats the genes separately rather than exploring their relation with each other. For a gene, the detailed relations between the levels of expression in the different conditions are neglected in this first-level analysis. Based on the assumption that expressional similarity (i.e. coexpression) implies some kind

Data Analysis and Visualization in Genomics and Proteomics Edited by Francisco Azuaje and Joaquin Dopazo

of regulatory or functional similarity of the genes (and vice versa), the challenge of finding genes that might be involved in the same biological process is thus transformed into the problem of clustering genes into groups based on their similarity in expression profiles.

The first generation of clustering algorithms applied to gene expression profiles (e.g. hierarchical clustering (Eisen *et al.*, 1998), k-means (Hartigan, 1975) and self-organizing maps (SOM; Kohonen, 1995)) were mostly developed outside biological research. Although encouraging results have been produced (Spellman *et al.*, 1998; Tavazoie *et al.*, 1999; Tamayo *et al.*, 1999), some of their characteristics (such as determination of the number of clusters, clustering of outliers and computational complexity) often complicate their use for clustering expression data (Sherlock, 2000).

For this reason, a second generation of clustering algorithms has started to tackle some of the limitations of the earlier methods. These algorithms include, among others, model-based algorithms (Yeungo *et al.*, 2001; McLachlan, Bean and Peel, 2000), the self-organizing tree algorithm (Herrero, Valencia and Dopazo, 2001), quality-based algorithms (Heyer, Kruglyak and Yooseph, 1999; De Smet *et al.*, 2002), and biclustering algorithms (Cheng and Church, 2000; Sheng, Moreau and De Moor, 2003). Also, some procedures have been developed to help biologists estimate some of the parameters needed for the first generation of algorithms, such as the number of clusters present in the data (Lukashin and Fuchs, 2000; Yeung *et al.*, 2001).

While it is impossible to give an exclusive survey of all the clustering algorithms that have been developed for gene expression data, we try here to illustrate some key issues. The selection of algorithms is based on their popularity, their ability to handle the specific characteristics of microarray data and inevitably some personal biases. This chapter is organized as follows.

In Section 10.2, we address a few common issues for the discussion of clustering algorithms. In particular, we first discuss the preprocessing of microarray data, which is needed to overcome some difficult artifacts before clustering. Then, we address the basic but necessary ideas of the orientation of clustering (clustering genes versus clustering experiments) and the distance metrics commonly used to compare gene expression profiles.

We discuss the application of classical clustering algorithms to microarray data in Sections 10.3–10.5, where hierarchical clustering, k-means clustering, and self-organization maps are respectively addressed. Then, in Section 10.6, we identify common drawbacks of the first-generation clustering algorithms and give a wish list of some desirable features that an ideal clustering algorithm should carry.

Next, we look at some second-generation clustering algorithms, such as the self-organizing tree algorithm (SOTA; Herrero, Valencia and Dopazo, 2001) in Section 10.7, the quality-based clustering algorithms (Heyer, Kruglyak and Yooseph, 1999; De Smet *et al.*, 2002) in Section 10.8, mixture models for microarray data (Yeung *et al.*, 2001; McLachlan, Bean and Peel, 2002) in Section 10.9, and biclustering algorithms (Sheng, Moreau and De Moor, 2003) in Section 10.10.

Changes in details such as the preprocessing procedures, the algorithm or even the distance metrics might lead to different clustering results. Thus, in Section 10.11, we discuss methods used to validate clustering results.

10.2 Some Preliminaries

Before going into clustering algorithms *per se*, there are a few issues worth recalling.

Preprocessing microarray data

A correct preprocessing strategy, which not only removes as much as possible of the systematic noise present in microarray data but also provides a basis for the comparison between genes, is truly essential to an effective cluster analysis (in accordance with the '*garbage in, garbage out*' principle). Common procedures for preprocessing include the following five steps (Moreau *et al.*, 2002).

1. *Normalization.* First, it is necessary to normalize hybridization intensities within a single experiment or across experiments by computing and removing the biases to correct the data, before one can compare the results from different microarray experiments (Quackenbush, 2001).

2. *Nonlinear transformation.* Expression ratios (e.g. from two-channel cDNA microarray experiments using a test and reference sample) are not symmetrical in the sense that upregulated genes have expression ratios between one and infinity, while downregulated genes have expression ratios squashed between one and zero (Quackenbush, 2001). Taking the logarithms of these expression ratios results in symmetry between expression values of up- and downregulated genes. Furthermore, the noise on a microarray measurement is multiplicative as a function of the intensity of the signal. Taking the logarithm of the expression values makes noise approximately additive, except for low-intensity signals. The generalized log transformation combines normalization and transformation to provide this property over the whole signal range (Durbin and Rocke, 2004).

3. *Missing value replacement.* Microarray experiments often contain missing values that need to be replaced for many cluster algorithms. Techniques of missing value replacement (e.g. using the k-nearest-neighbour method or the singular value decomposition, SVD) have been described (Troyanskaya *et al.*, 2001), taking advantage of the rich information provided by the expression patterns of other genes in the data set.

4. *Filtering.* For any microarray study, many genes do not contribute to the underlying biological progress and show little variation over the different experiments.

These genes will have seemingly random and meaningless profiles after standardization (see further). Another problem arises from the highly unreliable expression profiles containing many missing values. The quality of the cluster would significantly degrade if these data were passed to the clustering algorithms as such. Filtering removes such expression profiles typically by putting a minimum threshold for the standard deviation of the expression values in a profile and a maximum threshold on the percentage of missing values (Eisen *et al.*, 1998).

5. *Standardization or rescaling*. Biologists are mainly interested in grouping gene expression profiles that have the same relative behaviour, (i.e. genes that are up- and downregulated together). Genes showing the same relative behaviour but with diverging absolute behaviour (e.g. gene expression profiles with a different baseline or a different amplitude but going up and down at the same time) will have a relatively high Euclidean distance (see Section 10.2.3). Cluster algorithms based on this distance measure will therefore wrongfully assign the genes to different clusters. This effect can largely be prevented by applying standardization or rescaling to the gene expression profiles so that they have zero mean and unit standard deviation.

Clustering genes versus clustering experiments

Instead of clustering genes, we can also cluster experimental conditions, where the task is to find groups of experimental conditions (which can be, for example, tumour samples) across which all the genes behave similarly. This type of clustering can be helpful for problems such as the discovery of histopathological tumours. While most of the discussion will be oriented towards clustering genes, most of it can be applied *mutatis mutandis* to clustering conditions.

Distance metrics

Depending on the way we define a cluster, clustering methods can be divided into two types – model-based clustering methods and distance-based clustering methods. Model-based clustering algorithms assume that the data points in the high-dimensional space are generated by a mixture of probabilistic models with different parameters. Each of these models is thus defined as a cluster. We will talk about this type of clustering method in detail in Section 10.9.

Distance-based clustering methods (to which most of the classical clustering methods belong, such as hierarchical clustering, k-means and SOM), in contrast, cluster data points according to some function of their pairwise distances. Some common distance metrics for clustering microarray data are the following.

1. *Pearson correlation*. The Pearson correlation r is the dot product of two normalized vectors, or in other words, the cosine between two vectors. It measures the

similarity in the shapes of two profiles, while not taking the magnitude of the profiles into account, and therefore suits well the biological intuition of coexpression (Eisen *et al.*, 1998).

2. *Squared Pearson correlation*. This is the square of the Pearson correlation, which considers two vectors pointing in the exact opposite directions to be perfectly similar (i.e., in this case, $r = -1$ while $r^2 = 1$), which might also be interesting for biologists (because repression is a form of coexpression).

3. *Euclidean distance*. Euclidean distance measures the length of the straight line connecting the two points. It measures the similarity between the absolute behaviours of genes, while the biologists are more interested in their relative behaviours. Thus, a standardization procedure is needed before clustering using Euclidean distance. Importantly, after standardization, the Euclidean distance between two points x and y is related to the Pearson correlation by $|x - y|^2 = 2(1 - |r|)$ (Alon *et al.*, 1999).

4. *Jackknife correlation*. The jackknife correlation (Heyer, Kruglyak and Yooseph, 1999) is an improvement for the Pearson correlation (which is not robust to outliers). Jackknife correlation increases the robustness to single outliers by computing a collection of all the possible leave-one-(experiment-)out Pearson correlations between two genes and then selecting the minimum of the collection as the final measure for the correlation.

10.3 Hierarchical Clustering

The first introduction of hierarchical clustering to the world of biology was its application to the construction of phylogenetic trees. Early applications of the method to gene expression data analysis (Eisen *et al.*, 1998; Spellman *et al.*, 1998) have proved its usefulness.

Hierarchical clustering has almost become the *de facto* standard for gene expression data analysis, probably because of its intuitive presentation of the clustering results. The whole clustering process is presented as a tree called a dendrogram; the original data are often reorganized in a heat map demonstrating the relationships between genes or conditions.

In hierarchical (agglomerative) clustering (Eisen *et al.*, 1998), each expression profile is initially assigned to one cluster; at each step, the distance between every pair of clusters is calculated and the pair of clusters with the minimum distance is merged; the procedure is carried on iteratively until a single cluster is assembled.

After the full tree is obtained, the determination of the final clusters is achieved by cutting the tree at a certain level or height, which is equivalent to putting a threshold on the pairwise distance between clusters. Note that the final cluster positions is thus rather arbitrary.

Distance measure between two clusters

As we mentioned, in every step of agglomerative clustering, the two clusters that are closest to each other will be merged. Here comes the problem of how we define the distance between two clusters. There are four common options:

1. *Single linkage*. The distance between two clusters is the distance between the two closest data points in these clusters (each point taken from a different cluster).

2. *Complete linkage*. The distance between two clusters is the distance between the two furthest data points in these clusters.

3. *Average linkage*. Both single linkage and complete linkage are sensitive to outliers (Duda, Hart and Stork, 2001). Average linkage provides an improvement by defining the distance between two clusters as the average of the distances between all pairs of points in the two clusters.

4. *Ward's method*. At each step of agglomerative clustering, instead of merging the two clusters that minimize the pairwise distance between clusters, Ward's method (Ward, 1963) merges the two clusters that minimize the 'information loss' for the step. The 'information loss' is measured by the change in the sum of squared error of the clusters before and after the merge. In this way, Ward's method assesses the quality of the merged cluster at each step of the agglomerative procedure.

These methods yield similar results if the data consist of compact and well separated clusters. However, if some of the clusters are close to each other or if the data have a dispersed nature, the results can be quite different (Duda, Hart and Stork, 2001). Ward's method, although less well known, often produces the most satisfactory results.

Visualization of the results

A heat map presenting the gene expression data, with a dendrogram to its side indicating the relationship between genes (or experimental conditions), is the standard way to visualize the result of hierarchical cluster analysis on microrray data. The length of a branch in the dendrogram is proportional to the pairwise distance between the clusters. Importantly, the leaves of the dendrogram, and accordingly the rows of the heat map, can be swapped (without actually changing the information contained in the tree) so that the similarity between adjacent genes is maximized, and hence the patterns embedded in the data become obvious in the heat map. However, the time complexity of such an optimal organization of the dendrogram is $O(2^{N-1})$ (because for each of the $N-1$ merging steps there are two possible orders to arrange the concerned clusters). Yet, the structure of the dendrogram remains an important problem, because although the dendrogram itself does not determine the clusters for the users, a good ordering of the leaves can help the users to identify and interpret the clusters. A heuristic approach aiming to find a good solution was developed (Eisen *et al.*, 1998) by weighting genes using combined source of information, and then placing the genes with lower average weight earlier in the final ordering. Further, Bar-Joseph, Gifford and Jaakkola (2001) reported a dynamic programming method that helps to reduce the time and memory complexities for solving the optimal leaf-ordering problem.

10.4 *k*-Means Clustering

k-means clustering (Hartigan, 1975) is a simple and widely used partitioning method for data analysis. Tavazoie *et al.* (1999) provided an example for applying *k*-means clustering to microarray data.

The number of clusters k in the data is needed as an input for the algorithm. The algorithm then initializes the mean vector for each of the k clusters either by hard assignment (e.g. from the input), or by random generation. These initial mean vectors are called the seeds. Next, the *k*-means algorithm proceeds iteratively with the following two steps: (1) using the given mean vectors, the algorithm assigns each gene (or experiment) to the cluster represented by the closest mean vector; (2) the algorithm recalculates the mean vectors (which are the sample means) for all the clusters. The iterative procedure converges when all the mean vectors of the clusters remain stationary.

A significant problem associated with the *k*-means algorithm is the arbitrariness of predefining the number of clusters, since it is difficult to predict the number of clusters in advance. In practice, this implies the use of a trial-and-error approach where a comparison and biological validations of several runs of the algorithm with different parameter settings are necessary (Moreau *et al.*, 2002). Another parameter that will influence the result of *k*-means clustering is the choice of the seeds. The algorithm suffers from the problem of converging to local minima. This means that with different seeds the algorithm can yield very different result.

10.5 Self-Organizing Maps

SOM (Kohonen, 1995) is a technique to visualize the high-dimensional input data (in our case, the gene expression data) on an output map of neurons, which are sometimes also called nodes. The map is often presented in a two-dimensional grid (usually of hexagonal or rectangular geometry) of neurons. In the high-dimensional input space, the structure of the data is represented by prototype vectors (serving similar functions as the mean vectors in the *k*-means algorithm), each of which is related to a neuron in the output space.

As an input for the algorithm, the dimension of the output map (e.g. a map of 6×5 neurons) needs to be specified. After initializing the prototype vectors, the algorithm iteratively performs the following steps. (1) Every input vector (e.g. representing a gene expression profile) is associated with the closest prototype vector, and thus is also associated with the corresponding neuron in the output space. (2) The coordinates of a prototype vector are updated based on a weighted sum of all the input vectors that are assigned to it. The weight is given by the neighbourhood function applied in the output space. As a result, a prototype vector is pulled more towards input vectors that are closer to the prototype vector itself and is less influenced by the input vectors located further away. In the meantime, this adaption procedure of the prototype vectors is reflected on the output nodes – nodes associated with similar prototype vectors are pulled closer together on the output map. (3) The initial variance of the neighbourhood function is chosen so that the neighbourhood covers all the neurons, but then the variance decreases during every iteration so as to achieve a smoother mapping. The algorithm terminates when convergence of the prototype vectors is achieved or after completing a pre-defined number of training iterations.

Because of the advantage in visualization, choosing the geometry of the output map is not as crucial a problem as the choice of the number of clusters for a k-means method. Like the k-means method, the initial choice of prototype vectors remains a problem that influences the final clustering result of SOM clustering. A good way to seed the prototype vectors is to use the result from a principal component analysis (PCA) (Kohonen, 1995).

The usefulness of SOM on clustering microarray data is illustrated by Tamayo *et al.* (1999).

10.6 A Wish List for Clustering Algorithms

The limitations of the first-generation algorithms together with the specific characteristics of gene expression data call out for clustering methods tailored for microarray data analysis. Collecting the lessons from the first-generation algorithms and the demands defined by the specific characteristics of microarray data, we compose here a subjective wish list of the features of an ideal clustering method for gene expression data.

A problem shared by the first-generation algorithms is the decision on the number of clusters in the data. In k-means clustering and SOM clustering, this decision has to be made before the algorithms are executed, while in hierarchical clustering it is postponed until the full dendrogram is formed, where the problem then is to determine where to cut the tree.

Another problem of the first-generation algorithms is that they all assign every gene in the data set (even outliers) to a particular cluster. A proper filtering step in the preprocessing (see Section 10.2.1) helps to reduce the number of outliers, but is insufficient. Therefore, a clustering algorithm should be able to identify genes that are not relevant for any clusters and leave them as they are.

A third problem is robustness. For all the three clustering techniques addressed above, difference in the choice of distance metrics (either for the vectors or for the clusters) will result in different final clusters. In k-means clustering and SOM clustering, the choice of seeds for the mean vectors or the prototype vectors also greatly influences the result. Taking into account the noisy nature of microarray data, improving the robustness should be one of the goals when designing novel clustering algorithms for gene expression data.

A fourth problem is the high dimensionality of microarray data, which requires the clustering algorithm to be fast and not memory hungry (a major problem of hierarchical clustering where the full distance matrix should be computed).

Finally, the biological process under study in a microarray experiment is a complicated process, where genes interact with each other in different pathways. Consequently, a gene under study might be directly or indirectly involved in several pathways. With this idea in mind, clustering algorithms that allow a gene to belong to multiple clusters would be favourable.

The desirable properties here are not exhaustive, but they give a number of clear directions for the development of clustering algorithms tailored to microarray data.

10.7 The Self-Organizing Tree Algorithm

SOTA (Herrero, Valencia and Dopazo, 2001) combines both SOM and (divisive) hierarchical clustering. As in SOM, SOTA maps the original input gene profiles to an output space of nodes. However, the nodes in SOTA are in the topology (or geometry) of a binary tree instead of a two-dimensional grid. In addition, the number of nodes in SOTA is not fixed from the beginning (in contrast to SOM); the tree structure of the nodes grows during the clustering procedure. Starting from a binary tree with two leaves, the algorithm iterates between the following two steps (see Figure 10.1).

With the given tree structure fixed, the gene expression profiles are sequentially and iteratively presented to the nodes located at the leaves of the tree (these nodes are called cells). Subsequently, each gene expression profile is associated with the cell that maps closest to it. The prototype vector of this cell and its neighbouring nodes, including its parent node and its sister cell, are then updated based on some neighbourhood weighting parameters (which perform the same role as the neighbourhood function in SOM). Thus, a cell is moved into the direction of the expression profiles that are associated with it. This presentation of the gene expression profiles to the cells continues until convergence.

After convergence of the above procedure is reached, the cell containing the most variable population of expression profiles (the variation is defined here by the maximal distance between two profiles that are associated with the same cell) is replicated into two daughter cells (causing the binary tree to grow), whereafter the entire process is restarted.

The algorithm stops (the tree stops growing) when a threshold of variability is reached for each cell. In this way, the number of clusters does not need to be specified in advance. The threshold variability can be determined by means of permutation tests on the data set.

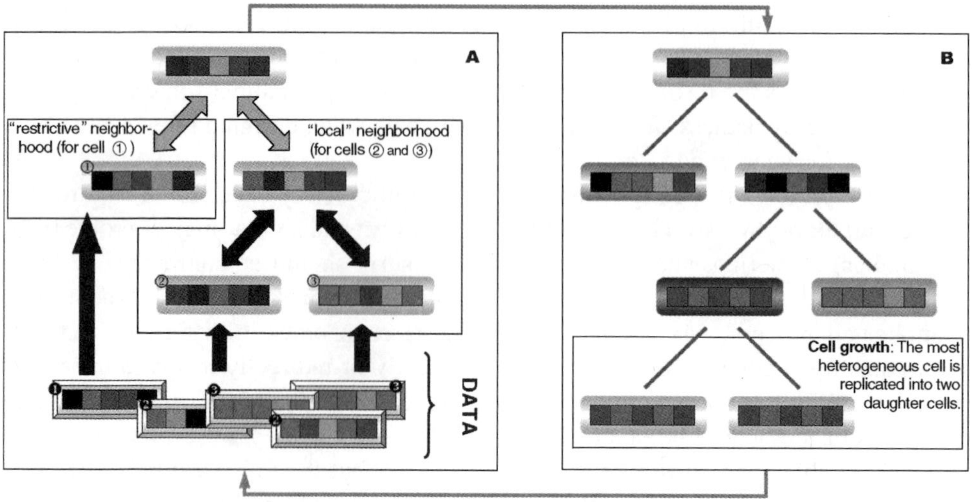

Figure 10.1 The iterative procedure of SOTA consists of two steps: (A) Each gene profile is associated with the cell whose prototype vector is located closest to it. Then the prototype vectors of the cells are updated based on the neighbourhood weighting parameters. (The black arrows between the nodes indicate where the updates take place, while the grey ones indicates where the updates are no longer performed.) This procedure iterates until convergence is reached. (B) The cell whose associated profile exhibits the largest variability is duplicated into two daughter cells (the darker the cell, the more heterogeneous it is)

10.8 Quality-Based Clustering Algorithms

Quality-based algorithms produce cluster with a quality guarantee that ensures that all members of a cluster are coexpressed.

QT_Clust

Heyer, Kruglyak and Yooseph, (1999) introduced the concept of quality-based clustering. Their implementation is called QT_Clust; it is a greedy procedure that finds one cluster at a time. It considers each expression profile in the data in turn. For each expression profile, it determines which other profiles are within the specified distance in its neighbourhood. This specified distance therefore serves as the quality guarantee. In this way, a candidate cluster is formed for every expression profile. The candidate cluster with the largest number of expression profiles is selected as an output of the algorithm. Then, the expression profiles of the selected cluster are removed, and the whole procedure starts again to find the next cluster. The algorithm stops when the number of profiles in the largest remaining cluster falls below a prespecified threshold.

By using a stringent quality guarantee, it is possible to find clusters with tightly related expression profiles (i.e. clusters containing highly coexpressed genes). Moreover, genes that are not really coexpressed with other members of the data set are not included in any of the clusters.

Adaptive quality-based clustering

Adaptive quality-based clustering (De Smet *et al.*, 2002) uses a heuristic two-step approach to find one cluster at a time. In the first step, a quality-based approach is performed to locate a cluster center. Using a preliminary estimate of the radius (i.e. the quality) of the cluster, a cluster centre is located in an area where the density (i.e. the number) of gene expression profiles is locally maximal. In the second step, the algorithm re-estimates the quality (i.e. the radius) of the cluster so that the genes belonging to the cluster are, in a statistical sense, significantly coexpressed. To this end, a bimodal and one-dimensional probability distribution (the distribution consists of two terms: one for the cluster and one for the rest of the data) describing the Euclidean distance between the data points and the cluster centre is fitted to the data using an expectation-maximization (EM) algorithm. The cluster is subsequently removed from the data and the whole procedure is restarted. Only clusters whose size exceed a predefined number are presented to the user.

In adaptive quality-based clustering, the users have to specify a significance level as the threshold for quality control. This parameter has a strict statistical meaning and is therefore much less arbitrary (in contrast to the case in QT_Clust). It can be chosen independently of a specific data set or cluster and it allows for a meaningful default value (95 per cent) that in general gives good results. This makes the approach user friendly without the need for extensive parameter fine-tuning. Second, with the ability to allow the clusters to have different radii, adaptive quality-based clustering produces clusters adapted to the local data structure.

10.9 Mixture Models

Model-based clustering (Hartigan, 1975) has already been used in the past for other applications outside bioinformatics, but its application to microarray data is comparatively recent (Yeung *et al.*, 2001; McLachlan, Bean and Peel, 2002).

Model-based clustering assumes that the data are generated by a finite mixture of underlying probability distributions, where each distribution represents one cluster. The problem, then, is to associate every gene (or experiment) with the best underlying distribution in the mixture, and at the same time to find out the parameters for each of these distributions.

Mixture model of normal distributions

When multivariate normal distributions are used, each cluster is represented by a hypersphere or a hyperellipse in the data space. The mean of the normal distribution gives the centre of the hyperellipse, and the covariance of the distribution specifies its orientation, shape and volume. The covariance matrix for each cluster can be represented by its eigenvalue decomposition, with the eigenvectors determining the orientation of the cluster, and the eigenvalues specifying the shape and the volume of the cluster. By using different levels of restrictions on the form of the covariance matrix (i.e. its eigenvectors and eigenvalues), one can control the trade-off between model complexity (the number of parameters to be estimated) and flexibility (the extent to which the model fits the data).

The choice of the normal distribution is partly based on its desirable analytic convenience. Moreover, the assumption for fitting a normal distribution to gene expression profiles is considered to be reasonable, especially when the standard preprocessing procedures (see Section 10.2.1) have been applied (Yeung *et al.*, 2001; Baldi and Brunak, 2001). Of course, other underlying distributions, such as gamma distributions or mixtures of Gaussian and gamma distributions, can also be used to describe expression profiles. So far, no precise conclusions have been made on what is the most suitable distribution for gene expression data (Baldi and Brunak, 2001).

Regardless of the choice of underlying distributions, a mixture model is usually learned by an EM algorithm. Given the microarray data and the current set of model parameters, the probability to associate a gene (or experiment) to every cluster is evaluated in the E step. Then, the M step finds the parameter setting that maximizes the likelihood of the complete data. The complete data refer to both the microarray data (observed data) and the assignment of the genes (or experiments) to the clusters (unobserved data). The likelihood of the model increases as the two steps iterate, and convergence is guaranteed.

The EM procedure is repeated for different numbers of clusters and different covariance structures. The result of the first step is thus a collection of different models fitted to the data and all having a specific number of clusters and specific covariance structure. Then, the best model with the most appropriate number of clusters and covariance structure in this group of models is selected. This model selection step involves the calculation of the Bayesian information criterion (BIC) for each model.

Yeung *et al.* (2001) reported good results of such analysis as described above using their MCLUST software on several synthetic and real expression data sets.

Mixture of factor analysis

For the clustering experiments (e.g. tissue samples), however, a problem arises in fitting a normal mixture to the data because the number of genes is much larger than the number of experiments. To solve this problem, McLachlan, Bean and Peel (2002) applied a mixture of factor analysis to the clustering of experiments (see Figure 10.2). The idea can be interpreted as follows. A single factor analysis performs a dimensional reduction in the gene space of a cluster. That is to say, in factor analysis, vectors of experiments located in the original n-dimensional hyperellipse (where n represents the number of genes) are projected onto their corresponding vectors of factors located in an m-dimensional unit sphere (usually $m \ll n$). By using a mixture

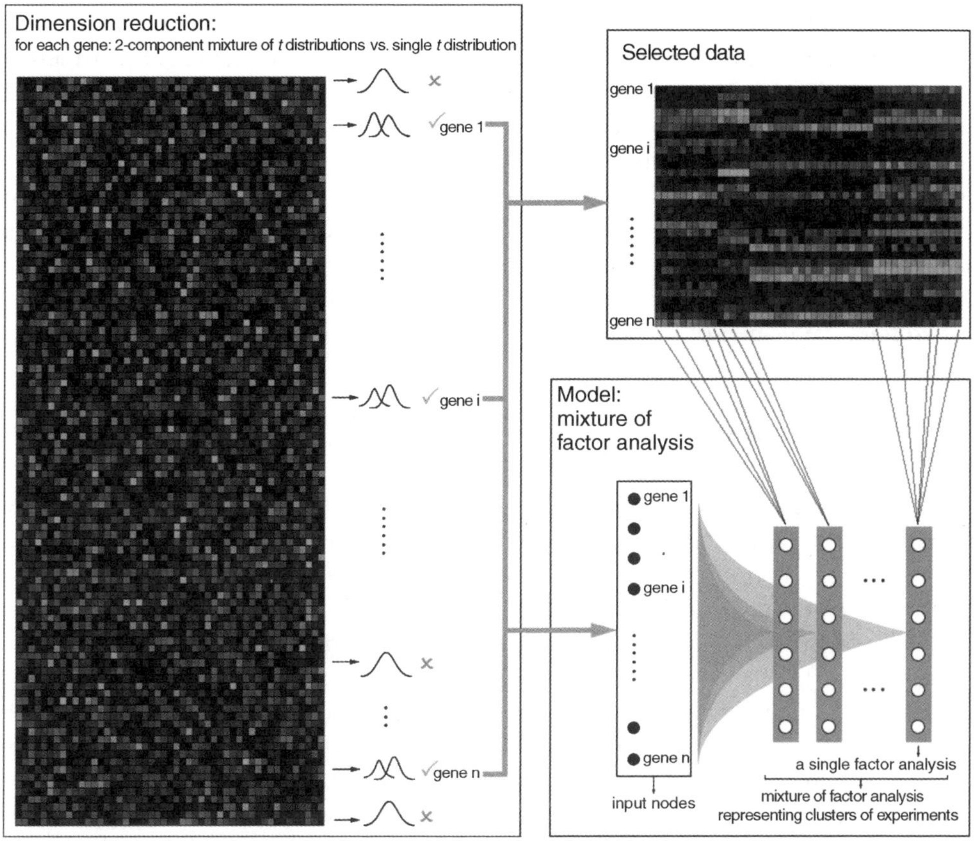

Figure 10.2 McLachlan, Bean and Peel (2002) use a two-component mixture model of *t* distributions to examine every gene expression profile against a single *t* distribution. Expression profiles to which the mixture models fit better (in terms of, for example, likelihood) are selected for further analysis. A mixture of factor analysis is applied on the selected data to cluster the experimental conditions

of factor analysis, clustering of the experiments is done on a reduced feature space (i.e. the m-dimensional factor space) instead of on the original huge-dimensional gene space. The EM algorithm is also used to learn the mixture of factor analysis model.

However, the choice for the number of factors in such a model remains a dilemma. If the number is too small, the full correlation structure of the genes cannot be captured; while if it is too large, the EM algorithm for the parametrization of the model can encounter computational difficulties. To alleviate the problem, McLachlan, Bean and Peel (2000) added another stage to reduce the dimension of the gene space before applying the mixture of factor analysis to the clustering of the experiements. In this stage, both a two-component mixture model of univariate t distributions (where the association of the experiments to the two components is unknown) and a single t distribution are fitted to the data for each gene. A threshold on the likelihood ratio between the two models is then applied to determine whether the gene is responsible for the clustering of experiments.

A *t* mixture model is more suitable for describing a gene expression profile than a normal mixture model because the former is more robust to outliers. A *t* distribution has an additional parameter called the degree of freedom compared with a normal distribution. The degree of freedom can be seen as a parameter for adjusting the thickness of the tail of the distribution. A *t* distribution with a relatively small degree of freedom will have a thicker tail than a normal distribution with the same mean and variance. However, as the degree of freedom goes to infinity, the *t* distribution approaches the normal distribution. Because of the thicker tail of a *t* distribution, the model learned for the *t* mixture is more robust to the outliers in gene profiles. Therefore, the degree of freedom can be viewed as a robustness tuning parameter.

10.10 Biclustering Algorithms

Biclustering means to cluster both the genes and the experiments at the same time. Among early papers on biclustering methods, clustering algorithms were applied (iteratively) to both dimensions of a microarray data set (Alon *et al.*, 1999; Getz, Levine and Domany, 2000). As a result, genes and experiments are reorganized so as to improve the manifestation of the patterns inherited in both the genes and the experiments. In other words, biclustering algorithms of this type divide the data into checkerboard units of patterns. More recently, other algorithms specifically designed for finding this kind of pattern have also been developed. An example is provided by Lazzeroni and Owen (2000), who used a plaid model – a specific form of mixture of normal distributions – to describe microarray data. EM was used for the parametrization of the model. For another example, the spectral biclustering method (Kluger *et al.*, 2003) applies SVD for solving the problem. However, this type of biclustering algorithm has limitations (Hastie *et al.*, 2000) when the expression profiles of some genes under study divide the samples by one biological explanation (say, tumour type) while some others divide the samples according to another biological process (e.g. drug response).

The second type of biclustering algorithm aims to find genes that are responsible for the classification of the samples. Examples are the gene shaving method (Hastie *et al.*, 2000), which searches for clusters of genes that vary as much as possible across the samples with the help of PCA; and a minimum description length method (Jörsten and Yu, 2003).

The third type of biclustering algorithm questions conventional clustering algorithms by the idea that genes that share functional similarities do not have to be coexpressed over all the experimental conditions under study. Instead of clustering genes based on their overall expressional behaviour, these algorithms look for patterns where genes share similar expressional behaviour over only a subset of experimental conditions. The same idea can be used for clustering the experimental conditions. Suppose a microarray study is carried out on tumour samples of different histopathological diagnoses. The problem then is to find tumour samples that have similar gene expression levels for a subset of genes (so as to obtain an expressional fingerprint for the tumour). To distinguish the two orientations for this type of biclustering problem, we will refer to the former case as biclustering genes, and the latter case as biclustering experiments. This type of biclustering algorithm was pioneered by Cheng and Church (2000), where a heuristic approach is proposed to find patterns as large as possible that have minimum mean squared residues, while allowing variance to be present across the experiments when biclustering genes (or across the genes when biclustering experiments). Model-based approaches have also been applied for this type of problem. Barash and Friedman (2002) used an EM algorithm for model parametrization, while Sheng *et al.* (2003) proposed a Gibbs sampling strategy for model learning.

The idea of applying Gibbs sampling to clustering was inspired by the success of the Gibbs sampling algorithm in solving the motif-finding problem (Thijs *et al.*, 2002). The model consists in associating a binary random variable (label) with each of the rows and each of the columns in the data set so that a value of 1 indicates that the row or the column belongs to the bicluster and a 0 indicates otherwise. Then the task of the algorithm is to estimate the value for each of these labels. The algorithm opts for Gibbs sampling, a Bayesian approach for the estimation, and examines the posterior distribution of the labels given the data (see Figure 10.3). Finally, a threshold is put on the posterior distribution and selects the rows and columns that have probabilities larger than the threshold as the positions of the bicluster. To find multiple biclusters in the data, the labels associated with the experiments for a found bicluster

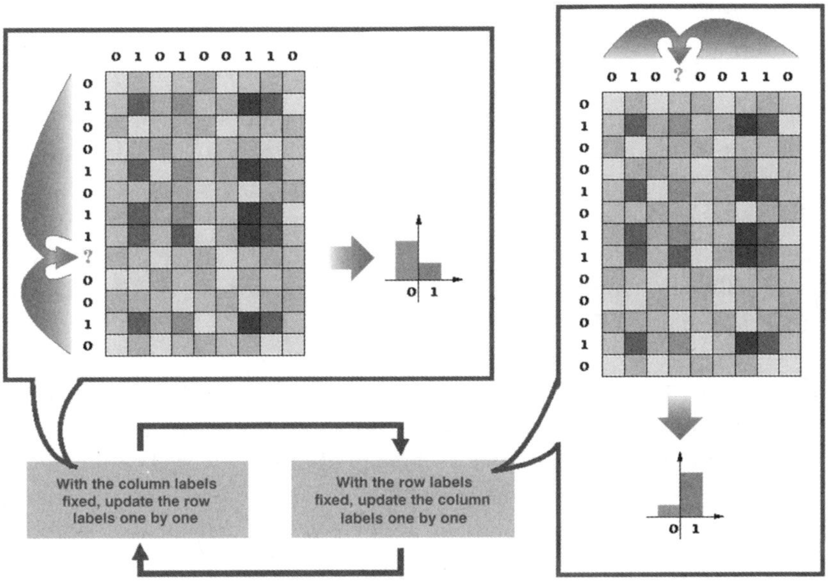

Figure 10.3 With all the other labels fixed, the Gibbs biclustering algorithm calculates the posterior conditional distribution of a label (indicating whether a gene or a condition belongs to the bicluster) at each iteration. Subsequently, a label is drawn from the obtained conditional distribution and is assigned to the gene or the experimental condition

are set permanently to zero when looking for further clusters. The masking of the experiments is chosen for both biclustering the genes and biclustering the experiments based on the idea that a gene should be allowed to belong to different clusters.

10.11 Assessing Cluster Quality

As mentioned before, different runs of clustering will produce different results, depending on the specific choice of preprocessing, algorithm, distance measure, and so on. Many methods often produce clusters even for random data. Therefore, validation of the relevance of the cluster results is of utmost importance. Validation can be either statistical or biological. Statistical cluster validation can be done by assessing cluster coherence, by examining the predictive power of the clusters, or by testing the robustness of a cluster result against the addition of noise.

Alternatively, the relevance of a cluster result can be assessed by a biological validation. Of course it is hard, not to say impossible, to select the best cluster output, since 'the biologically best' solution will be known only if the studied biological system is completely characterized. Although some biological systems have been described extensively, no such completely characterized benchmark system is now available. A common method to biologically validate cluster outputs is to search for enrichment of functional categories within a cluster. Detection of regulatory motifs is also an appropriate biological validation of the cluster results (Tavazoie *et al.*, 1999). Some of the recent methodologies described in the literature to validate clustering results are discussed as follows.

1. *Testing cluster coherence.* Based on biological intuition, a cluster result can be considered reliable if the within-cluster distance is small (i.e., all genes retained are tightly coexpressed) and the cluster has an average profile well delineated from the remainder of the data set (i.e. a maximal inter-cluster distance). Such criteria can be formalized in several ways, such as the sum-of-squared-error criterion of *k*-means, silhouette coefficients (Kaufman and Rousseeuw, 1990) or Dunn's validity index (Azuaje, 2002).

2. *Figure of merit.* The FOM (Yeung, Haynor and Ruzzo, 2001) is a simple quantitative data-driven methodology that allows comparisons between outputs of different clustering algorithms in terms of their predictive power. The methodology is related to the jackknife approach and the leave-one-out cross-validation. The clustering algorithm (for the genes) is applied to all experimental conditions (the data variables) except for one left-out condition. If the algorithm performs well, we expect that if we look at the genes from a given cluster their values for the left-out condition will be highly coherent. Therefore, for each cluster, the sum of squared deviations is computed for the expression levels under the left-out condition and over all the genes in the cluster. With the left-out condition fixed, the

FOM is subsequently calculated as the root mean of these sums obtained for all the clusters. The aggregate FOM is further computed as the sum of the FOMs over all the experimental conditions so as to compare different clustering algorithms.

3. *Sensitivity analysis*. Gene expression levels are the superposition of real biological signals and experimental errors. A way to assign confidence to a cluster membership of a gene consists in creating new in silico replicas of the microarray data by adding to the original data a small amount of artificial noise and clustering the data of those replicas. If the biological signal is stronger than the experimental noise in the measurements of a particular gene, adding small artificial variations (in the range of the experimental noise) to the expression profile of this gene will not drastically influence its overall profile and therefore will not affect its cluster membership. Through some robustness statistics (Bittner *et al.*, 2000), sensitivity analysis lets us detect which clusters are robust within the range of experimental noise and therefore trustworthy for further analysis.

   The main issue in this method is to choose the noise level for sensitivity analysis. Bittner *et al.* (2000) perturbed the data by adding random Gaussian noise with zero mean and a standard deviation that is estimated as the median standard deviation for the log-ratios for all genes across the experiments.

   The bootstrap analysis methods described by Kerr and Churchill (2001) use the residual values of a linear analysis of variance (ANOVA) model as an estimate of the measurement error. By using an ANOVA model, non-consistent measurement errors can be separated from variations caused by alterations in relative expression or by consistent variations in the data set. The residuals are subsequently used to generate new replicates of the data set by bootstrapping (adding residual noise to estimated values).

4. *Use of different algorithms*. Just as clustering results are sensitive to adding noise, they are sensitive to the choice of clustering algorithm and to the specific parameter settings of a particular algorithm. Many clustering algorithms are available, each of them with different underlying statistics and inherent assumptions about the data. The best way to infer biological knowledge from a clustering experiment is to use different algorithms with different parameter settings. Clusters detected by most algorithms will reflect the pronounced signals in the data set. Again, statistics similar to those of Bittner *et al.* (2000) are used to perform these comparisons. (See Chapter 11 for a further discussion of the use of different algorithms.)

5. *Enrichment of functional categories*. One way to biologically validate results from clustering algorithms is to compare the gene clusters with existing functional classification schemes. In such schemes, genes are allocated to one or more functional categories (Tavazoie *et al.*, 1999; Segal *et al.*, 2001) representing their biochemical properties, biological roles and so on. Finding clusters that have been

significantly enriched for genes with similar function is a proof that a specific clustering technique produces biologically relevant results.

Using the cumulative hypergeometric probability distribution, we can measure the degree of enrichment by calculating the probability or P-value of finding by chance at least k genes in this specific cluster of n genes from this specific functional category that contains f genes out of the whole g annotated genes:

$$P = 1 - \sum_{i=0}^{k-1} \frac{\binom{f}{i}\binom{g-f}{n-i}}{\binom{g}{n}} = \sum_{i=k}^{\min(n,f)} \frac{\binom{f}{i}\binom{g-f}{n-i}}{\binom{g}{n}}.$$

These P-values can be calculated for each functional category in each cluster. Note that these P-values must be corrected for multiple testing according to the number of functional categories.

10.12 Open Horizons

When research on clustering of microarray data started, a common opinion was that clustering was a 'closed' area of statistical research where little innovation was possible. Dozens of papers about clustering microarray data have now been published, demonstrating time and again significant improvements over classical methods. Yet, classical methods (in particular hierarchical clustering) remain dominant in biological applications, despite real shortcomings. The conclusion most probably is that new methods have not demonstrated sufficient added value to overcome the *status quo* established by a few pioneering works. As an example, Table 10.1 provides a summary of how well the second-generation clustering algorithms described in this paper meet our wish list presented in Section 10.6.

Lack of benchmarking significantly impairs the demonstration of major improvements. This situation is itself created by the subjectivity of interpreting clustering results in many situations, and weak benchmarks (such as the yeast cell cycle data set by Cho *et al.*, 1998) have only added to the confusion. The most likely way out is the production of a large, carefully designed set of microarray experiments, specifically dedicated to the evaluation of clustering algorithms.

Another major open problem is the limited connection between clustering and biological knowledge. Clustering does not stand by itself but is tightly linked to the biological interpretation of its results and the subsequent use of these results. Cluster methods that incorporate functional, regulatory and pathway information directly in the algorithm are highly desirable. Also, clustering is only the starting point for further analysis, so strategies that integrate clustering tightly with its downstream analysis (e.g. regulatory sequence analysis, guilt by association) will improve on the final biological predictions (Moreau *et al.*, 2002). Probabilistic relational models and

Table 10.1 How well do the second-generation clustering algorithms meet our wish list?

	Decision on no. of clusters	Assign every gene to a particular cluster?	Robustness	Time complexity	Allow a gene in multiple clusters?
SOTA	By putting a threshold on the variability of the cells	Yes	Comparable to that of SOM	Linear in no. of expression profiles	No
QT_clust	By putting a threshold on the quality of a cluster	No	Global solution	Quadratic in no. of expression profiles	No
Adap. qual. based	By specifying a significance level	No	Global solution	Linear in no. of expression profiles	No
Model based	By model comparison in terms of BIC	No	The use of EM leads to local minimum solutions	Depends on the implementation	Yes
Gibbs biclustering	Automatic decision	No	The chance for finding local minima is reduced (comparing with EM)	Linear performance can be achieved depending on the implementation	Yes

their variants, such as biclustering algorithms, hold a great potential in this regard, as already demonstrated in some applications (Segal *et al.*, 2001, 2003).

References

Alon, U., Barkai, N., Notterman, D. A. *et al.* (1999) Broad patterns of gene expression revealed by clustering analysis of tumor and normal colon tissues probed by oligonucleotide arrays. *Proc Natl Acad Sci USA*, **96**, 6745–6750.

Azuaje, F. (2002) A cluster validity framework for genome expression data. *Bioinformatics*, **18** (2), 319–320.

Baldi, P. and Brunak, S. (2001) *Bioinformatics: the Machine Learning Approach, Adaptive Computation and Machine Learning*, 2nd edn. MIT Press, Cambridge, MA.

Barash, Y. and Friedman, N. (2002) Context-specific Bayesian clustering for gene expression data. *J. Comput. Biol.*, **9**, 169–191.

Bar-Joseph, Z., Gifford, D. K. and Jaakkola, S. (2001) Fast optimal leaf ordering for hierachical clustering. *Bioinformatics*, **17** (Suppl. 1), S22–S29.

Bittner, M., Meltzer, P., Chen, Y. *et al.* (2000) Molecular classfication of cutaneous malignant melanoma by gene expression profiling. *Nature*, **406**, 536–540.

Cheng, Y. and Church, G. M. (2000) Biclustering of expression data. In *ISMB 2000 Proceedings*, 93–103.

Cho, R. J., Campbell, M. J., Winzeler, E. A. *et al.* (1998) A genome-wide transcriptional analysis of mitotic cell cycle. *Mol Cell*, **2**, 65–73.

De Smet, F., Mathys, J., Marchal, K., Thijs, G., De Moor, B. and Moreau, Y. (2002) Adaptive quality-based clustering of gene expression profiles. *Bioinformatics*, **18** (5), 735–746.

Duda, R. O., Hart, P. E. and Stork, D. G. (2001) *Pattern Classification*, 2nd edn. Wiley, New York.

Durbin, B. P. and Rocke, D. M. (2004) Variance-stabilizing transformations for two-color microarrays. *Bioinformatics*, **20** (5), 660–667.

Eisen, M. B., Spellman, P. T., Brown, P. O. and Botstein, D. (1998) Cluster analysis and display of genome-wide expression patterns. *Proc. Natl. Acad. Sci. USA*, **95**, 14 863–14 868.

Getz, G., Levine, E. and Domany, E. (2000) Coupled two-way clustering analysis of gene microarray data. *Proc. Natl. Acad. Sci. USA*, **97** (22), 12 079–12 084.

Hartigan, J. A. (1975) *Clustering Algorithms (Wiley Series in Probability)*. Wiley, New York.

Hastie, T., Tibshirani, R., Eisen, M. B. *et al.* (2000) 'Gene shaving' as a method for identifying distinct sets of genes with similar expression patterns. *Genome Bio.*, **1** (2), research0003.1–0003.21.

Herrero, J., Valencia, A. and Dopazo, J. (2001) A hierachical unsupervised growing neural network for clustering gene expression patterns. *Bioinformatics*, **17** (2), 126–136.

Heyer, L. J., Kruglyak, S. and Yooseph, S. (1999) Exploring expression data: identification and analysis of coexpressed genes. *Genome Res*, **9**, 1106–1115.

Jöorsten, R. and Yu, B. (2003) Simultaneous gene clustering and subset selection for sample classification via MDL. *Bioinformatics*, **19** (9), 1100–1109.

Kaufman, L. and Rousseeuw, P. J. (1990) *Finding Groups in Data: an Introduction to Cluster Analysis*, Wiley, New York.

Kerr, M. K. and Churchill, G. A. (2001) Bootstrapping cluster analysis: Assessing the reliability of conclusions from microarray experiments. *Proc. Natl. Acad. Sci. USA*, **98** (16), 8961–8965.

Kluger, Y., Basri, R., Chang, J. T. and Gerstein, M. (2003) Spectral biclustering of microarray data: Coclustering genes and conditions. *Genome Res.*, **13**, 703–716.

Kohonen, T. (1995) *Self-Organizing Maps (Springer Series in Information Sciences)*. Springer.

Lazzeroni, L. and Owen, A. (2000) *Plate Models for Gene Expression Data, technical report. Department of Statistics, Stanford University.*

Lukashin, A. V. and Fuchs, R. (2000) Analysis of temporal gene expression profiles: clustering by simulated annealing and determining the optimal number of clusters. *Bioinformatics*, **17** (5), 405–414.

McLachlan, G. J., Bean, R. W. and Peel, D. (2002) A mixture model-based approach to the clustering of microarray expression data. *Bioinformatics*, **18** (3), 413–422.

Moreau, Y., De Smet, F., Thijs, G., Marchal, K. and De Moor, B. (2002) Functional bioinformatics of microarray data: From expression to regulation. *Proc. IEEE*, **90** (11), 1722–1743.

Quackenbush, J. (2001) Computational analysis of microarray data. *Nature Rev.*, **2**, 418–427.

Segal, E., Shapira, M., Regev, A. *et al.* (2003) Module networks: identifying regulatory modules and their condition-specific regulators for gene expression data. *Nature Genetics*, **34** (2), 166–176.

Segal, E., Taskar, B., Gasch, A., Friedman, N. and Koller, D. (2001) Rich probabilistic models for gene expression. *Bioinformatics*, **17** (Suppl. 1), S243–S252.

Sheng, Q., Moreau, Y. and De Moor, B. (2003) Biclustering microarray data by Gibbs sampling. *Bioinformatics*, **19** (Suppl. 2), II196–II205.

Sherlock, G. (2000) Analysis of large-scale gene expression data. *Curr. Opin. Immunol*, **12**, 201–205.

Spellman, P. T., Sherlock, G., Zhang, M. Q. *et al.* (1998) Comprehensive identification of cell cycle-regulated genes of the yeast *Saccharomyces cerevisiae* by microarray hybridization. *Mol. Bio. Cell*, **9**, 3273–3297.

Tamayo, P., Slonim, D., Mesirov, J. *et al.* (1999) Interpreting patterns of gene expression with self-organizing maps: Methods and application to hematopoietic differentiation. *Proc. Natl. Acad. Sci. USA*, **96**, 2907–2912.

Tavazoie, S., Hughes, J. D., Campbell, M. J., Cho, R. J. and Church, G. M. (1999) Systematic determination of genetic network architecture. *Nature Genetics*, **22**, 281–285.

Thijs, G., Marchal, K., Lescot, M. *et al.* (2002) A Gibbs sampling method to detect overrepresented motifs in the upstream regions of coexpressed genes. *J. Comput. Biol.*, **9**, 447–464.

Troyanskaya, O., Cantor, M., Sherlock, G. *et al.* (2001) Missing value estimation methods for DNA microarrays. *Bioinformatics*, **17** (6), 520–525.

Ward, J. H. (1963) Hierarchical grouping to optimize an objective function. *J. Am. Stat. Assoc.*, **58**, 239–244.

Yeung, K., Fraley, C., Murua, A., Raftery, A. and Ruzzo, W. (2001) Model-based clustering and data transformations for gene expression data. *Bioinformatics*, **17** (10), 977–987.

Yeung, K., Haynor, D. and Ruzzo, W. (2001) Validating clustering for gene expression data. *Bioinformatics*, **17** (4), 309–318.

11

Unsupervised Machine Learning to Support Functional Characterization of Genes: Emphasis on Cluster Description and Class Discovery

Olga G. Troyanskaya

Abstract

In recent years, multiple types of high-throughput functional genomic data have become available to facilitate rapid functional annotation of sequenced genomes. However, such data often sacrifice specificity for scale, and thus sophisticated analysis methods are necessary to make accurate predictions of gene function based on large-scale datasets. This chapter presents an overview of unsupervised analysis of microarray data followed by an in-depth discussion of integrated analysis of heterogeneous biological data for accurate gene function prediction. This discussion focuses on a general probabilistic method for such integration, called MAGIC, and provides an overview of the methodology, application and evaluation of this technology.

Keywords

data integration, function prediction, genomic data analysis, Bayesian network

11.1 Functional Genomics: Goals and Data Sources

The availability of complete genomic sequences of several eukaryotic organisms, including the human genome (Goffeau *et al.*, 1996; Wood *et al.*, 2002; Adams *et al.*, 2000; Consortium, 1998; Lander *et al.*, 2001; Venter *et al.*, 2001), has brought

Data Analysis and Visualization in Genomics and Proteomics Edited by Francisco Azuaje and Joaquin Dopazo

molecular biology into a new era of systematic functional understanding of cellular processes. The sequences themselves provide a wealth of information, but functional annotation is a necessary step toward comprehensive description of genetic systems of cellular controls, including those whose malfunctioning becomes the basis of genetic disorders such as cancer (Kitano, 2002; Steinmetz and Deutschbauer, 2002; Ideker, Galitski and Hood, 2001). High-throughput functional technologies, such as genomic (Lipshutz *et al.*, 1999; Schena *et al.*, 1995) and soon proteomic microarrays (Cahill and Nordhoff, 2003; Sydor and Nock, 2003; Oleinikov *et al.*, 2003; Huang, 2003; Cutler, 2003), allow one to rapidly assess general functions and interactions of proteins in the cell. While classical genetic and cell biology techniques continue to play an important role in the detailed understanding of cellular mechanisms, the combination of rapid functional annotation with targeted exploration by traditional methods will facilitate fast and accurate identification of causal genes and key pathways affected in disease.

Increasing amounts of high-throughput data are available for functional annotation of eukaryotic genomes. In the model organism yeast *Saccharomyces cerevisiae*, these datasets include protein–protein interaction studies (affinity precipitation, Larsson and Mosbach, 1979; two-hybrid techniques, Fields and Song, 1989), synthetic rescue (Novick, Osmond and Botstein, 1989) and lethality (Novick, Osmond and Botstein, 1989; Bender and Pringle, 1991) experiments, and microarray analysis (Lipshutz *et al.*, 1999; Schena *et al.*, 1995). The most commonly available data are coexpression datasets due to their relatively low cost and easily accessible technology. For example, the recently established NCBI Gene Expression Omnibus database (Edgar, Domrachev and Lash, 2002) currently contains over 350 gene expression datasets, 53 of which are yeast, and close to 100 human datasets. This increase in functional data is also reflected in the rise of multiple functional databases, especially for yeast, including the Biomolecular Interaction Network Database (Bader, Betel and Hogue, 2003a), the Database of Interacting Proteins (Xenarios *et al.*, 2002), the Molecular Interactions Database (Zanzoni *et al.*, 2002), the General Repository for Interaction Datasets (Breitkreutz, Stark and Tyers, 2003), the MIPS Comprehensive Yeast Genome Database (Mewes *et al.*, 2002) and the model organism database for yeast – Saccharomyces Genome Database (SGD) (Issel-Tarver *et al.*, 2002).

The goal of these high-throughput data is rapid functional annotation of the sequenced genomes. Even in yeast, the best-studied eukaryote, 1481 of 5788 open reading frames (ORFs) are still unnamed, and functional annotation is unknown for 1865 ORFs. High-throughput functional data is important for rapid functional annotation of these unknown genes, but it is important to recognize that high-throughput methods sacrifice specificity for scale in the quality to coverage trade-off, yielding to many false positives in the datasets (Grunenfelder and Winzeler, 2002; Steinmetz and Deutschbauer, 2002; Chen and Xu, 2003; Bader *et al.*, 2003; von Mering *et al.*, 2002; Deane *et al.*, 2002). Recent work has highlighted this problem, showing that two of the yeast two-hybrid datasets share few overlaps (Ito *et al.*, 2001) and different cDNA microarrays exhibit between 10 and 30 per cent variation among corresponding microarray elements (Yue *et al.*, 2001). For gene function annotation, and later pathway and network analysis, an increase in accuracy is essential, even if it comes at the cost of some sensitivity (Bader *et al.*, 2003).

This chapter presents an overview of unsupervised analysis methods for gene function prediction, highlighting the challenges of creating methods that achieve appropriate specificity and sensitivity. The first section provides a brief discussion of gene expression microarray analysis followed by presentation of methods based on integrated analysis of diverse functional genomic data. Such integrated analysis provides higher accuracy of predictions. The last section provides a detailed discussion of MAGIC (Multi-source Association of Genes by Integration of Clusters), a Bayesian network-based method for integrated analysis of functional genomic data for gene function prediction (Troyanskaya *et al.*, 2003).

11.2 Functional Annotation by Unsupervised Analysis of Gene Expression Microarray Data

Currently, gene expression microarray datasets are the most commonly available functional genomic data due to their relatively low cost and easily accessible technology. The recently established NCBI Gene Expression Omnibus database (Edgar, Domrachev and Lash, 2002) at time of press contained over 350 gene expression datasets, 53 of which are yeast, and close to 100 human datasets, and other databases throughout the world provide additional gene expression data. This data can be used to identify groups of coexpressed genes, and such groups, or clusters, can facilitate function prediction for unknown proteins.

Many clustering algorithms have been proposed for unsupervised identification of groups of coexpressed genes from gene expression microarray data (Chapter 10). The general goal of such algorithms is to find biologically relevant groupings of genes from microarray data, so that each resulting cluster includes genes that are functionally related. It is difficult to evaluate cluster quality, or to assess whether a given cluster is a 'biologically relevant' grouping of genes, because no true gold standard exists for biological data. However, we can see whether a cluster is enriched for a particular functional attribute, for example genes involved in DNA damage repair. This can be done by calculating how many genes with each biological function a cluster contains, and comparing that number to how many genes with such a function would be expected by chance in a cluster of this size. The significance of this enrichment is usually assessed using the hypergeometric distribution (Robinson *et al.*, 2002).

Hierarchical clustering using average or complete linkage is probably the most widely applied method (Eisen *et al.*, 1998), and self-organizing maps (SOMS) are another commonly used technique (Tamayo *et al.*, 1999). Other authors have suggested using mutual information relevance networks (Butte and Kohane, 2000), clustering by simulated annealing (Lukashin and Fuchs, 2001), model-based clustering (McLachlan, Bean and Peel, 2002; Yeung *et al.*, 2001; Ghosh and Chinnaiyan,

2002) and graph-theoretic approaches (Sharan and Shamir, 2000), as well as other methods (Sherlock, 2000).

More recently, several groups have introduced methods based on two-dimensional clustering. These methods take into account the fact that functionally related genes may be coexpressed only under certain conditions, not necessarily spanning the entire range of experiments included in each dataset. In fact, a gene may participate in two different pathways under two sets of experimental conditions, and thus should belong to more than one cluster. Therefore, the clustering problem for microarray data can be restated to identify submatrices in the gene expression matrix that correspond to groups of genes coexpressed over a range of experiments. Algorithms that have addressed this two-way clustering problem include a two-sided clustering algorithm plaid model (Lazzeroni and Owen, 2000), a two-way hierarchical clustering method by Alon *et al.* (1999), a biclustering method by Cheng and Church (2000), in which low-variance submatrices of the complete data matrix are found, and a graph-theoretic-based method, CLICK (Sharan, Maron-Katz and Shamir, 2003). Another promising trend in clustering algorithms has been the emergence of methods that are probabilistic in nature, thus allowing one gene to be a member of more than one cluster (Yeung *et al.*, 2001; Sasik *et al.*, 2001; Ghosh and Chinnaiyan, 2002; Lazzeroni and Owen, 2000; Cheng and Church, 2000). Such algorithms do not necessarily perform biclustering, but do address the problem of accounting for proteins that participate in more than one pathway.

With a plethora of clustering methods available and new ones proposed regularly, the issue of choosing the most suitable method still remains open. Each clustering algorithm offers some advantages, but also has some drawbacks; each makes different assumptions, and each can be more or less successfully applied to different types of data (Fasulo *et al.*, 1999). When considering the challenging problem of microarray data analysis, the issue of choosing the most appropriate clustering method or the most biologically sound clustering output becomes even more important. Finding reliable groupings of genes is especially hard due both to the dimensionality of the data (thousands of genes measured in tens or at best few hundreds of experiments) and to the lack of reliable external validation, or gold standard.

Thus, microarray clustering evaluation methods most often rely on internal evaluation standards, such as cluster homogeneity, as opposed to external gold standards. Yeung, Haynr and Ruzzo use a concept similar to the error-sum-of-squares criterion in model selection to show that performance of the microarray clustering algorithm depends on the specific data, the number of clusters formed and the figure of merit used to assess the quality of clustering (Yeung, Haynr and Ruzzo, 2001). Chen *et al.* compare performances of clustering algorithms by homogeneity and separation characteristics of resulting clusters (Chen *et al.*, 2002), and Kerr and Churchill used a linear model and resampling (Kerr and Churchill, 2001). Some methods do rely on external standards; for example, gene ontology annotations can be used to assess functional coherence of clusters of genes in organisms with well annotated genomes (Gibbons and Roth, 2002).

11.3 Integration of Diverse Functional Data for Accurate Gene Function Prediction

As described above, microarray analysis can provide gene function predictions by assessing coexpression relationships in a high-throughput fashion. However, while gene coexpression data is an excellent tool for hypothesis generation, microarray data alone often lacks the degree of specificity needed for accurate gene function prediction. For such purposes, an increase in accuracy is needed, even if it comes at the cost of some sensitivity. This improvement in specificity can be achieved through incorporation of heterogeneous functional data in an integrated analysis.

Bioinformatics methods for effective integration of high-throughput heterogeneous data can provide the improvement in specificity necessary for accurate gene function annotation and network analysis based on high-throughput data (Marcotte *et al.*, 1999a; Ideker, Galitski and Hood, 2001; Steinmetz and Deutschbauer, 2002; Troyanskaya *et al.*, 2003). While the exact amount of overlap and correlation among functional datasets is unclear (Deane *et al.*, 2002; Edwards *et al.*, 2002; Kemmeren *et al.*, 2002; Werner-Washburne *et al.*, 2002), data integration has been shown to increase the accuracy of gene function prediction compared with a single high-throughput method (Marcotte *et al.*, 1999a, 1999b; Schwikowski, Uetz and Fields, 2000; Bader and Hogue, 2002; von Mering *et al.*, 2002; Gerstein, Lan and Jansen, 2002; Ge *et al.*, 2001). Von Mering *et al.* showed that using more than one type of functional data for interaction predictions increased accuracy (von Mering *et al.*, 2002). Another group demonstrated that integrating more heterogeneous information increases the number of protein–protein interactions identified (Gerstein, Lan and Jansen, 2002). This potential of data integration recently led to several groups proposing methods for heterogeneous data integration.

A simple scheme for increasing accuracy in function prediction based on heterogeneous data is to consider the intersection of interaction maps for different high-throughput datasets (Tong *et al.*, 2002). While this scheme reduces the false positives, it has the drawback that the lowest-sensitivity dataset will limit sensitivity of the entire analysis. As published large-scale interaction studies are not comprehensive even in model organisms, this strict sensitivity limitation is too restrictive for large-scale and general function prediction.

Several other groups suggested approaches that provide increased sensitivity of function prediction from the intersection scheme above. In the first study of this type, Marcotte *et al.* predicted a number of potential protein functions for *S. cerevisiae* based on a heuristic combination of different types of data (Marcotte *et al.*, 1999a, 1999b). Schwikowski, Uetz and Fields assigned putative protein function based on the number of interactions an unknown protein has with proteins from different functional categories (with no weighting for quality of experimental method) (Schwikowski, Uetz and Fields, 2000). These studies combine the information from different sources in a heuristic fashion, where confidence levels for protein–protein links are defined on a case-by-case basis. This approach is successful in these studies and served as a clear proof of concept, but it may be hard to generalize to new datasets, data types or organisms because each approach is developed with specific data and application goal in mind and therefore lacks a general scheme or representation.

More recently, several computational methods have been suggested that combine datasets in a confidence-dependent manner (Pavlidis *et al.*, 2001; Friedman *et al.*, 2000; Segal *et al.*, 2003; Imoto, Goto and Miyano, 2002; Ihmels *et al.*, 2002). Most of these methods focus on modelling one or several particular data types, such as gene expression data combined with phylogenetic profiles via support vector machines (Pavlidis *et al.*, 2002) or gene expression combined with transcription factor binding sites in a Bayesian system (Segal *et al.*, 2003). These methods provide innovative and useful ways of modelling the particular data type combinations for a particular goal. For example, Segal *et al.* designed a Bayesian framework for identifying sets of coregulated genes based on known regulatory genes and gene expression data (Segal *et al.*, 2003).

11.4 MAGIC – General Probabilistic Integration of Diverse Genomic Data[1]

To address the need for a generalizable method for comprehensive data integration, an approach should perform confidence-based combination of a variety of data types in an algorithmic fashion and should easily adapt to new data sources. Recently, several such approaches have been introduced, including the first probabilistic approach of this type – a Bayesian network-based method called MAGIC (Multi-source Association of Genes by Integration of Clusters) (Troyanskaya *et al.*, 2003). MAGIC is a flexible probabilistic framework for integrated analysis of high-throughput biological data. The current version of the system is implemented for *S. cerevisiae*, for which multiple useful data sources exist. The system is based on a Bayesian network (Pearl, 1988) that combines evidence from diverse data sources (including microarray analysis methods) to predict whether two proteins are functionally related (involved in a common biological process). The network essentially performs a probabilistic 'weighting' of data sources, thus avoiding double-counting evidence and allowing for formal representation of expert knowledge about the methods. Each predicted functional relationship is assigned a posterior belief, allowing the user to vary the level of stringency of the predictions.

The advantage of a probabilistic approach is its generality and adaptability. MAGIC uses the Bayesian network architecture, which can easily incorporate new data sources, datasets, and analysis methods. It readily incorporates expert knowledge in the prior probability parameters in the Bayesian framework, thus formally integrating relative accuracies of different experimental and computational techniques in

[1]Parts of this work were originally published in *PNAS*, Copyright 2003 National Academy of Sciences.

the analysis and minimizing potential bias toward well studied areas in its reasoning. In addition, Bayesian networks are generally robust to noise in prior probabilities and in training data. These characteristics of Bayesian networks yield high accuracy of gene function predictions produced by MAGIC, and the probabilistic nature of the system provides confidence levels for each output.

MAGIC's system design

The MAGIC system has a distributed design that promotes flexibility for adding new input methods and datasets. MAGIC provides a general framework that can incorporate a number of data types and microarray analysis methods. The framework includes yeast protein–protein interactions from the General Repository of Interaction Datasets (GRID) (Breitkreutz, Stark and Tyers, 2002) and pairs of genes that have experimentally determined binding sites for the same transcription factor, derived from the *Saccharomyces cerevisiae* Promoter Database (SCPD) (Zhu and Zhang, 1999). In addition, MAGIC incorporates gene expression data analyses by the three most widely used microarray analysis methods: *k*-means clustering (K-means), self-organizing maps (SOM) and hierarchical clustering (Hier).

The inputs for the system are groupings (or clusters) of genes based on coexpression or other experimental data (e.g. transcription factor binding sites). MAGIC's main component, its Bayesian network, combines evidence from input groupings and generates a posterior belief for whether each gene i – gene j pair has a functional relationship. For each pair of genes, MAGIC essentially asks the following question: 'What is the probability, based on the evidence presented, that products of gene i and gene j have a functional relationship (i.e. are involved in the same biological process)?'.

The Bayesian network receives as input gene–gene relationship matrices, each representing one data source, where element $s_{i,j} \neq 0$ if gene i and gene j have a functional relationship and $s_{i,j} = 0$ if they do not. As each different method (or a different set of parameters of the same method) creates each matrix, the definition of criteria for functional relationship for each input matrix relies on the method used to create the particular matrix (e.g. genes that are in the same cluster for clustering algorithms). The score $s_{i,j}$ corresponds to the strength of each method's belief in the existence of relationship between gene i and gene j. This score can be a binary (e.g. results of co-immunoprecipitation experiments), continuous or discrete variable (for example $-1 \leq s \leq 1$ for Pearson correlation).

The flexible input format allows genes to be members of more than one group or cluster, and thus does not exclude biclustering or fuzzy clustering methods. The output format is the same as the input format. The flexibility of input and output formats ensures that MAGIC can incorporate any type of gene–gene grouping, including protein–protein interaction data, outputs of clustering methods and sequence-based data (for example shared transcription factor binding sites).

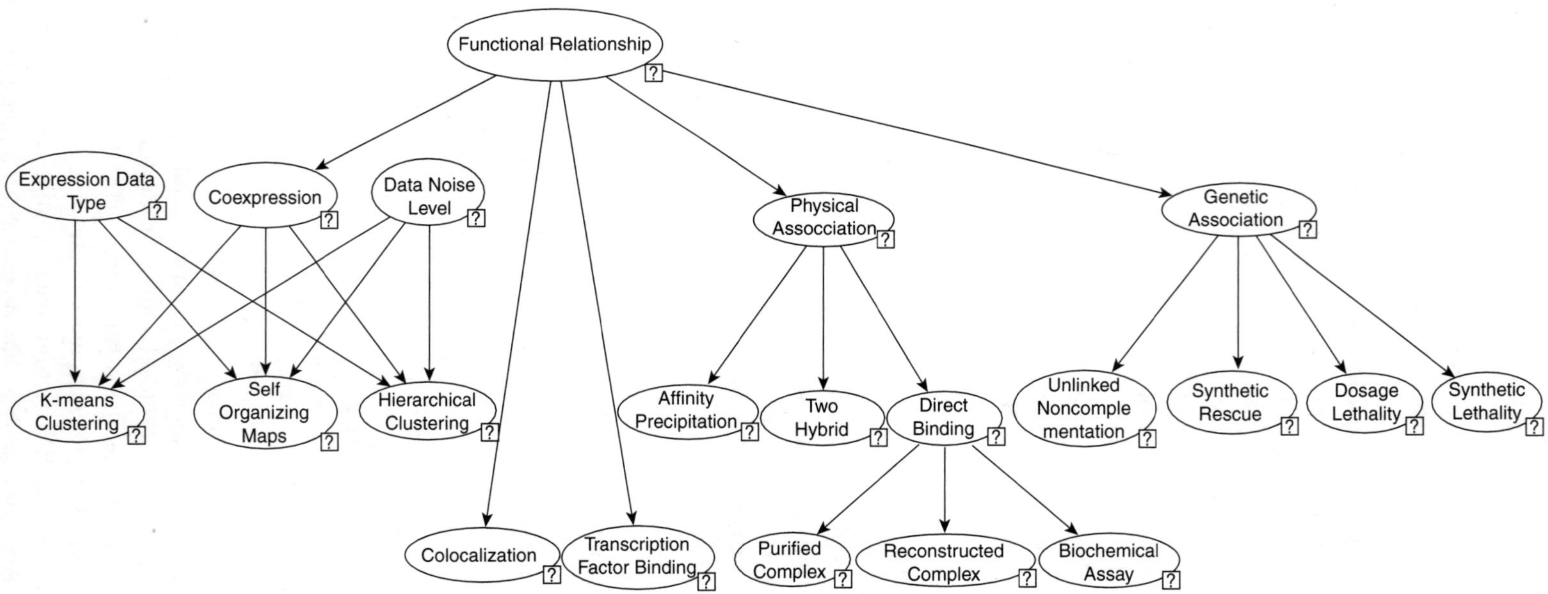

Figure 11.1 General architecture of MAGIC's Bayesian network

MAGIC's Bayesian network structure was determined through consultation with experts in yeast genomics and microarray analysis. The resulting structure (Figure 11.1) adequately reflects relationships between evidence from different data types for the purpose of ensemble analysis and avoids double counting of evidence. A separate network is instantiated for each pair of genes by initializing bottom level nodes with evidence. Conditional probability tables (CPTs) for each connection between nodes were assessed formally by yeast genetics experts. Probabilistic combination performed by the network essentially represents a formal probabilistic 'weighting' of each type of evidence based on the knowledge about each data source encoded in the network's CPTs. Combination of clustering methods is performed through a single 'Coexpression' node, which allows all of the expression analysis methods outputs for one dataset to be combined based on each method's characteristics, such as robustness to noise level in data or optimality for a specific data type (e.g. temporal data). Non-expression-based data consists of colocalization data, experimentally identified transcription factor binding sites and experimental evidence for physical or genetic associations of two proteins. The genetic and physical relationship data is divided into experimental evidence types according to the GRID database (http://biodata.mshri.on.ca/grid/servlet/HelpHtmlPages?pageID=3).

MAGIC[2] combines inputs based on the type of relationship they detect (for example coexpression for microarray clustering methods). It makes some independence assumptions that allow for a more accurate population of the conditional probability tables based on information elicited from yeast experts. Given the relatively sparse nature of non-microarray experimental data, these independence assumptions are unlikely to affect the results. In addition, the different underlying principles of the methods represented in the network make their combination robust for functional inference (Marcotte and Date, 2001; Marcotte *et al.*, 1999b; Pavlidis *et al.*, 2002).

The prior probabilities for the Bayesian network were formally assessed by seven experts in the field of yeast molecular biology (SGD curators) through detailed formal questionnaires. The experts were questioned independently, and displayed substantial agreement in their prior beliefs. The method of constructing Bayesian networks based on probabilities provided by experts in the field has been successfully used previously, for example in the PATHFINDER network for pathology diagnosis (the network structure and prior probabilities for PATHFINDER were based on consultations with one pathology expert) (Heckerman, 1991). If a sufficient amount of functional data is available, the network priors and structure could be automatically learned (Heckerman, 1999).

MAGIC was implemented in C++ under Linux, and a web-based user interface is under development and will be available from http://function.cs.princeton.edu/. The implementation used SMILE library and the GeNIe modelling environment

[2]Naming of protein–protein interaction detection methods included in MAGIC follows GRID.

developed by the Decision Systems Laboratory of the University of Pittsburgh (http://www.sis.pitt.edu/~dsl).

Evaluation: assessing accuracy of gene function prediction

To evaluate the quality of a gene grouping, one needs to measure the biological relevance or accuracy of gene–gene functional pairs belonging to that gene grouping. Biological relevance is the key criterion in evaluating pairs of genes with predicted functional relationships, yet it is a difficult metric to assess. If gene i and gene j are predicted to have a functional relationship, but no prior biological knowledge links their functionality, is that an erroneous clustering, experimental error, or a novel biological discovery? While no perfect gold standard for gene groupings exists, the curator-controlled annotation of the *S. cerevisiae* genome with Gene Ontology (GO) terms (Ashburner *et al.*, 2000; Dwight *et al.*, 2002) provides a reflection of the current biological knowledge and thus a reasonable biological standard for evaluation of functional pairs of *S. cerevisiae* genes (also see the Gene Ontology discussion in Chapter 7).

Gene Ontology contains three types of term: (1) molecular function, (2) biological process and (3) cellular component. GO has a hierarchical structure with multiple inheritance, and each gene (or protein) can be annotated with one or more GO terms from disparate parts of the GO tree. This evaluation focuses on the biological process part of GO, which is the most relevant part of the ontology for evaluation of gene groupings based on the presence of functional relationships because genes annotated to the same GO term from the biological process ontology are believed (in current biological literature) to be involved in the same biological process.

The hierarchical nature of GO and multiple inheritance in the GO structure can lead to problems in evaluation if we consider only the particular GO term that a gene is annotated with. For example, gene i may be annotated with term g, while gene j with g's immediate ancestor g' (e.g. gene i is annotated with 'GO:0007216: metabotropic glutamate receptor signalling pathway' and gene j is annotated with 'GO:0007215: glutamate signalling pathway' – a parent node of GO:0007216). Although genes i and j are functionally similar based on their GO annotation, they are technically annotated with different GO terms. To alleviate this problem, this evaluation considers any gene annotated with GO term g to be also implicitly annotated with every ancestor of g, up to level 3 of the GO tree (with 'Gene_Ontology' considered level 1). This evaluation is robust to changes of the exact level of cut-off.

This evaluation reflects the biological relevance of gene groupings by using Gene Ontology as a gold standard. This evaluation approach is not flawless: Gene Ontology may have annotation errors, and the functions of many genes in the yeast genome are unknown. The evaluation is conservative: a false positive (FP) pair of genes could represent a true error or a novel discovery. There may be some biases in the subsets of genes that are or are not currently annotated by GO terms, but there is no reason to believe these biases would affect clustering methods differently. This method therefore provides a reasonable and biologically grounded comparative evaluation framework for gene groupings.

Due to the cost of follow-up experimental investigation, the key problem in creating biologically relevant gene groupings tends to be specificity, not sensitivity. Unfortunately, calculating specificity and sensitivity requires knowledge of the total of number of true positives (TP) and negatives in *S. cerevisiae*, numbers that are currently impossible to assess accurately. Therefore, we assess the accuracy of each method through the proportion of TP pairs in its predictions, where TP pairs are defined as pairs of (gene *i*, gene *j*) such that gene *i* and gene *j* have an overlapping (explicit or implicit) GO term annotation:

$$\text{proportionTP}_{\text{method}} = \frac{\text{No. of pairs that share GO term assignment}}{\text{total no. of pairs predicted by method}}$$

The predicted pairs for each input method are available from adjacency matrices representing gene groupings, as described above.

MAGIC integrates various gene groupings in a systematic fashion, yielding posterior probabilities for functional relationship between every pair of genes in the yeast genome. Because the stringency of MAGIC's predictions can be controlled by varying a cut-off for the posterior beliefs sufficient to consider two genes functionally related, MAGIC's performance depends on different levels of stringency applied to confidence scores in its output. The stringency of the input clustering methods can be varied as well by varying the cut-off of $s_{\text{A, B}}$, the average correlation of two genes (A, B) to the centroid of the cluster they are both members of: $s_{\text{A,B}} = \frac{1}{2}\sum_{g=\text{A, B}} \frac{\text{Cov}(g, \text{centroid}_c)}{\sigma_g \sigma_{\text{centroid}_c}}$. Such cut-off optimization is not performed when these clustering methods are used routinely for microarray analysis, which is unfortunate, as evidenced by the ROC curves of these clustering methods performance in Figure 11.2.

Application of MAGIC to *S. cerevisiae* data

To illustrate the utility of MAGIC for integrated analysis of heterogeneous biological data, MAGIC was applied to *S. cerevisiae* functional genomic data, including protein–protein interactions, transcription factor binding sites, and gene expression microarray data from a stress response microarray dataset (Gasch *et al.*, 2000). MAGIC incorporates gene groupings based on microarray analysis with the often more accurate non-expression-based data sources, and MAGIC consistently increases the proportion of TP pairs when compared with its input methods (Figure 11.2(A)). In gene function prediction, high specificity is key for creating biologically relevant gene groupings. When we consider predictions with the highest proportion of TP

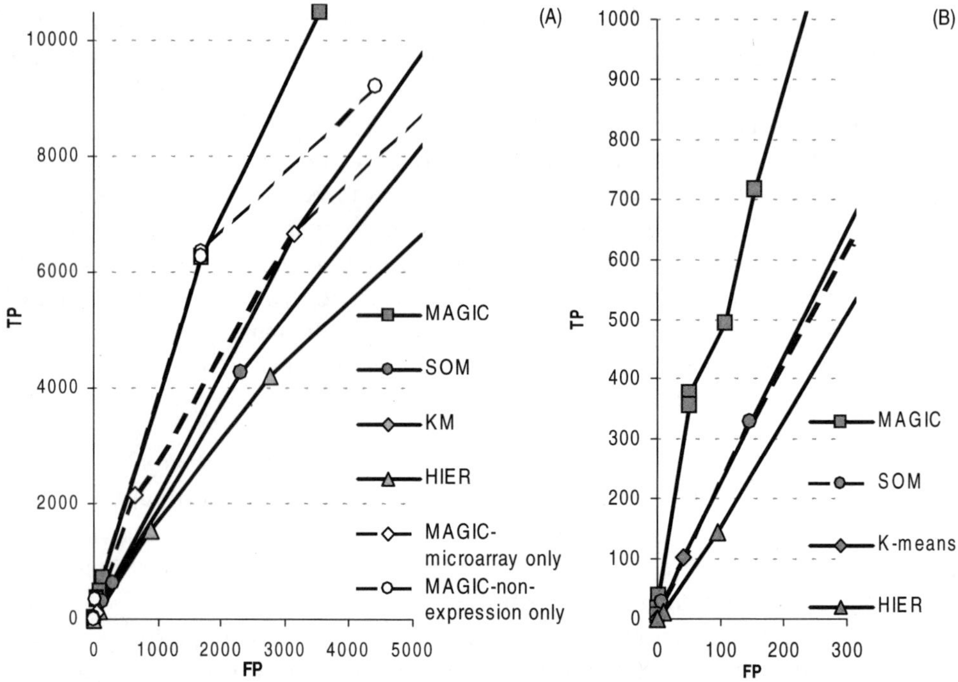

Figure 11.2 Trade-off between the number of TP and FP pairs for each method. (A) MAGIC increases the proportion of TP pairs in a broad high-specificity region compared to expression-based clustering methods, MAGIC based on purely microarray data (MAGIC – microarray only) or purely on non-expression data (MAGIC – non-expression only). (B) Comparison in the region of highest accuracy (<1000 TP pairs). MAGIC predicts more TP pairs for each number of FPs than its input methods

pairs made by each method (when at least 100 TP pairs are predicted), MAGIC, which uses the non-optimized inputs, performs better than the optimized clustering methods, with a 17 percent increase in proportion of TP pairs over the best of the input methods and the largest number of TP pairs predicted (Figure 11.2(B)). This difference in performance declines at very large numbers of predicted pairs (40 000 and higher), where the proportion of TP rates for all methods are around or below 50 percent and thus at levels not suitable for accurate gene function prediction. Thus, by combining heterogeneous data in an integrated analysis, MAGIC creates more biological relevant gene groupings, with the highest improvement in the high-specificity region.

MAGIC's output is pairs of proteins with a score that reflects the confidence that the two protein in the pair are functionally related. Groupings of genes (clusters) can be constructed based on MAGIC's pairwise output by considering all genes with functional relationship to the same gene as a group. Clusters are defined around each gene i or each row of the adjacency matrix ($i = 1 \ldots$ total number of genes). For example, in the adjacency matrix A_m for output of method m, the cluster around gene i includes any gene j for which $A_m(i,j) > 0$ (or $A_m(i,j) >$ cut-off). Other, more complex algorithms can be used here as well, but this simple method directly addresses the issue of gene function prediction by creating gene groupings around each gene with unknown biological process.

Clusters created based on integrated data include those identified by Gasch *et al.* in microarray data (Gasch *et al.*, 2000), but MAGIC separates these clusters into smaller, more functionally specific groups. In addition, through integration of diverse data sources, the system provides a coherent summary of all functional data associated with a particular pair of proteins. For example, genes involved in protein degradation are induced during the response to environmental stress. MAGIC identifies a cluster of genes involved in ubiquitin-dependent protein catabolism, provides potential functional annotation for an open reading frame (orf) present in that cluster (YGL004C), and confirms the recently added annotation for YNL311C (Figure 11.3). The cluster contains 12 genes. In the version of SGD annotations used for evaluation in this study, nine of the proteins are annotated to ubiquitin-dependent protein catabolism, one (Rad23, not shown) to 'nucleotide excision repair', and YNL311C and YGL004C do not have a known biological process assignment. MAGIC predicted that YNL311C and YGL004C are probably involved in ubiquitin-dependent protein catabolism. In the most recent release of the annotation (February 2003), YNL311C has been annotated to this process. The other unknown orf, YGL004C, has been assigned an SGD reserved name RPN14. This example illustrates the utility of MAGIC as a tool to aid gene function annotation.

The group includes Rad23, though its current GO annotation is to 'nucleotide-excision repair, DNA damage recognition'. Based on current literature, Rad23's

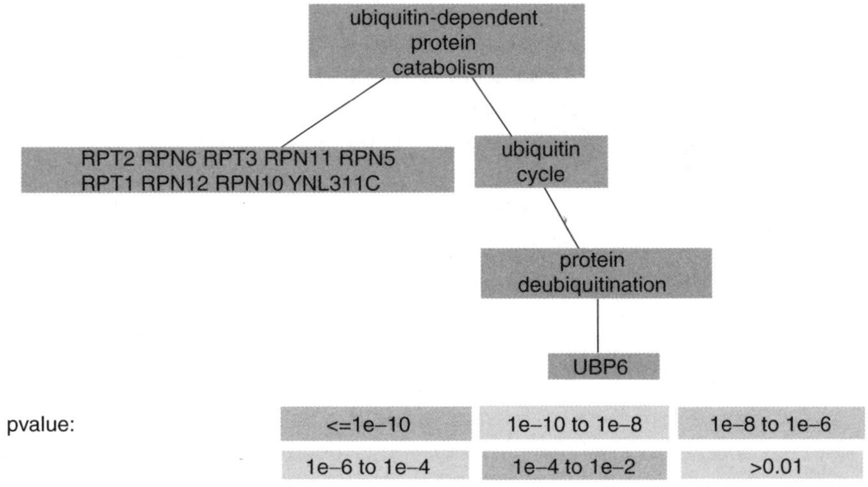

Figure 11.3 Ubiquitin-dependent protein catabolism cluster represented using GO Term Finder (http://genome-www4.stanford.edu/cgi-bin/SGD/GO/goTermFinder).

involvement in DNA repair is probably due to its inhibition of the degradation of repair proteins in response to DNA damage (van Laar, van der Eb and Tevleth, 2002). It has been shown that Rad23 physically interacts with the 26S proteasome and may also be involved in other protein degradation pathways (van Laar, van der Eb and Tevleth, 2002). The grouping generated by MAGIC identifies outdated and potentially misleading annotation of Rad23.

Thus, in addition to predicting gene function of unknown genes that are found in groups with well characterized ones, MAGIC also provides a means of quality control for the existing functional annotations of partially characterized genes. Another such example is a group that consists of three genes: BUD31, CEF1 and PRP8. Both CEF1 and PRP8 are well characterized splicing factors (Will and Luhrmann, 1997; Tsai *et al.*, 1999). BUD31 is currently annotated to bud site selection based on a genome-wide screen for mutants defective in the bipolar budding pattern (Ni and Snyder, 2001). However, Ni and Snyder found that several nuclear proteins, including genes involved in RNA processing, also exhibit defects in bud site selection, most probably as an indirect effect of the processing of RNA for genes directly involved in budding (Ni and Snyder, 2001). In addition, BUD31 has a putative nuclear localization signal[3] (Boeckmann *et al.*, 2003). Thus, BUD31 might be involved in RNA processing rather than directly playing a role in bud site selection. By searching for genes with annotations that do not fit with the other annotations of genes in a group, one can target particular genes that may be associated with spurious or incomplete functional information.

11.5 Conclusion

A key challenge in present-day molecular biology is functional annotation of unknown genes in sequenced genomes. Classical functional annotation techniques are too labour intensive and slow to accomplish this task in the near future; therefore, we must rely on the high-throughput experimental methods to direct more traditional experimentation. However, these large-scale techniques sacrifice specificity for scale, and thus computational analysis is required to provide accurate gene function predictions based on high-throughput functional genomic data. Such techniques often focus on gene expression microarray data, but as other sources of functional data become available, integrated analysis of diverse data becomes possible.

Such integrated analysis increases accuracy of gene function prediction as compared to methods based on gene expression data alone and provides a coherent view of functional information derived from diverse types of high-throughput data. It allows for formal probabilistic reasoning and predictions based on heterogeneous data sources, and is generalizable to new data sources as they become available.

[3]http://us.expasy.org/cgi-bin/sprot-ft-details.pl?P25337@DOMAIN@2@11

However, these integration-based methods are still limited by the coverage of functional genomics datasets and the quality of high-throughput data available. Future development of more accurate integrative methodologies and their expansion to multi-cellular organisms complemented by the development of high-throughput experimental technologies is critical for complete functional annotation of model organism and human genomes.

References

Adams, M. D. *et al.* (2000) The genome sequence of *Drosophila melanogaster Science*, **287** (5461) 2185–2195.

Alon, U., Barkai, N. *et al.* (1999) Broad patterns of gene expression revealed by clustering analysis of tumor and normal colon tissues probed by oligonucleotide arrays. *Proc Atal Acad Sci USA* **96** (12): 6745–50.

Ashburner, M., C. A. Ball, *et al.* (2000) Gene ontology: tool for the unification of biology. The Gene Ontology Consortium. *Nat Genet* **25** (1): 25–9.

Bader, G. D., Betel, D. and Hague, C. W. (2003) BIND: the Biomolecular Interaction Network Database. *Nucleic Acids Res*, **31** (1): 248–250.

Bader, G. D., Heilbut, A. *et al.* (2003) Functional genomics and proteomics: charting a multi-dimensional map of the yeast cell. *Trends Cell Biol* **13** (7): 344–356.

Bader, G. D. and Hogue, C. W. (2002) Analyzing yeast protein–protein interaction data obtained from different sources. *Nat Biotechnol* **20** (10): 991–997.

Bender, A. and Pringle, J. R. (1991) Use of a screen for synthetic lethal and multicopy suppressee mutants to identify two new genes involved in morphogenesis in *Saccharmoyces cerevisiae Mol Cell Biol* **11** (3): 1295–1305.

Boeckmann, B., Bairoch, A. *et al.* (2003) The SWISS-PROT protein knowledgebase and its supplement TrEMBL in 2003. *Nucleic Acids Res* **31** (1): 365–370.

Breitkreutz, B. J., Stark, C. and Tyers, M. (2002) The GRID: The General Repository for Interaction Datasets. *Genome Biol* **3** (12): PREPRINT0013.

Breitkreutz, B. J., Stark, C. and Tyers, M. (2003) The GRID: The General Repository for Interaction Datasets. *Genome Biol* **4** (3).

Butte, A. J. and Kohane, I. S. (2000) Mutual information relevance networks: functional genomic clustering using pairwise entropy measurements. *Poc Symp Biocomput*, 418–429.

Cahill, D. J. and Nordhoff, E. (2003) Protein arrays and their role in proteomics. *Adv Biochem Eng Biotechnol* **83**: 177–187.

Chen, G., Jaradat, S. A. *et al.* (2002) Evaluation and comparison of clustering algorithms in analyzing ES cell gene expression data. *Statistica Sinica* **12**: 241–262.

Chen, Y. and Xu, D. (2003) Computational analyses of high-throughput protein–protein interaction data. *Curr Protein Pept Sci* **4** (3): 159–181.

Cheng, Y. and Church, G. M. (2000) Biclustering of expression data. *Proc Int Conf Intell Syst Mol Biol* **8**: 93–103.

Consortium, T. C. e. S. (1998) Genome sequence of the nematode *C. elegans*: a platform for investigating biology. *Science* **282** (5396): 2012–2018.

Cutler, P. (2003) Protein arrays: the current state-of-the-art. *Preteomics* **3** (1): 3–18.

Deance, C. M., L. Salwinski, *et al.* (2002) Protein interactions: two methods for assessment of the reliability of high throughput observations. *Mol Cell Proteomics* **1** (5): 349–356.

Dwight, S. S., M. A. Harris, *et al.* (2002) *Saccharomyces* Genome Database (SGD) provides secondary gene annotation using the Gene Ontology (GO) *Nucleic Acids Res* **30** (1): 69–72.

Edgar, R., Domrachev, M. and Lash, A. E. (2002) Gene Expression Omnibus: NCBI gene expression and hybridization array data repository. *Nucleic Acids Res* **30** (1): 207–10.

Edwards, A. M., Kus, B. *et al.* (2002) Bridging structural biology and genomics: assessing protein interaction data with known complexes. *Trends Genet* **18** (10): 529–36.

Eisen, M. B., Spellman, P. T. *et al.* (1998) Cluster analysis and display of genome-wide expression patterns. *Proc Nat Acad Sci* USA **95** (25): 14, 863–8.

Fasulo, D., T. Jiang, *et al.* (1999) An algorithmic approach to multiple complete digest mapping. *J Comput Biol* **6** (2): 187–207.

Fields, S. and Song, O. (1989) A novel genetic system to detect protein–protein interactions. *Nature* **340** (6230): 245–6.

Friedman, N., M. Linial, *et al.* (2000) Using Bayesian networks to analyze expression data. *J Comput Biol* **7** (3–4): 601–20.

Gasch, A. P., P. T. Spellman, *et al.* (2000) Genomic expression programs in the response of yeast cells to environments changes. *Mol Biol Cell* **11** (12): 4241–57.

Ge, H., Z. Liu *et al.* (2001) Correlation between transcriptome and interactome mapping data from *Saccharomyces cerevisiae. Nat Genet* **29** (4): 482–6.

Gerstein, M., Lan, N. and Jansen, R. (2002) Proteomics. Integrating interactomes. *Science* **295** (5553): 284–7.

Ghosh, D. and Chinnaiyan, A. M. (2002) Mixture modelling of gene expression data from microarray experiments. *Bioinformatics* **18** (2): 275–86.

Gibbons, F. D. and Roth, F. P. (2002) Judging the quality of gene expression-based clustering methods using gene annotation. *Genome Res* **12** (10): 1574–81.

Goffeau, A., Barrell, B. G. *et al.* (1996) Life with 6000 genes. *Science* **274** (5287): 546–567.

Grunenfelder, B. and Winzeler, E. A. (2002) Treasures and traps in genome-wide data sets: case examples from yeast. *Nat Rev Genet* **3** (9): 653–61.

Heckerman, D. (1991) *Probabilistic Similarity Networks*. MIT Press, Cambridge, MA.

Heckerman, D. (1999) A tutorial on learning with Bayesian networks. *Learning in Graphical Models*, M. I. Jordan (ed.) MIT Press, Cambridge, MA, 301–354.

Huang, R. P. (2003) Protein arrays, an excellent tool in biomedical research. *Front Biosci* **8**: d559–76.

Ideker, T., Galitski, T. and Hood, L. (2001) A new approach to decoding life: systems biology. *Annu Rev Genomics Hum Genet* **2**: 343–372.

Ihmels, J., G. Friedlander, *et al.* (2002) Revealing modular organization in the yeast transcriptional network. *Nat Genet* **31** (4): 370–7.

Imoto, S., Goto, T. and Miyam, S. (2002) Estimation of genetic networks and functional structures between genes by using Bayesian networks and nonparametric regression. *Pac Symp Biocomput.* 175–86.

Issel-Tarver, L., K. R. Christie, *et al.* (2002) *Saccharomyces* Genome Database. *Methods Enzymol* **350**: 329–46.

Ito, T., T. Chiba, *et al.* (2001) A comprehensive two-hybrid analysis to explore the yeast protein interacetone. *Proc Natl Acad Sci USA* **98** (8): 4569–74.

Kemmeren, P., N. L. van Berkum, *et al.* (2002) Protein interaction verification and functional annotation by integrated analysis of genome-scale data. *Mol Cell* **9** (5): 1133–43.

Kerr, M. K. and Churchill, G. A. (2001) Bootstrapping cluster analysis: assessing the reliability of conclusions from microarray experiments. *Proc Natl Acad Sci USA* **98** (16): 8961–5.

Kitano, H. (2002) Looking beyond the details: a rise in system-oriented approaches in genetics and molecular biology. *Curr Genet* **41** (1): 1–10.

Lander, E. S., L. M. Linton *et al.* (2001) Initial sequencing and analysis of the human genome. *Nature* **409** (6822): 860–921.

Larsson, P. O. and Mosbach, K. (1979) Affinity precipitation of enzymes. *FEBS Lett* **98** (2): 333–338.

Lazzeroni, L. and Owen, A. B. (2000) Plaid models for gene expression data. *Technical Report.*

Lipshutz, R. J., S. P. Fodor, *et al.* (1999) High density synthetic oligonucleotide arrays. *Nat Genet* **21** (Suppl 1): 20–4.

Lukashin, A. V. and R. Fuchs (2001) Analysis of temporal gene expression profiles: clustering by simulated annealing and determining the optimal number of clusters. *Bioinformatics* **17** (5): 405–14.

Marcotte, E. and Date, S. (2001) Exploiting big biology: integrating large-scale biological data for function inference. *Brief Bioinform* **2** (4): 363–74.

Marcotte, E. M., M. Pellegrini, *et al.* (1999a) Detecting protein function and protein–protein interactions from genome sequences. *Science* **285** (5428): 751–3.

Marcotte, E. M., M. Pellegrini, *et al.* (1999b) A combined algorithm for genome-wide prediction of protein function. *Nature* **402** (6757): 83–6.

McLachlan, G. J., Bean, R. W. and Peel, D. (2002) A mixture model-based approach to the clustering of microarray expression data. *Bioinformatics* **18** (3): 413–22.

Mewes, H. W., D. Frishman, *et al.* (2002) MIPS: a database for genomes and protein sequences. *Nucleic Acids Res* **30** (1): 31–4.

Ni, L. and Snyder, M. (2001) A genomic study of the bipolar bud site selection pattern in *Saccharomyces cerevisiae. Mol Biol Cell* **12** (7): 2147–70.

Novick, P., Osmond, B. C. and Botstein, D. (1989) Suppressors of yeast actin mutations. *Genetics* **121** (4): 659–74.

Oleinikov, A. V., M. D. Gray, *et al.* (2003) Self-assembling protein arrays using electronic semiconductor microchips and in vitro translation. *J Proteome Res* **2** (3): 313–9.

Pavlidis, P., J. Weston, *et al.* (2001) Gene functional classification from heterogeneous data. *Proc Int Conf Intell Syst Mol Biol* **5**: 242–248.

Pavlidis, P., J. Weston, *et al.* (2001) Gene functional classification from heterogeneous data. *Proc Int Conf Intell Syst Mol Biol* **5**: 242–248.

Pavlidis, P., J. Weston, *et al.* (2002) Learning gene functional classifications from multiple data types. *J Comput Biol* **9** (2): 401–11.

Pearl, J. (1988) *Probabilistic Reasoning in Intelligent Systems: Networks of Plausible Inference.* Kaufmann, San Mateo, CA.

Robinson, M. D., J. Grigull, *et al.* (2002) FunSpec: a web-based cluster interpreter for yeast. *BMC Bioinformatics* **3** (1): 35.

Sasik, R., T. Hwa, *et al.* (2001) Percolation clustering: a novel approach to the clustering of gene expression patterns in *Dictyostelium* development. *Pac Symp Biocomput*, 335–47.

Schena, M., D. Shalon, *et al.* (1995) Quantitative monitoring of gene expression patterns with a complementary DNA microarray. *Science* **270** (5235): 467–70.

Schwikowski, B., Uetz, P. and Fields, S. (2000) A network of protein–protein interactions in yeast. *Nat Biotechnol* **18** (12): 1257–61.

Segal, E., M. Shapira, *et al* (2003) Module networks: identifying regulatory modules and their condition-specific regulators from gene expression data. *Nat Genet* **34** (2): 166–76.

Sharan, R., Maron-Katz, A. and Shamir, R. (2003) CLICK and EXPANDER: a system for clustering and visualizing gene expression data. *Bioinformatics* **19** (14): 1787–99.

Sharan, R. and Shamir, R. (2000) CLICK: a clustering algorithm with applications to gene expression analysis. *Proc Int Conf Intell Syst Mol Biol*, **8**: 307–16.

Sherlock, G. (2000) Analysis of large-scale gene expression data. *Curr Opin Immunol* **12** (2): 201–5.

Steinmetz, L. M. and Deutschbauer, A. M. (2002) Gene function on a genomic scale. *J Chromatogr B Analyt Technol Biomed Life Sci* **782** (1/2): 151–63.

Sydor, J. R. and Nock, S. (2003) Protein expression of profiling arrays: tools for the multiplexed high-throughput analysis of proteins. *Proteome Sci* **1** (1): 3.

Tamayo, P., D. Slonim, *et al.* (1999) Interpreting patterns of gene expression with self-organizing maps: methods and application to hematopoietic differentiation. *Proc Natl Acad Sci USA* **96** (6): 2907–12.

Tong, A. H., B. Drees, *et al.* (2002) A combined experimental and computational strategy to define protein interaction networks for peptide recognition modules. *Science* **295** (5553): 321–4.

Troyanskaya, O. G., K. Dolinski, *et al.* (2003) A Bayesian framework for combining heterogenous data sources for gene function prediction (in *Saccharomyces cerevisiae*) *Proc Natl Acad Sci USA* **100** (14): 8348–53.

Tsai, W. Y., Y. T. Chow, *et al.* (1999) Cef1p is a component of the Prp 19p-associated complex and essential for pre-mRNA splicing. *J Biol Chem* **274** (14): 9455–62.

van Laar, T., van der Eb, A. J. and Tevleth, C. (2002) A role for Rad23 proteins in 26S proteasome-dependent protein degradation? *Mutat Res* **499** (1): 53–61.

Venter, J. C., M. D. Adams, *et al.* (2001) The sequence of the human genome. *Science* **291** (5507): 1304–51.

von Mering, C., R. Krause, *et al.* (2002) Comparative assessment of large-scale data sets of protein–protein interactions. *Nature* **417** (6887): 399–403.

Werner-Washburne, M., B. Wylie, *et al.* (2002) Comparative analysis of multiple genome-scale data sets. *Genome Res* **12** (10): 1564–73.

Will, C. L. and Luhrmann, R. (1997) Protein functions in pre-mRNA splicing. *Curr Opin Cell Biol* **9** (3): 320–8.

Wood, V., R. Gwilliam, *et al.* (2002) The genome sequence of *Schizosaccharomyces pombe. Nature* **415** (6874): 871–80.

Xenarios, I., L. Salwinski, *et al.* (2002) DIP, the Database of Interacting Proteins: a research tool for studying cellular networks of protein interactions. *Nucleic Acids Res* **30** (1): 303–5.

Yeung, K. Y., C. Fraley, *et al.* (2001) Model-based clustering and data transformations for gene expression data. *Bioinformatics* **17** (10): 977–87.

Yeung, K. Y., Haynor, D. R. and Ruzzo, W. L. (2001) Validating clustering for gene expression data. *Bioinformatics* **17** (4): 309–18.

Yue, H., P. S. Eastman, *et al.* (2001) An evaluation of the performance of cDNA microassays for detecting changes in global mRNA expression. *Nucleic Acids Res* **29** (8): E41–1.

Zanzoni, A., L. Montecchi-Palazzi, *et al.* (2002) MINT: a Molecular INTeraction database. *FEBS Lett* **513** (1): 135–40.

Zhu, J. and Zhang, M. Q. (1999) SCPD: a promoter database of the yeast *Saccharomyces cerevisiae. Bioinformatics.* **15** (7/8): 607–11.

12

Supervised Methods with Genomic Data: a Review and Cautionary View

Ramón Díaz-Uriarte

Abstract

We review well accepted methods to address questions about differential expression of genes and class prediction from gene expression data. We highlight some new topics that deserve more attention: testing of differential expression of specific groups of genes, intra-group heterogeneity and class prediction, gene interaction in predictors, visualization, difficulties in the biological interpretation of predictor genes and molecular signatures, and the use of ROC (receiver operating characteristic curve)-based statistics for evaluating predictors and differential expression. We end with a review of some serious problems that can limit the potential of these methods; we focus specially on inadequate assessment of the performance of new methods (due to inadequate estimation of error rates and to the use of few and 'easy' data sets) and failure to recognize observational studies and include needed covariates. A final comment is made about the need for freely available source code.

Keywords

differential expression, prediction, prognostic, microarrays, multiple testing, molecular signatures, software, statistics, machine learning, observation study

12.1 Chapter Objectives

Reviews of the analysis of gene expression data (e.g., Dăghici, 2002; Parmigiani *et al.*, 2003; Simon *et al.*, 2003a; Slonim, 2002; Speed, 2003; Tumor Analysis Best Practices

---

Data Analysis and Visualization in Genomics and Proteomics Edited by Francisco Azuaje and Joaquin Dopazo

Working Group, 2004) often mention three objectives: (a) class comparison, or finding/ranking of differentially expressed genes; (b) class prediction or prognostic prediction; (c) class discovery, also known as clustering or unsupervised analyses. We will not discuss class discovery or clustering here (it is discussed elsewhere in this book) and will concentrate on class comparison and class prediction. For the remaining two broad types of problem, this chapter has three main objectives: (a) to bring a statistician, computer scientist, or computational biologist quickly up to speed by providing pointers to the literature on well accepted and standard methods;[1] (b) to emphasize some topics that deserve more attention and are open to additional theoretical, empirical and computational contributions; (c) to alert editors, reviewers and general practitioners to several serious problems that can undermine the full potential of these techniques.

12.2 Class Prediction and Class Comparison

Class comparison asks whether different classes of subjects (e.g., lung cancer and prostate cancer patients) differ in their gene expression; the result is often a list of genes ranked by their degree of differential expression between classes; this objective can alternatively be to examine whether other non-categorical variables (such as expression of certain proteins or survival) are associated with gene expression. *Class prediction* or prognostic prediction tries to predict the class membership (or survival or protein expression or any prognostic variable) of a set of subjects given their gene expression data. Although related, these are different objectives that answer different biological questions and require different methods (unfortunately, this difference is not always recognized in empirical work). Ranking genes often precedes trying to use genes for class prediction (see also Sackett and Haynes, 2002), but genes that show large expression differences are not necessarily good predictors (see, e.g., p. 299 of Whitfield *et al.*, 2003).

12.3 Class Comparison: Finding/Ranking Differentially Expressed Genes

The most common procedures analyse each and all of the genes of the array, 'asking the same question' (e.g., 'is this gene differentially expressed between prostate and lung cancer patients?') for *each gene of the array*. In contrast, when there are *prespecified groups* of genes, one can ask whether that subset of genes, as a whole, shows evidence of differential expression (e.g., 'are genes X, Y, Z,

[1]Lack of space precludes a full review; other lists of references can be found in http://www.biostat.umn.edu/~weip/course/ge/sy11. html and http://biosun01.biostat.jhsph.edu/~gparmigi/688/readings.html, from two well known statisticians.

which are involved in cell cycle, differentially expressed between prostate and lung cancer patients?').

Especially when asking the same question for each gene of the array, there are often two different objectives: to obtain a list of genes for which 'their expression is statistically significantly different' and to rank genes based on some measure of how distant is the expression level between conditions (and this measure can be the p-value computed before) or how likely they are to differ. These objectives are related, but measuring the likelihood of differential expression requires additional assumptions, and obtaining p-values is more delicate than simply ranking. Even when p-values are obtained, however, they are used as informal rules of inference and to guide future experiments, rather than to provide 'black or white' answers.

Asking the same question for each gene of the array

Widely accepted methods, with available software, involve the use of standard statistical tests (e.g., t-test for two-class comparisons, ANOVA for multi-class comparisons, Cox models for survival data etc.), where analyses are carried out gene by gene (reviews in Cui and Churchill, 2003; Dudoit *et al.*, 2002; Yekutieli, Reiner and Benjamini, 2003; Simon *et al.*, 2003a, ch. 7). These analyses, although conducted gene by gene, need to take into account that thousands of null hypotheses are being tested (one for each gene): if we were to consider any of the genes with a 'rejected null' as differentially expressed, we would end up with many false rejections. Appropriate correction for *multiple testing* is often conducted using either control of the *family wise error rate* or the *false discovery rate*. Controlling the family wise error rate refers to controlling the probability of making one or more false discoveries, or falsely rejecting the null, over the whole family of tests; this approach was detailed by Westfall and Young (1993) and its application to microarrays was pioneered by Dudoit *et al.* (2002). In contrast, the false discovery rate approach controls the expected proportion of erroneously rejected nulls among the rejected hypotheses; FDR control has been worked on mainly by Yoav Benjamini, Daniel Yekutieli, and their collaborators (see http://www.math.tau.ac.il/~roee/index.htm for lists of references and links); a recent review and applications to microarrays is given by Reiner, Yekutieli and Benjamini, (2003); other approaches related to, or variations of, FDR are given by Storey (2002), Storey and Tibshirani (2003) and references therein; Ge, Dudoit, and Speed (2003) compare and discuss most of these different approaches. Detailed discussion of whether control of FWER or FDR is the most appropriate for a given situation is beyond the scope of this chapter; however, in many exploratory studies control of FDR is probably what most researchers need. In addition, methods for control of FDR do not require the subset pivotality assumption (Westfall and Young, 1993) to hold, and therefore are applicable to a wider range of tests; in addition, although control of FDR, as originally proposed by Benjamini and Hochberg (1995), works only for independent (or positively regression dependent) test statistics, the results of Reiner, Yekutieli and Benjamini (2003) show that violation of this assumption is generally inconsequential and there are also resampling-based FDR approaches that account for the dependence of the tests statistics.

Most gene-by-gene approaches, when computing the statistic for each gene, do not use the information contained in the rest of the genes, which could be wasteful; hierarchical Bayes or *empirical Bayes* methods allow us to *'borrow information' from all of the genes* in the array when making inferences about each of the genes (see Smyth, 2004).[2] Although not as well known as the above methods, Parmigiani and colleagues (Garrett and Parmigiani, 2003; Parmigiani *et al.*, 2002) model gene expression using *latent categories* that are interpreted as a gene being over-expressed, under-expressed or at baseline expression;[3] these models allow for denoising of the expression data, can enhance interpretability and help with visualization, and ease comparisons among platforms. Finally, Bickel (2004) has argued for testing *customized null hypotheses* that redefine differential expression in a biologically meaningful way (e.g., any non-zero difference is not necessarily biologically relevant), and use ROC-based statistics[4] (see below, section 12.5).

Asking questions about prespecified groups of genes

Among the tens of thousands of genes in an array, there might be prespecified sets of genes (e.g., those involved in cell cycle, or those found as relevant in a previous study) about which we might want to ask whether, as a whole, these subsets of genes show evidence of differential expression between groups of patients (or whether the expression of the whole set of genes is related to some other clinical variable, such as survival). Goeman *et al.* 2004 have proposed a method to test whether the expression pattern of a group of genes is related to some outcome of interest (be it class membership, survival, or a non-censored continuous variable). Their approach exploits the connection between differential expression among groups and predictability of clinical outcome, and the problem of number of genes being much larger than the

[2]Another review of 'moderated' or 'modified' t and F statistics is that by Cui and Churchill (2003). The approach developed by Gordon Smyth (Smyth, 2004) is applicable to a wide range of linear models (in contrast to some earlier approaches, that were only suited for specific comparisons), and an R (http://www.R-project.org) package, limma, is available from Bioconductor (http://www.bioconductor.org), and also incorporates accounting for multiple testing. However, although applicable to linear models, borrowing strength from all other genes is not as yet implemented in an easy to use tool for problems such as censored data, often analysed with Cox models.

[3]They use a Beyesian hierarchical mixture model – with uniform distributions for abnormally high and abnormally low expression, and normal distribution for baseline expression, and the model returns, for each gene and sample, the probability that it is over-, under- or baseline expressed. Software – R code – is available from http://astor.som.jhmi.edu/poe/. See also the work of Newton *et al.* (2004), who use a semiparametric hierarchical mixture model for a somewhat similar problem.

[4]R code is available from http://www.davidbickel.com

number of samples is overcome using penalized regression models.[5] This method constitutes a very promising way of conducting tests of differential expression of subsets of genes.[6]

A different approach has been suggested by Mootha *et al.* (2003), who examine if the members of a set of genes are enriched (i.e., a proportion larger than expected) among the most differentially expressed genes between two classes. This method should be applicable to any other type of comparison, such as multiclass comparisons (via ANOVA) or survival data. The main differences between the approaches of Mootha *et al.* (2003) and Goeman *et al.* (2004) are listed in Table 12.1. Although with a different objective, a method similar to that of Mootha *et al.* (2003) was proposed by Díaz-Uriarte, Al-Shahrour and Dopazo (2003) (see also Al-Shahrour *et al.*, 2004); as in Mootha *et al.*, (2003), the approach of Díaz-Uriarte, Al-Shahrour and Dopazo (2003) only works if genes with similar ranking or order belong to the same set, but in contrast to Mootha *et al.* (2003), the approach of Díaz-Uriarte, Al-Shahrour and

Table 12.1 Comparison of methods of Goeman *et al.* (2004) and Mootha *et al.* (2003) for testing hypotheses about pre-specified sets of genes

	Goeman *et al.* (2004)	Mootha *et al.* (2003)
Testing	If the set of genes that belongs to set *S* shows differential expression between classes *A* and *B*.	If the 'most differentially expressed' genes are mainly of one of the sets.
Statistic	Multivariate: all genes in the set fitted simultaneously using a generalized linear model.[1]	Univariate (gene-by-gene).
Ease of application	Requires development of maths for different cases (already done for two-class, multiclass and censored data).	Only needs ordering of genes with criteria of our choice.
Assumes equal behaviour of genes in set	No.	Genes in the set(s) of interest must have a similar ranking of the statistic.[2]
Application to different sets	Need to carry out different tests for each of different sets of genes.	Can be applied at once over different sets, and a permutation test carried out to test the single null hypothesis that no gene set is associated with the class distinction.

[1]In general, for multivariate hypotheses ('are the genes of set *S* differentially expressed between groups *A* and *B*?') we should prefer procedures that are fully multivariate (Krzanowski, 1988, pp. 235 ff.).
[2]Requiring the set of genes to have a similar ranking of the statistic does not by itself guarantee that the set of genes will be made of genes that are co-expressed.

[5]Penalized regression models are related to shrinkage methods, such as ridge regression, and models with random effects, and will drive many coefficients towards zero; they allow the fitting of models even when the number of samples (i.e., arrays) is smaller than the number of variables (i.e., genes).

[6]The code is available as the package 'globaltest' from Bioconductor.

Dopazo (2003) will detect sets of genes that are not extreme in their statistic of differential expression; however, it is a method targeted towards exploratory purposes rather than for statistical testing of prespecified hypotheses.

12.4 Class Prediction and Prognostic Prediction

Overview

As explained above, the goal here is to predict the clinically relevant characteristic of a subject (be it class membership, survival, prognosis or any other variable of interest) given the genetic profile of this subject. This is also an area of extremely active research, where the disciplines of statistics and machine learning have contributed much; Table 12.2 shows widely accepted methods and references.

Available reviews (see Table 12.2) show that relatively simple and well known methods such as *k*-nearest neighbour (KNN) and diagonal linear discriminant analysis (DLDA), together with support vector machines (SVMs), perform very well in most classification tasks in microarray data. Because of their performance and

Table 12.2 Well known and good-performing class prediction methods. Because classification has been much more studied than prediction of survival, the methods listed for survival data are not as well known

Method	References
Classification	
Diagonal linear discriminant analysis (DLDA)	Dudoit, Fridlyand and Speed (2002), Simon *et al.* (2003a), Romualdi *et al.* (2003), Huang and Pan (2003), Duda, Hart and Stork (2001) and Hastie, Tibshirani and Friedman (2001)[1]
k nearest neighbour	Dudoit, Fridlyand and Speed (2002a), Simon *et al.* (2003a), Romualdi *et al.* (2003), Duda, Hart and Stork (2001) and Hastie, Tibshirani and Friedman (2001)
Support vector machines (SVM)	Guyon *et al.* (2002), Lee and Lee (2003), Simon *et al.* (2003a), Romualdi *et al.* (2003), Duda, Hart and Stork (2001) and Hastie, Tibshirani and Friedman (2001).
Partial least squares	Stone and Brooks (1990), Garthwaite (1994), Ghosh (2003), Gusnanto, Pawitan and Ploner (2003), Huang and Pan (2003), Nguyen and Rocke (2002)
Random forests	Breiman (2001a), Liaw and Wiener (2002), Bureau *et al.* (2003), Gunther *et al.* (2003)
Survival data	
Partial least squares	Park, Tian and Kohane (2002)
Penalized Cox regression	Pawitan *et al.* (2004)

[1]Dudoit, Fridlyand and Speed (2002), Simon *et al.* (2003a) and Romualdi *et al.* (2003) are general reviews that include reviews and results from different data sets. Huang and Pan (2003) show the relationships between several of these (and other) methods. Duda, Hart and Stork (2001) and Hastie, Tibshirani and Friedman (2001) are general overviews, with additional background material in statistics and machine learning.

free availability[7] in quality implementations, DLDA, KNN and SVM should probably be used routinely as benchmarks when proposing new methods.

Five specific issues

We will discuss five issues that probably deserve more attention. First, for the user it quickly becomes evident that many methods yield non-unique solutions (see also Section 12.6), or, in other words, can return different solutions of very similar quality (e.g., prediction error rate), which itself leads to the question of how to choose among solutions. A direct way of approaching this problem is via *model combination and model averaging*. Model averaging is well known among Bayesians (e.g., Hoeting *et al.*, 1999; Wasserman, 2000), and theory shows that a (weighted) average of predictions from several models should perform better (at least no worse) than predictions from any single model. The Bayesian model averaging approach is not without problems, however, especially selection of priors and computation, and model definition. Model averaging is also available outside the Bayesian camp; stacking was initially proposed by Wolpert (1992) in the machine learning community, and later developed by Breiman (1996) and Ting and Witten (1999) (see also Hastie, Tibshirani and Friedman, 2001; Ripley, 1996, for short accounts). AIC-based model averaging has been developed by Buckland, Burnham and Augustin (1997) and Burnham and Anderson (2002). Somorjai *et al.* (2002) show successful examples of stacking applied to MR and IR spectra.[8] Finally, random forests perform a kind of model averaging by using an ensemble of trees.

Regardless of which model(s) are used, two general problems can affect all models/algorithms. First, most of the available methods assume additive effects of genes. Non-additive relationships or interactions, also called synergistic (or antagonistic) effects, are present when the outcome (e.g., being of class A) depends not just on the sum of the independent contributions of X and Y, but on their combined effects. Non-additive relationships are likely both between genes (e.g., the snail [NM_005985] gene) and between genes and other factors (Section 12.6). Random forests (Breiman, 2001a; Liaw and Wiener, 2002) implicitly incorporate interactions as they are an ensemble of classification trees, but the actual interactions are not easy to see. Boulesteix and Tutz (2004) and Boulesteix, Tutz and Strimmer (2003) have attempted to explicitly search for *patterns of interactions and use them in predictive models*. Second, the predictive capacity of many models can be hampered by *unrecognized heterogeneity within classes* that are regarded as homogeneous. Not much work has been done in this area. This problem, for instance, was recognized in the past (e.g., Rosenwald *et al.*, 2003) and is dealt with by Munagala, Tibshirani and Brown (2004).[9]

[7]For instance, in R, DLDA is available in package 'sma', KNN in package 'class' (part of the VR bundle) and SVM in package 'e1071', the latter from the libsvm library of Chang and Lin (2003).

[8]However, the present author has attempted, without success, both stacking and AIC-based model combination of logistic and multiresponse linear regression with genomic data.

[9]Unfortunately, their code depends on non-free software.

A final set of problems involves the *biological interpretation of class prediction models* (together with making sense of information for potentially tens of thousands of coefficients). Most methods for building predictors tend not to return models that allow for easy biological interpretation of why and how those predictors are used, and how the genes in the predictors affect and relate to the class prediction. These problems are detailed by Díaz-Uriarte (2004) and an example is methods that use dimension reduction via PCA or PLS, where all genes have loadings on all the components, making it virtually impossible to interpret the biological meaning, if any, of the components.[10]

Visualization methods can help with biological interpretation in this task. For microarray data the *biplot*, as extended by Pittelkow and Wislon (2003),[11] is particularly useful, specially use of the GE-biplot both before and after selecting genes according to different criteria of relevance.

In addition, *'molecular signatures' or 'gene expression signatures'* are key features in many studies in cancer research (Alizadeh *et al.*, 2000; Golub *et al.*, 1999; Pomeroy *et al.*, 2002; Rosenwald *et al.*, 2002; Shaffer *et al.*, 2001; Shipp *et al.*, 2002) and seem to imply the idea of coordinate expression of subsets of genes, so that some of these sets with coordinate expression would be related to some criterion of interest (e.g., cancer type, or survival) (for a near definition of a signature see p. 375 of Shaffer *et al.*, 2001). Recently Stegmaier *et al.* (2004) have provided a very interesting example of a high-throughput, generic, method for screening of compounds that induce differentiation of leukaemia cells, based on the gene expression signature of five genes; so gene expression signatures work as a surrogate for a biological state. In spite of their apparent relevance, however, there seems to be no approach for identifying molecular signatures. Recently, we proposed a method that is explicitly designed to try to identify molecular signatures: it finds sets of genes that are tightly coexpressed and that can be used as successful predictors (Díaz-Uriarte, 2004). This method could also help uncover situations that are inconsistent with the assumptions underlying the existence of a few, easily interpretable, signature components of coexpressed genes. However, there are several unsolved issues. On the one hand, the implicit model underlying the work of Díaz-Uriarte (2004) is one where most of the genes are not relevant for prediction, relevant genes are involved in one and only one 'signature component' (i.e., non-overlapping signature components) and the signature components are common, and behave similarly, in different groups; there are, however, richer biological models for biological signatures. In addition, there are related issues regarding differences in patterns of gene coexpression within and among groups and potential instability concerns (see also Section 12.6) about some results (see Sections 3.2 and 3.3 of Díaz-Uriarte, 2004). Some of these issues might be solved with extensions to the method, and some might require completely different

[10]Naively interpreting components using loadings or eliminating genes with small loadings is often not justified and can lead to unexpectedly suboptimal solutions (Cadima and Jolliffe, 2001; Jolliffe, 2002).

[11]R code is available from Y. Pittelkow on request (see http://cbis.anu.edu.au/software.html).

approaches. For example, modifications of the Plaid model of Lazzeroni and Owen (2002) (see also Turner, Bailey and Krzanowski, 2004), which might allow a more principled, model-based, approach of the problem, within a richer class of models; or an extension of the simultaneous clustering and classification approach of Jörnsten and Yu (2003), where we could add normal mixture models with restrictions on the covariance matrix for clustering; or an approach based on the latent class methods of Parmigiani and colleagues (Garrett and Parmigiani, 2003; Parmigiani *et al.*, 2002), where signature components are based on under-, over- or baseline expression (instead of expression levels), and potentially non-overlapping sets of genes for different classes. Work along these lines is currently in progress in our group. In any case, regardless of the exact method used, it is also relevant that the search for molecular signatures highlights that finding a few sets of genes with biological interpretability can be worthwhile even if it leads to small losses in predictive performance (see also Somorjai, Dolenko and Baumgartner, 2003) because good classification performance, per se, does not shed any light on the underlying biological or clinical phenomena.

12.5 ROC Curves for Evaluating Predictors and Differential Expression

Particularly for the two-class setting, common measures of performance (e.g., Baker, Kramer and Srivastava, 2002; Hastie, Tibshirani and Friedman, 2001; Pepe, 2003) are *sensitivity*, or true positive rate, the probability of predicting a positive outcome when the true state is positive (i.e., $\frac{\text{TP}}{\text{TP+FN}}$ in Table 12.3) and *specificity*, the probability of predicting a negative outcome when the true state of a case is negative (i.e., $\frac{\text{TN}}{\text{TN+FP}}$).[12]

Table 12.3 Confusion matrix for a two-class classification problem, with an indication of the usual labels for the four types of outcome

	Predicted	
True	Diseased	Healthy
Diseased	True positive (TP)	False negative (FN)
Healthy	False positive (FP)	True negative (TN)

[12]Lemon, Liyanarachchi and You (2003) have argued that the *positive-predictive value* (PPV), '[...] the likelihood that a positive test result indicates a true positive' (i.e., $\frac{\text{TP}}{\text{TP+FP}}$) can be more relevant than sensitivity and specificity; however, this needs to be done carefully. In fact, for cancer screening the *predictive value positive* (PVP) (similar in spirit to the PPV) and the *predictive value negative* (PVN) are probably more important than the sensitivity and specificity, but they must be computed taking into account the prevalence, and not just the entires from Table 12.3, as explained by Baker, Kramer and Srivastava (2002), Pepe (2003) and van Belle (2002). This caveat is particularly important for very low-prevalence diseases.

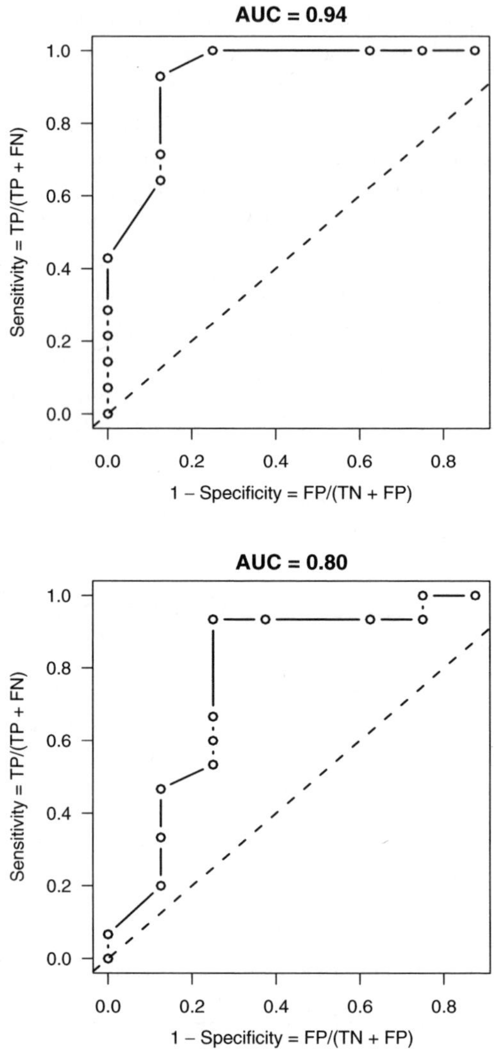

Figure 12.1 Two ROC curves from real microarray data; on top of each we indicate the area under the ROC curve

Sensitivity and specificity are often used to construct a receiver operating characteristic (*ROC*) curve.[13] An ROC curve (see, e.g., Figure 12.1) (e.g. Pepe, 2003; Pepe *et al.*, 2001; van Belle, 2002, ch. 4) is a plot of sensitivity in the ordinate against one minus specificity or the *false positive rate* (i.e., $\frac{\text{FP}}{\text{TN+FP}}$) in the abscissa: in other words, a plot of the probability of a hit against the probability of false alarm (Duda, Hart and Stork, 2001). This shows us how the sensitivity and the false positive rate change as

[13]The package ROC in Bioconductor offers several utilities for building and using ROC curves.

we modify the threshold that classifies a subject as a member of one class or the other. In addition, we can use as a statistic the 'area under the curve' for an ROC curve, which is '[. . .] an overall measure of classification accuracy over all possible decision thresholds' (Bickel, 2004; Pepe, 2003).

ROC curves and ROC-based statistics are widely (and successfully) used to evaluate the diagnostic utility of medical tests (e.g., X-rays, ultrasound, biochemical tests etc., as reviewed in the excellent book by Pepe, 2003). It seems reasonable that similar approaches could be used with microarray data, specially since ROC-based statistics are very flexible devices that allow us, for example, to model covariate effects on the ROC curves, and to combine multiple test results (see Pepe, 2003, for a review). As mentioned above (Section 3), Bickel (2004) and Pepe *et al.* (2003) have argued for the use of ROC-based statistics to rank genes. These authors (see also Xu and Li, 2003) argue that ranking genes using ROC-based statistics is more meaningful than using t- and F-based statistics or p-values. Using the area under the ROC curve for two groups is a measure of differential expression that also provides information on the discriminatory capacities of genes: the empirical area under the ROC curve is equal to the probability that a randomly selected patient from one of the groups will have a larger expression value than a randomly selected patient from the other group (Bickel, 2004; Pepe, 2003), and this summary, from the clinical or biological perspective, is often much more meaningful than a t-statistic or a p-value. In addition, the area under the ROC curve is equivalent to the Wilcoxon rank sum statistic (≡Mann–Whitney U statistic), and thus it is a distribution-free rank statistic (Pepe, 2003; Pepe *et al.*, 2003). Besides the area under the whole curve, Pepe *et al.* (2003) suggest using the empirical estimates of the ROC at a given false positive rate, t_0, ROC(t_0), and the partial area under the ROC at t_0, pAUC(t_0), as measures of differential expression. These statistics do depend on t_0, and a reasonable t_0 could be the false positive rate that is acceptable in practice: when screening asymptomatic people, where prevalence of cancer is very low in average risk populations, it is important to keep the false positive rate extremely low because otherwise there would be large numbers of people undergoing expensive and invasive procedures (Baker, Kramer and Srivastava, 2002; Pepe *et al.*, 2003).

12.6 Caveats and Admonitions

Estimating the error rate of the predictor

To evaluate the performance of a predictor, it is common to provide the error rate of the predictions. However, many papers, including 'high-profile' ones, report error rates that are severely biased, leading to overoptimistic claims about the performance of different methods. This is a most unfortunate situation because lack of appropriate rigour in the application and adherence to appropriate rules of evidence undermines trust in the promises of these technologies. These severe problems were addressed in the bioinformatics literature in Ambroise and McLachlan (2002) and Simon *et al.* (2003b). In spite of the seriousness of the problem, the practice of reporting severely biased error rates is still common, and this has prompted a recent review (Ransohoff, 2004) that tries, once again, to alert users, reviewers and editors against computing, reporting and accepting overly optimistic error rates. We will review here the two most common problems, remembering that our objective when providing an estimate of the error rate is to provide an estimate of the likely error rate we will make when we apply our classifier to new data sets from the same population.

One possible problem is reporting the '*resubstitution rate*', the error rate computed from the very same observations as were used to build the classifier, because the resubstitution error rate is severely biased down due to overfitting: if we fit a classifier to a data set, we can expect it to 'adapt to' some peculiarities of the data, which will make it work well with those data, but might lead it to work poorly with data not yet seen by the classifier or learner. This problem is even more serious with microarray data, where there are thousands of genes that can be part of a predictor. With so many variables, and so few samples, it is very easy to find a predictor that works perfectly in a completely random data set (see, for example, Figure 8.4 in Simon *et al.*, 2003a). To solve this problem either cross-validation or bootstrap have been used; both methods build the predictor using a subset of the data, and then predict the values for the remaining data, thus insuring that the predictions are from data not used for the training.

A second common problem is to carry out the cross-validation *after* the gene selection: all samples are used for gene selection, and the cross-validation process does not include gene selection. This leads to very optimistic estimates of the error rate, as shown by Ambroise and McLachlan (2002) and Simon *et al.* (2003a,b) because we incur a problem similar to overfitting when the gene selection is carried out. The solution is to perform cross-validation or bootstrap so that all steps of the analysis (including gene selection, but also other potential steps such as imputation) are included in the cross-validation.[14] Whether cross-validation (and what size of folds) or bootstrap (and what type of bootstrap) should be used is beyond the scope of this review (see Ambroise and McLachlan, 2002; Braga-Neto and Dougherty, 2004; Davison and Hinkley, 1997; Efron and Gong, 1983; Efron and Tibshirani, 1993, 1997; Simon *et al.*, 2003a,b).

Reinventions of the wheel and comparisons among methods

There are two related problems that slow the development of the field simply by overwhelming researchers with new publications and algorithms. On the one hand,

[14]Of course, all these comments apply to other approaches, such as stepwise, forward and backward selection methods in linear or logistic regression; in addition, these selection methods are well known for their instability and their leading to biased p-values (see, e.g., Section 4.3 of Harrell, 2001). In any case, these variable selection methods ought to be subject, too, to cross-valdation or bootstrap.

there is a fair amount of 'repeated reinventions of the wheel', or ignorance of previously dealt with problems (many of them with solutions by now). In addition, many new methods that are published are not evaluated against 'standard' competing methods (see also Section 4), or are evaluated using only data sets regarded as 'easy' (e.g., the leukaemia data set of Golub *et al.*, 1999) making it hard to assess how new methods really perform (in sharp contrast, for example, Dettling and Bühlmann, 2004, use six different data sets and three competing predictors). Hopefully, stricter standards for evaluation of proposed methods (together with the requirements of a freely available 'reference implementation' – Section 7) will decrease the number of new proposed methods, will shorten the 'to-read' pile and will allow researchers to carry out wider and more exhaustive searches for more mature solutions to similar problems from other fields.

Stability of results, or which set of candidate genes is biologically relevant?

Suppose a predictor has been built that includes 20 genes. How far can we take biological interpretation on the relevance of these genes? A paper by Somorjai, Dolenko and Baumgartner (2003) suggests that often not very far; the problem is the instability or non-uniqueness of results, a phenomenon called the 'Rashomon effect' by L. Breiman (2001b). It is very common that, if we re-run a given procedure with only minor changes or using bootstrap samples, we end up with very different sets of models, suggesting that there are many different 'optimal' subsets of genes (because there are many different descriptions that give approximately the same minimum error rate; Breiman, 2001b). Somorjai, Dolenko and Baumgartner (2003) show how this can arise because of small sample sizes and an extremely small sample per feature ratio (i.e., very small number of arrays relative to the number of genes). Somorjai, Dolenko and Baumgartner (2003) suggest using a variety of classifiers or predictors and finding whether the same features are selected; if the same set of genes is repeatedly selected, we would be more confident that the set is reasonably robust. Of course, this way of examining robustness to selection methods cannot be used if feature selection is carried out using the same filter method for different classifiers (e.g., finding the 200 genes with largest F-ratio, and then using those 200 genes with DLDA, KNN and SVM). Additionally, the bootstrap can be used to examine variation in solutions achieved. The multiplicity problem deserves much more careful attention and prompts for cautious interpretation of results.

Recognizing observational studies and the need of including covariates

Although microarray studies are often referred to as 'experiments', they are frequently observational studies. The differences between observational and experimental studies are well known in statistics and epidemiology, and affect both analyses and interpretation of results. Observational studies present several potential problems, particularly the following.

- Background differences between groups and presence of potential confounding variables; confounding is a pervasive problem. Potter (2003) illustrates it with examples of the relation between vegetable consumption and cancer being confounded by differences in smoking associated with vegetable consumption (smokers also tend to eat fewer vegetables) and differences in expression profiles between cancer types being related to the unmeasured confounding of age and sex. A related problem is interaction, such as when the degree of association between an exposure factor (e.g., expression of gene A) and the disease is different for different levels of the confounding variable, such as sex (Collett, 2003); there is evidence that this might be the case in lung cancer (Patel, Bach and Kris, 2004). The problems of confounding and interaction are discussed in more detail below.

- Biases arising from handling of units (e.g., case samples are frozen several hours after collection whereas control samples are frozen immediately; Potter, 2003) or from biases during the selection of subjects for the study or from informative patterns of missingness.

- Samples too small to allow for generalizations to the populations of interest, and problems of reproducibility.

These issues are well known in epidemiology, which studies patterns of disease and possible factors that affect these patterns of disease by using mainly observational data (Collett, 2003; Potter 2003). However, as indicated by Potter (2003), concerns related to microarrays being often observational studies are mostly absent from standard papers and textbooks on microarray design and analysis (Churchill, 2002; Drăghici, 2002; Simon and Dobbin, 2003; Simon *et al.*, 2003b; Speed, 2003; Tumor Analysis Best Practices Working Group, 2004; Yang and Speed, 2002). In particular, it is surprising that confounding and interaction have not been given more consideration (see also Ntzani and Ioannidis, 2003, who show that an alarmingly large number of predictive studies with DNA arrays do not include adjustments for other known, and potentially competing, predictors). Confounding and interaction can be addressed, at least partially, by appropriately using relevant covariates in the statistical models.[15]

How is this relevant for microarray data? As Potter (2003) illustrates, many of the differences seen in expression profiles between different types of cancer can be the result of confounding by age and sex. Another example is provided by Patel, Bach

[15]Harrell (2001, pp. 3, 390) emphasizes the importance of multivariable modelling in observational studies because they allow us to control (hold constant mathematically the effect of) variables that might differ between groups because the study is observational.

and Kris (2004), who have reviewed evidence that clearly indicates that there are sex-specific differences in susceptibility to, and biology and progression of, lung cancer. Some of these sex-specific differences could be related to differential expression of certain genes, decreased DNA repair capacity in women, increased incidence of certain mutations and oestrogen signalling. All of these factors and differences make it extremely likely that both confounding and interaction will occur related to sex in studies of the relationship between gene expression and cancer,[16] and in the development of predictive models. However, the good news is that sex and age of patients are often known for each microarray sample; these two variables, thus, should routinely be included in the analysis as covariates and to examine possible interactions. (Interestingly, Patel, Bach and Kris, 2004, call for undertaking sex-specific research in lung cancer.) Of course, comments regarding sex and age are extensive to other potential confounders (e.g., diet, exercise, region of origin etc.), for which information might be available. Controlling for the effect of confounders with strong effects (and, from the biology we know, sex and age are likely to be confounders with strong effects in many cases), can lead to increases in statistical power, because a source of variation is being taken into account rather than being thrown into the error term.[17] Thus, by controlling the effects of covariates we can be more likely to detect differential expression between conditions. On the other hand, if differences between groups are mainly due to confounders (e.g., because of a disproportionate presence of one sex in one of the groups), only after controlling for the confounder can we trust that differential expression of certain genes or the predictive ability of our model is not due to confounding. With respect to interactions (e.g., that the effects of changes in the expression of certain genes depend strongly on, say, sex), their presence can be an important finding in itself, as is the case of sex differences and lung cancer biology (Patel, Bach and Kris, 2004). Finally, if there are interactions with, say, sex, we will obtain lower error rates if we develop different predictive models for men and women than if we use a model that makes predictions independently of sex.

Collaboration between statisticians and biologists and the use of software 'magic bullets'

Successful use of microarrays to answer biologically relevant questions will require close collaboration between biologists and statisticians during the complete process of the study. The need for statisticians' advice during the experimental design has been discussed before (Churchill, 2002; Simon and Dobbin, 2003; Yang and Speed, 2002) and is not the subject of this chapter; however, it should be remembered that full details of the experimental set-up are necessary for the use of appropriate statistical

[16]Interactions are very likely, given the complex mappings between transcript levels and protein levels (O'Neill, Catchpoole and Golemis, 2003).

[17]This is the idea behind blocking in experimental design: controlling a known source of variation.

methods. In the context of this chapter, statisticians need to realize that there are often many subtleties in the interpretation of microarray results that preclude simple mappings from RNA expression data to phenotypes (O'Neill, Catchpoole and Golemis, 2003). At the same time, statistical help is needed to insure that the statistical model and test being used is addressing the biological questions of interest. What in any case is unrealistic is to expect that if the biologist sends a file with 15 000 rows by 200 columns (genes by subject) to the statistician, the statistician will return to the biologist the list of, say, 30 genes that are the answer to the biological question. But that is, in fact, what some users often expect from software tools or statistical consulting, and what some statisticians might believe is possible/desirable. This also means that the questions asked are sometimes reformulated to accommodate the available software.

The problem of these expectations and procedures is that they lack key ingredients often needed to provide an answer to the underlying biological question. Table 12.4 lists some typical questions that a statistician might ask.[18] Only after these (and other) questions have been answered is it time to search for the appropriate tool, which might be a web tool, a GUI-based stats program, or might require the competent use of command-driven programs or the development of new programs to carry out the customized required analyses.

Table 12.4 Some relevant questions statisticians and biologists should engage in a dialogue about

Are genes grouped in families, and are we interested in the overall responses of groups of genes, or should we look at individual genes?
Are certain genes or spots in the array more relevant biologically, maybe because they are easier to measure reliably with other assays?
Is there additional information on which genes are likely to be differentially expressed?
Do you really need the best possible predictor that statistical computing will get you, or do you want a small list of genes very likely to be differentially expressed?
In what stage of the scientific discovery process is this study, and how tight control do you require over the type I error rate?
What other information and variables about the patients, besides the microarray data, do you have available?
What population do you expect the results of these studies to be relevant for?
Are these the original, complete data, and are these the original biological questions, or have the data and questions gone through an already long run of analyses which has already filtered data and reoriented hypotheses?
What is the next stage of this study, or what do you want to do with these results?
What additional studies could be done to confirm the results from these analyses?

[18] van Belle (2002) provides a very accessible account for the reasons behind these, and many other, questions statisticians ask.

12.7 Final Note: Source Code Should Be Available

Many new method papers are published every month, and biologists and applied statisticians do not have the time to implement each and every idea that is published, nor to deal with the complications associated with patented algorithms. Sometimes, however, when researchers ask for software from authors of method papers they face answers such as '...my method is straightforward to implement from the explanations in my paper', '...the method will soon be available as part of program XYZ (which is proprietary)', or '...I am not in the business of providing software to anyone'.

In the opening lecture of the Royal Statistical Society meeting of 2002, titled 'Statistical methods *need* software', Brian Ripley (2002) proposed '[...] a reference implementation, some code which is warranted to give the authors intended answers in a moderately-sized problem. It need not be efficient, but it should be available to anyone and everyone'. Calls for availability of software, including source code, in bioinformatics research have also been made in other settings (see, e.g., Dudoit, Gentleman and Quackenbush, 2003; Marshall, 2003), and the Open Bioinformatics Foundation (http://www.open-bio.org/) is 'focused on supporting open source programming in bioinformatics'. The Free Software Foundation (http://www.fsf.org) and the Open Source Initiative (http://www.opensource.org/) explain free and open source software. The reasons for making source code available in bioinformatics and microarray research are summarized by Dudoit, Gentleman and Quackenbush (2003, p. 46) and are reproduced in Table 12.5.

Table 12.5 Reasons why source code should be available in bioinformatics, from p. 46 of Dudoit, Gentleman and Quackenbush (2003)

- full access to the algorithms and their implementation, which allows users to understand what they are doing when they run a particular analysis
- the ability to fix bugs and extend and improve the supplied software
- encouraging good scientific computing and statistical practice by providing appropriate tools, instruction and documentation
- providing a workbench of tools that allow researchers to explore and expand the methods used to analyse biological data
- ensuring that the international scientific community is the owner of the software tools needed to carry out research
- promoting reproducible research by providing open and accessible tools with which to carry out that research (reproducible research as distinct from independent verification)

In this review, and following the above spirit, we have been highly biased towards methods for which software, including source code, is available; besides the philosophical issues involved, this is also a pragmatic decision.

Acknowledgements

J. Goeman and V. Mootha for answers about the workings of their procedures. C. Lázaro-Perea provided detailed and careful comments on the manuscript that forced me to rewrite it and greatly improved it. The editors and reviewers for very helpful comments on the manuscript. The author was partially supported by the Ramón y Cajal programme of the Spanish MCyT (Ministry of Science and Technology); funding partially provided by project TIC2003-09331-C02-02 of the Spanish MCyT.

References

Al-Shahrour, F., Herrero, J., Mateos, Á., Santoyo, J., Díaz-Uriarte, R. and Dopazo, J. (2004) Using gene outology on genome-scale studies to find significant associations of biologically relevant terms to groups of genes. *Neural Networks for Signal Processing XIII.*

Alizadeh, A. A., Eisen, M. B., Davis, R. E., Ma, C., Lossos, I. S., Rosenwald, A., Boldrick, J. C., Sabet, H., Tran, T., Yu, X., Powell, J. I., Yang, L., Marti, G. E., Moore, T., Hudson, J. J., Lu, Lewis, D. B., Tibshirani, R., Sherlock, G., Chan, W. C., Greiner, I. C., Weisenburger, D. D., Armitage, J. O., Warnke, R. and Staudt, L. M., (2000) Distinct types of diffuse large b-cell lymphoma identified by gene expression profiling. *Nature*, **403**, 503–511.

Ambroise, C. and McLachlan, G. J. (2002) Selection bias in gene extraction on the basis of microarray gene-expression data. *Proc Natl Acad Sci USA*, **99** (10), 6562–6566.

Baker, S. G., Kramer, B. S. and Srivastava, S. (2002) Markers for early detection of cancer: statistical guidelines for nested case–control studies. *BMC Med Res Methodol*, **2**, 4.

Benjamini, Y. and Hochberg, Y. (1995) Controlling the false discovery rate: a practical and powerful approach to multiple testing. *J R Stat Soc B*, **57**, 289–300.

Bickel, D. R. (2004) Degrees of differential gene expression: detecting biologically significant expression differences and estimating their magnitudes. *Bioinformatics*, **20**, 682–688.

Boulesteix, A. and Tutz, G. (2004) Identification of interaction patterns and classification with applications to microarray data. *SFB386 Discussion Paper, 369.*

Boulesteix, A. L., Tutz, G. and Strimmer, K. (2003) A CART-based approach to discover emerging patterns in microarray data. *Bioinformatics*, **19**, 2465–2472.

Braga-Neto, U. M. and Dougherty, E. R. (2004) Is cross-validation valid for small-sample microarray classification? *Bioinformatics*, **20**, 374–380.

Breiman, L. (1996) Stacked regressions. *Machine Learning*, **24**, 49–64.

Breiman, L. (2001a) Random forests. *Machine Learning*, **45**, 5–32.

Breiman, L. (2001b) Statistical modeling: the two cultures (with discussion). *Stat Sci*, **16**, 199–231.

Buckland, S. T., Burnham, K. P. and Augustin, N. H. (1997) Model selection: an integral part of inference. *Biometrics*, **53**, 603–618.

Bureau, A., Dupuis, J., Hayward, B., Falls, K. and Van Eerdewegh, P. (2003) Mapping complex traits using Random Forests. *BMC Genet*, **4** Suppl 1, S64.

Burnham, K. P. and Anderson, D. R. (2002) *Model Selection and Multimodel Inference: a Practical Information-Theoretic Approach*, 2nd ed. Springer, New York.

Cadima, J. F. C. L. and Jolliffe, I. T. (2001) Variable selection and the interpretation of principal subspaces. *Journal of Agricultural, Biological, and Environmental Statistics*, **6**, 62–79.

Chang, C.-C. and Lin, C.-J. (2003) *Libsvm: a Library for Support Vector Machines*, technical report. Available from http://www.csie.ntu.edu.tw/~cjlin/libsvm/[accessed 23/09/04].

Churchill, G. A. (2002) Fundamentals of experimental design for cDNA microarrays. *Nat Genet*, **32** (Suppl.), 490–495.

Collett, D. (2003) *Modelling Binary Data*, 2nd edn. Chapman and Hall, London.

Cui, X. and Churchill, G. A. (2003) Statistical tests for differential expression in cDNA microarray experiments. *Genome Biol*, **4**, 210.

Davison, A. C. and Hinkley, D. V. (1997) *Bootstrap Methods and their Application*. Cambridge University Press, Cambridge.

Dettling, M. and B̈hlmann, P. (2004) Finding predictive gene groups from microarray data. *J Multivariate Anal*, **90**, 106–131.

Díaz-Uriarte, R. (2004) *Molecular Signatures from Gene Expression Data*, technical report. Available from http://www.arxiv.org/abs/q-bio.QM/0401043 [accessed 23/09/04].

Díaz-Uriarte, R., Al-Shahrour, F. and Dopazo J. (2003) In *Methods of Microarray Data Analysis III, papers from Camda '02*. Kluwer, Dordrecht, The use of GO terms to understand the biological significance of microarray differential gene expression data. 233–247.

Drăghici, S. (2002) *Data Analysis for DNA Microarrays*. Chapman and Hall, London.

Duda, R. O., Hart, P. E. and Stork, D. G. (2001) *Pattern Classification*, 2nd edn. Wiley, New York.

Dudoit, S., Fridlyand, J. and Speed, T. P. (2002) Comparison of discrimination methods for the classification of tumors using gene expression data. *J Am Stat Assoc*, **97** (457), 77–87.

Dudoit, S., Gentleman, R. C. and Quackenbush, J. (2003) Open source software for the analysis of microarray data. *Biotechniques Suppl*, 45–51.

Dudoit, S., Yang, Y. Callow, M. and Speed, T. (2002) Statistical methods for identifying differentially expressed genes in replicated cDNA experiments. *Statistica Sinica*, **12**, 111–139.

Efron, B. and Gong, G. (1983) A leisurely look at the bootstrap, the jacknife, and cross-validation. *Am Stat*, **37** (1), 36–48.

Efron, B. and Tibshirani, R. J. (1993) *An Introduction to the Bootstrap*. Chapman and Hall, London.

Efron, B. and Tibshirani, R. J. (1997) Improvements on cross-validation: the 632+ bootstrap method. *J Am Stat Assoc*, **92**, 548–560.

Garrett, E. and Parmigiani, G. (2003) POE: statistical methods for qualitative analysis of gene expression. In *The Analysis of Gene Expression Data: Methods and Software*. Springer, New York, 362–387.

Garthwaite, P. H. (1994) An intepretation of partial least squares. *J Am Stat Assoc*, **89** (425), 122–127.

Ge, Y., Dudoit, S. and Speed, T. (2003) Resampling-based multiple testing for microarray data analysis (with discussion). *TEST*, **12**, 1–77.

Ghosh, D. (2003) Penalized discriminant methods for the classification of tumors from gene expression data. *Biometrics*, **59**, 992–1000.

Goeman, J. J., van de Geer, S. A., de Kort, F. and van Houwelingen, H. C. (2004) A global test for groups of genes: testing association with a clinical outcome. *Bioinformatics*, **20**, 93–99.

Golub, T. R., Slonim, D. K. Tamayo, P. Huard, C. Gaasenbeek, M. Mesirov, J. P. Coller, H. Loh, M. L. Downing, J. R. Caligiuri, M. A. Bloomfield, C. D. and Lander, E. S. (1999) Molecular classification of cancer: class discovery and class prediction by gene expression monitoring. *Science*, **286**, 531–537.

Gunther, E. C., Stone, D. J. Gerwien, R. W. Bento, P. and Heyes, M. P. (2003) Prediction of clinical drug efficacy by classification of drug-induced genomic expression profiles in vitro. *Proc Natl Acad Sci USA*, **100**, 9608–9613.

Gusnanto, A., Pawitan, Y. and Ploner, A. (2003) *Variable Selection in Gene and Protein Expression Data*, technical report, Department of Medical Epidemiology and Biostatistics, Karolinska Institutet, Stockholm.

Guyon, I., Wecton, J., Bainhill, S. and Vapnik, V. (2002) Gene selection for cancer classification using support vector machines. *Machine Learning*, **46**, 389–422.

Harrell, J. F. E. (2001) *Regression Modeling Strategies*. Springer, New York.

Hastie, T., Tibshirani, R. and Friedman, J. (2001) *The Elements of Statistical Learning*. Springer, New York.

Hoeting, J. A., Madigan, D. Raftery, A. E. and Volinsky, C. T. (1999) Bayesian model averaging: a tutorial. *Stat Sci*, **14**, 382–417.

Huang, X. and Pan, W. (2003) Linear regression and two-class classification with gene expression data. *Bioinformatics*, **19**, 2072–2078.

Jolliffe, I. T. (2002) *Principal Component Analysis*, 2nd edn. Springer, New York.

Jörnsten, R. and Yu, B. (2003) Simultaneous gene clustering and subset selection for sample classification via MDL. *Bioinformatics* **19**, 1100–1109.

Kvzanowski, W. J. (1998) *Principle of Multvariate Analysis*. Oxford University Press, Oxford.

Lazzeroni, L. and Owen, A. (2002) Plaid models for gene expression data. *Statistica Sinica*, **12**, 61–86.

Lee, Y. and Lee, C.-K. (2003) Classification of multiple cancer types by multicategory support vector machines using gene expression data. *Bioinformatics*, **19**, 1132–1139.

Liaw, A. and Wiener, M. (2002) Classification and regression by randomforest. *Rnews*, **2**, 18–22.

Marshall, E. (2003) The upside of good behavior: make your data freely available. *Science*, **299**, 990.

Mootha, V. K., Lindgren, C. M., Eriksson, K. F., Subramanian, A., Sihag, S., Lehar, J., Puigserver, P., Carlsson, E., Ridderstrale, M., Laurila, E., Houstis, N., Daly, M. J., Patterson, N., Mesirov, J. P., Golub, T. R., Tamayo, P., Spiegelman, B., Lander, E. S., Hirschhorn, J. N., Altshuler, D. and Groop, L. C. (2003) PGC-1alpha-responsive genes involved in oxidative phosphorylation are coordinately downregulated in human diabetes. *Nat Genet*, **34**, 267–273.

Munagala, K., Tibshirani, R. and Brown, P. O. (2004) Cancer characterization and feature set extraction by discriminative margin clustering. *BMC Bioinformatics*, **5**, 21.

Newton, M. A., Noueiry, A., Sarkar, D. and Ahlquist, P. (2004) Detecting differential gene expression with a semiparametric hierarchical mixture method. *Biostatistics*, **5**, 155–176.

Nguyen, D. V. and Rocke, D. M. (2002) Multi-class cancer classification via partial least squares with gene expression profiles. *Bioinformatics*, **18** (9), 1216–1226.

Ntzani, E. E. and Ioannidis, J. P. (2003) Predictive ability of DNA microarrays for cancer outcomes and correlates: an empirical assessment. *Lancet*, **362**, 1439–1444.

O'Neill, G. M., Catchpoole, D. R. and Golemis, E. A. (2003) From correlation to causality: microarrays, cancer, and cancer treatment. *Biotechniques Suppl*, 64–71.

Park, P. J., Tian, L. and Kohane, I. S. (2002) Linking gene expression data with patient survival times using partial least squares. *Bioinformatics*, **18** (S1), S120–S127.

Parmigiani, G., Garrett, E. Anbazhaghan, R. and Gabrielson, E. (2002) A statistical framework for expression-based molecular classification in cancer. *J. R Stat Soc B*, **64**, 717–736.

Parmigiani, G., Garrett, E., Irizarry, R. and SL. Z. (2003) The Analysis of gene expression data: an overview of methods and software. In *The Analysis of Gene Expression Data: Methods and Software*. Springer, New York, 1–45.

Patel, J. D., Bach, P. B. and Kris, M. G. (2004) Lung cancer in US women: a contemporary epidemic. *JAMA*, **291**, 1763–1768.

Pawitan, Y., Bjöhle, J., Wedren, S., Humphreys, K., Skoog, L., Huang, F., Amler, L., Shaw, P., Hall, P. and Bergh, J. (2004) Gene expression profiling for prognosis using Cox regression. *Stat Med*, **23** (11), 1767–1780.

Pepe, M. S. (2003) *The Statistical Evaluation of Medical Tests for Classification and Prediction*. Oxford University Press, Oxford.

Pepe, M. S., Etzioni, R., Feng, Z., Potter, J. D., Thompson, M. L., Thornquist, M., Winget, M. and Yasui, Y. (2001) Phases of biomarker development for early detection of cancer. *J Natl Cancer Inst*, **93**, 1054–1061.

Pepe, M. S., Longton, G., Anderson, G. L. and Schummer, M. (2003) Selecting differentially expressed genes from microarray experiments. *Biometrics*, **59**, 133–142.

Pittelkow, Y. E. and Wislon, S. R. (2003).Visualisation of gene expression data – the Ge-biplot, the chip-plot and the gene-plot. *Stat Applications Genetics Mol Biol*, **2**, article 6.

Pomeroy, S., Tamayo, P., Gaasenbeek, M., Sturla, L., Angelo, M., McLaughlin, M., Kim, J., Goumnerova, L., Black, P., Lau, C., Allen, J., Zagzag, D., Olson, J., Curran, T., Wetmore, C., Biegel, J., Poggio, T., Mukherjee, S., Rifkin, R., Califano, A., Stolovitzky, G., Louis, D., Mesirov, J., Lander, E. and Golub, T. (2002) Prediction of central nervous system embryonal tumour outcome based on gene expression. *Nature*, **415**, 436–442.

Potter, J. D. (2003) Epidemiology, cancer genetics and microarrays: making correct inferences, using appropriate designs. *Trends Genet*, **19**, 690–695.

Ransohoff, D. F. (2004) Opinion: rules of evidence for cancer molecular-marker discovery and validation. *Nat Rev Cancer*, **4**, 309–314.

Reiner, A., Yekutieli, D. and Benjamini, Y. (2003) Identifying differentially expressed genes using false discovery rate controlling procedures. *Bioinformatics*, **19**, 368–375.

Ripley, B. D. (1996) *Pattern Recognition and Neural Networks*. Cambridge University Press, Cambridge.

Ripley, B. D. (2002) Statistical methods need software: a view of statistical computing, Opening lecture, *RSS 2002*. http://www.stats.ox.ac.uk/~ripley/RSS2002.pdf [accessed 23/09/04].

Romualdi, C., Campanaro, S., Campagna, D., Celegato, B., Cannata, N., Toppo, S., Valle, G. and Lanfranchi, G. (2003) Pattern recognition in gene expression profiling using DNA array: a comparative study of different statistical methods applied to cancer classification. *Hum Mol Genet* **12** (8), 823–836.

Rosenwald, A., Wright, G., Chan, W. C., Connors, J. M., Campo, E., Fisher, R. I., Gascoyne, R. D., Muller-Hermelink, H. K., Smeland, E. B., Giltnane, J. M., Hurt, E. M., Zhao, H., Averett, L., Yang, L., Wilson, W. H., Jaffe, E. S., Simon, R., Klausner, R. D., Powell, J., Duffey, P. L., Longo, D. L., Greiner, T. C., Weisenburger, D. D., Sanger, W. G., Dave, B. J., Lynch, J. C., Vose, J., Armitage, J. O., Montserrat, E., Lopez-Guillermo, A., Grogan, T. M., Miller, T. P., LeBlanc, M.Ott, G., Kvaloy, S., Delabie, J., Holte, H., Krajci, P., Stokke, T., Staudt, L. M. and the Lymphoma/Leukemia Molecular Profiling Project (2002) The use of molecular profiling to predict survival after chemotherapy for diffuse large-b-cell lymphoma. *N Engl J Med*, **346** (25), 1937–1947.

Rosenwald, A., Wright, G., Leroy, K., Yu, X., Gaulard, P., Gascoyne, R. D., Chan, W. C., Zhao, T., Haioun, C., Greiner, T. C., Weisenburger, D. D., Lynch, J. C., Vose, J., Armitage, J. O., Smeland, E. B., Kvaloy, S., Holte, H., Delabie, J., Campo, E., Montserrat, E., Lopez-Guillermo, A., Ott, G., Muller-Hermelink, H. K., Connors, J. M., Braziel, R., Grogan, T. M., Fisher, R. I., Miller, T. P., LeBlanc, M., Chiorazzi, M., Zhao, H., Yang, L., Powell, J. Wilson, W. H., Jaffe, E. S., Simon, R., Klausner, R. D. and Staudt, L. M., (2003) Molecular diagnosis of primary mediastinal B cell lymphoma identifies a clinically favorable subgroup of diffuse large B cell lymphoma related to Hodgkin lymphoma. *J Exp Med*, **198**, 851–862.

Sackett, D. L. and Haynes, R. B. (2002) The architecture of diagnostic research. *BMJ*, **324**, 539–541.

Shaffer, A., Rosenwald, A., Hurt, E., Giltnane, J., Lam, L., Pickeral, O. and Staudt L. (2001) Signatures of the immune response. *Immunity*, **15**, 375–385.

Shipp, M. A., Ross, K. N., Tamayo, P., Weng, A. P., Kutok, J. L., Aguiar, R. C. T., Gaasenbeek, M., Angelo, M., Reich, M., Pinkus, G. S., Ray, T. S., Koval, M. A., Last, K. W. Norton, A., Lister, T. A., Mesirov, J., Neuberg, D. S., Lander, E. S., Aster, J. C., and Golub, T. R. (2002) Diffuse large b-cell lymphoma outcome prediction by gene-expression profiling and supervised machine learning. *Nature Med*, **8** (1), 68–74.

Simon, R. M. and Dobbin, K. (2003) Experimental design of DNA microarray experiments. *Biotechniques Suppl*, 16–21.

Simon, R. M., Korn, E. L., McShane, L. M., Radmacher, M. D., Wright, G. W. and Zhao, Y. (2003a) *Design and Analysis of DNA Microarray Investigations*. Springer, New York.

Simon, R., Radmacher, M. D., Dobbin, K. and McShane, L. M. (2003b) Pitfalls in the use of DNA microarray data for diagnostic and prognostic classification. *J Natl Cancer Inst*, **95** (1), 14–18.

Slonim, D. K. (2002) From patterns to pathways: gene expression data analysis comes of age. *Nat Genet*, **32** Suppl., 502–508.

Smyth, G. K. (2004) Linear models and empirical Bayes methods for assessing differential expression in microarray experiments. *Stat Applications Genetics Mol Biol*, **3**, article 3.

Somorjai, R. L., Dolenko, B. and Baumgartner, R. (2003) Class prediction and discovery using gene microarray and proteomics mass spectroscopy data: curses, caveats, cautions. *Bioinformatics*, **19**, 1484–1491.

Somorjai, R. L., Dolenko, B., Nikulin, A., Nickerson, P., Rush, D., Shaw, A., Glogowski, M., Rendell, J. and Deslauriers, R. (2002) Distinguishing normal from rejecting renal allografts: application of a three-stage classification strategy to MR and IR spectra of urine. *Vibrational Spectroscopy*, **28**, 97–102.

Speed, T. E. (2003) *Statistical Analysis of Gene Expression Microarray Data.* Chapman and Hall, London.

Stegmaier, K., Ross, K. N., Colavito, S. A., O'Malley, S., Stockwell, B. R. and Golub, T. R. (2004) Gene expression-based high-throughput screening(GE-HTS) and application to leukemia differentiation. *Nat Genet*, **36**, 257–263.

Stone, M. and Brooks, R. J. (1990) Continuum regression: cross-validated sequentially constructed prediction embracing ordinary least squares, partial least squares and principal components regression (with discussion). *J R Stat Soc B*, **52** (2), 237–269.

Storey, J. (2002) A direct approach to false discovery rates. *J R Stat Soc B*, **64**, 479–498.

Storey, J. D. and Tibshirani, R. (2003) Statistical significance for genomewide studies. *Proc Natl Acad Sci USA*, **100**, 9440–9445.

Ting, K. M. and Witten, I. H. (1999) Issues in stacked generalization. *J Artif Intell Res*, **10**, 271–289.

Tumor Analysis Best Practices Working Group (2004) Guidelines: expression profiling – best practices for data generation and interpretation in clinical trials. *Nat Rev Genet*, **5**, 229–237.

Turner, H., Bailey, T. and Krzanowski, W. (2004) Improved biclustering of microarray data demonstrated through systematic performance tests. *Comput Stat Data Anal*, in press.

van Belle, G. (2002) *Statistical Rules of Thumb.* Wiley, New York.

Wasserman, L. (2000) Bayesian model selection and model averaging. *J Math Psychol* **44**, 92–107.

Westfall, P. H. and Young S. S. (1993). *Resampling-Based Multiple Testing. Examples and Methods for p-Value Adjustment.* Wiley, New York.

Whitfield, C. W., Cziko, A. M. and Robinson, G. E. (2003) Gene expression profiles in the brain predict behavior in individual honey bees. *Science*, **302**, 296–299.

Wolpert, D. H. (1992) Stacked generalization. *Neural Networks*, **5**, 241–259.

Xu, R. and Li, X. (2003) A comparison of parametric versus permutation methods with applications to general and temporal microarray gene expression data. *Bioinformatics*, **19**, 1284–1289.

Yang, Y. H. and Speed, T. (2002) Design issues for cDNA microarray experiments. *Nat Rev Genet*, **3**, 579–588.

13

A Guide to the Literature on Inferring Genetic Networks by Probabilistic Graphical Models

Pedro Larrañaga, Iñaki Inza and **Jose L. Flores**

Abstract

In this chapter we discuss the advantages of the use of probabilistic graphical models for modelling molecular networks at different levels. We also provide an overview to the literature on inferring genetic networks by probabilistic graphical models. Different types of probabilistic graphical model – Bayesian networks, Gaussian networks – are introduced and methods for learning these models from data are presented. These models provide a concise language for describing joint probability distributions by means of local distributions. This fact and the possibility of reasoning inside the model, apart from their declarative nature, provide an advantage to inferring molecular networks and to transforming heterogeneous data sets into biological insights about the underlying mechanisms.

Keywords

molecular networks, genetic networks, gene interaction, probabilistic graphical models, Bayesian networks, Gaussian networks, learning from data

13.1 Introduction

In recent years research in molecular biology has been living through a revolution. Nowadays, it is possible to measure molecular networks and their components at multiple levels. These include mRNA transcript quantities, protein–protein and

Data Analysis and Visualization in Genomics and Proteomics Edited by Francisco Azuaje and Joaquin Dopazo

protein–DNA interactions, chromatin structure and protein quantities, localization and modifications. This huge amount of data offers much promise for novel insights about cellular processes (Lander, 1999).

One of the main challenges of computational biology is to develop tools and methods able to transform all this heterogeneous data into biological knowledge about the underlying mechanism. These tools and methods should allow us to go beyond a mere description of the data and provide knowledge in the form of testable models. By this simplifying abstraction that constitutes a model, we will be able to obtain prediction of the system behaviour under different conditions. The model can also be used to learn the function of some of the components of the system (Friedman, 2004).

In this chapter we present a guide to the literature on inferring genetic regulatory networks – also known as the reverse engineering process – by means of probabilistic graphical models. We can see a genetic network as a set of genes in which individual genes influence the activity of other genes. By inferring genetic regulatory networks we aim to discover the nature of regulation between genes. This regulation between two genes can be directed or indirected (mediated by a third gene).

Probabilistic graphical models represent joint probability distributions by means of a product of local distributions, each of them only involving a few variables. In order to represent genetic networks by using probabilistic graphical models we associate each gene with a random variable. The values of this random variable are determined by the expression level of the gene. These types of stochastic model have proved to be perfectly adequate – as we explain in Section 13.4 – to represent the regulation between genes.

The chapter is organized as follows. In Section 13.2 genetic networks are shown as central elements for molecular biology and the problem of constructing genetic networks from data is presented as a major challenge in computational biology. In this section we also review previous work in modelling genetic networks with different mathematical tools. In Section 13.3, some general concepts about probabilistic graphical models are presented. Several methods for modelling Bayesian and Gaussian networks from data are also reviewed. Section 13.4 presents an overview of the literature and the different approaches for inferring molecular networks by means of probabilistic graphical models are classified. Finally, closing conclusions and future work possibilities are analysed in Section 13.5.

13.2 Genetic Networks

The vast quantity of data generated by genomic expression arrays affords researchers a significant opportunity to transform biology, medicine and pharmacology using systematic computational methods. The availability of genomic (and eventually proteomic) expression data promises to have a profound impact on the understanding of basic cellular processes, the diagnosis and treatment of disease and the efficacy of designing and delivering targeted therapeutics. Particularly relevant to these objectives is the development of a deeper understanding of the various mechanisms by which cells control and regulate the transcription of their genes.

While the potential utility of expression data is immense, some obstacles are needed to overcome before significant progress can be realized. First, data from expression arrays is inherently noisy. Second, our knowledge regarding genetic regulatory networks is extremely limited. Third, gene expression is regulated in a complex and seemingly combinatorial manner. For the previous reasons the modelling of genetic networks by means of stochastic systems, as well as probabilistic graphical models, seems to be a reasonable approach.

Genetic network modelling was pioneered by a handful of scientists (Kauffman, 1971) several decades ago. A classical classification of methods for modelling genetic networks (Szallasi, 2001) can be made taking these two criteria into account: the size of the genetic network and the nature of the regulatory interactions.

From the viewpoint of the *size*, genetic networks can be studied on several complexity levels:

- small-scale (Becskei and Serrano, 2000) networks, when only a few genes are studied,
- intermediate-level (Hill, Tomasi and Sethna, in preparation) networks, when the interaction of a couple of tens of genes are studied, and
- large-scale modelling (Kauffman, 1993), which deals with realistic sized networks of thousands of interacting genes, clearly related to phenomena that appear in nature.

The consideration of the second criterion, that is, the *nature of regulatory interactions*, is more relevant for the content of this chapter. According to this criterion, different approaches can be grouped taking the function or mathematical tool that describes the regulatory interaction between the genes into account.

- *Boolean rules* (Kauffman, 1993; Somogyi and Sniegoski, 1996), where a gene can be modelled as a binary element and may receive one or several inputs from other genes or itself. In this type of model the output at time $t+1$ is deterministically computed from the input at time t according to logical or Boolean rules. The gene expression state at a given time point and the regulatory interactions between the genes unambiguously determine the gene expression state at the next time point. The REVerse Engineering ALgorithm (REVEAL) (Liang, Fuhrman and Somogyi, 1998) based on the systematic analysis of the mutual information between input and output states, is an example of an algorithm that infers Boolean nets from data.
- *Differential equations* can also be used as a more accurate approach to the modelling of regulatory interactions. Unfortunately, this leads to serious computational

problems that could possibly be circumvented by various methods (Glass and Kauffman, 1973).

- *Stochastic models* (Arkin, Ross and McAdams, 1998), which, in contrast to both previous approaches, starting from a given gene expression state, are able to generate more than one successive gene expression value. The crucial fact that in reality genetic networks are stochastic is supported by theoretical considerations and experimental results, both in prokaryotes (McAdams and Arkin, 1997) and eukaryotes (Abkowitz, Catlin and Guttorp, 1996).

- The so-called *time-shifted* (D'haeseleer, Liang and Somogyi, 2000, Herrero, Díaz-Uriarte and Dopazo, 2003) approach makes use of clustering of co-expression profiles, allowing us to infer shared regulatory inputs and functional pathways. This coarse resolution shows groups of genes under common transcriptional control.

Machine learning techniques have also been investigated for identifying gene interactions. For instance, Becquet *et al.* (2002) and Creighton and Hanash (2003) apply association rules for this task, while Wu *et al.* (2003) use k-way interaction log-linear modelling.

13.3 Probabilistic Graphical Models

Probabilistic graphical models represent multivariate joint probability distributions via a product of terms, each of which involves only a few variables. The structure of the product is represented by a graph that relates variables that appear in a common term. This graph specifies the product form of the distribution and also provides tools for reasoning about the properties entailed by the product. For a sparse graph, the representation is compact and in many cases allows effective inference and learning.

In this section we will introduce two types of probabilistic graphical model known as Bayesian networks and Gaussian networks that have been used during the last decade for reasoning in domains with an intrinsic uncertainty. Both types of probabilistic graphical model are well suited for inferring genetic regulatory networks from data, as we will discuss in Section 13.4.

Notation and semantics

We use X_i to represent a random variable. A possible instance of X_i is denoted x_i. $\rho(X_i = x_i)$ (or simply $\rho(x_i)$) represents the *generalized probability distribution* (DeGroot, 1970) over the point x_i. Similarly, we use $\boldsymbol{X} = (X_1, \ldots, X_n)$ to represent an n-dimensional random variable, and $\boldsymbol{x} = (X_1, \ldots, X_n)$ to represent one of its possible instances. The *joint generalized probability distribution* of $\boldsymbol{X}$ is denoted $\rho(\boldsymbol{X} = \boldsymbol{x})$ (or simply $\rho(\boldsymbol{x})$). The *generalized conditional probability distribution* of the variable X_i given the value x_j of the variable X_j is represented as $\rho(X_i = x_i|X_j = x_j)$ (or simply as $\rho(x_i|x_j)$). We use D to represent a data set, i.e. a set of N instances of the variables $(X_1, \ldots, X_n)$.

If the variable X_i is discrete, $\rho(X_i = x_i) = p(X_i = x_i)$ (or simply $p(x_i)$) is called the *mass probability* for the variable X_i. If all the variables in $\boldsymbol{X}$ are discrete, $\rho(\boldsymbol{X} = \boldsymbol{x}) = p(\boldsymbol{X} = \boldsymbol{x})$ (or simply $p(\boldsymbol{x})$) is the *joint probability mass*, and $\rho(X_i = x_i|X_j = x_j) = p(X_i = x_i|X_j = x_j)$ (or simply $p(x_i|x_j)$) is the *conditional mass probability* of the variable X_i given that $X_j = x_j$.

In the case that X_i is continuous, $\rho(X_i = x_i) = f(X_i = x_i)$ (or simply $f(x_i)$) is the *density function* of X_i. If all the variables in $\boldsymbol{X}$ are continuous, $\rho(\boldsymbol{X} = \boldsymbol{x}) = f(\boldsymbol{X} = \boldsymbol{x})$ (or simply $f(\boldsymbol{x})$) is the *joint density function*, and $\rho(X_i = x_i|X_j = x_j) = f(X_i = x_i|X_j = x_j)$ (or simply $f(x_i|x_j)$) is the *conditional density function* of the variable X_i given that $X_j = x_j$.

Let $\boldsymbol{X} = (X_1, \ldots, X_n)$ be a vector of random variables. We use x_i to denote a value of X_i, the ith component of $\boldsymbol{X}$, and $\boldsymbol{y} = (x_i)_{X_i \in \boldsymbol{Y}}$ to denote a value of $\boldsymbol{Y} \subseteq \boldsymbol{X}$. A *probabilistic graphical model* for $\boldsymbol{X}$ is a graphical factorization of the joint generalized probability distribution, $\rho(\boldsymbol{X} = \boldsymbol{x})$ (or simply $\rho(\boldsymbol{x})$). The representation consists of two components: a structure and a set of local generalized probability distributions. The structure S for $\boldsymbol{X}$ is a directed acyclic graph (DAG) that represents a set of conditional (in) dependence (Dawid, 1979)[1] assertions on the variables in $\boldsymbol{X}$.

The structure S for $\boldsymbol{X}$ represents the assertions that, for all $i = 1, \ldots, n$, X_i and its non-descendent are independent given $\boldsymbol{Pa}_i^S$.[2] Thus, the factorization is as follows:

$$\rho(\boldsymbol{x}) = \rho(x_1, \ldots, x_n) = \prod_{i=1}^{n} \rho(x_i|\boldsymbol{pa}_i^S). \tag{13.1}$$

The local generalized probability distributions associated with the probabilistic graphical model are precisely those in the previous equation.

In this presentation, we assume that the local generalized probability distributions depend on a finite set of parameters $\boldsymbol{\theta}_S \in \Theta_S$. Thus, we rewrite the previous equation as follows:

$$\rho(\boldsymbol{x}|\boldsymbol{\theta}_S) = \prod_{i=1}^{n} \rho(x_i|\boldsymbol{pa}_i^S, \boldsymbol{\theta}_i) \tag{13.2}$$

where $\boldsymbol{\theta}_S = (\boldsymbol{\theta}_1, \ldots, \boldsymbol{\theta}_n)$. Taking both components of the probabilistic graphical model into account, this will be represented by $M = (S, \boldsymbol{\theta}_S)$.

[1]Given $\boldsymbol{Y}, \boldsymbol{Z}, \boldsymbol{W}$ three disjoints sets of variables, we say that $\boldsymbol{Y}$ is *conditionally independent* of $\boldsymbol{Z}$ given $\boldsymbol{W}$ if for any $\boldsymbol{y}, \boldsymbol{z}, \boldsymbol{w}$ we have $\rho(\boldsymbol{y}|\boldsymbol{z}, \boldsymbol{w}) = \rho(\boldsymbol{y}|\boldsymbol{w})$.

[2]$\boldsymbol{Pa}_i^S$ represents the set of parents – variables from which an arrow that ends in X_i comes out – of the variable X_i in the probabilistic graphical model with structure given by S.

Bayesian networks

Bayesian networks have been surrounded by a growing interest in recent years, as shown by the large number of dedicated books and the wide range of theoretical and practical publications in this field. Textbooks include the classic work of Pearl (1988). Neapolitan (1990) explains the basics of propagation algorithms and these are studied in detail by Shafer (1996). Jensen (1996) is a recommended tutorial introduction while in the work of Castillo, Gutiérrez and Hadi (1997) another sound introduction with many worked examples can be found. Lauritzen (1996) provides a mathematical analysis of graphical models, and more recently Cowell *et al.* (1999), Jensen (2001) and Neapolitan (2003) have provided excellent compilations of material covering recent advances in the field.

The Bayesian network paradigm is used mainly to reason in domains with an intrinsic uncertainty. The reasoning inside the model, that is, the propagation of the evidence through the model, depends on the structure reflecting the conditional (in) dependences between its variables. Cooper (1990) proved that this task is NP-hard in the general case of multiply connected Bayesian networks. The most popular algorithm to accomplish this task was proposed by Lauritzen and Spiegelhalter (1988) – later improved by Jensen, Olesen and Anderson (1990) – and is based on a manipulation of the Bayesian network structure.

Notation

In the particular case of each variable $X_i \in \boldsymbol{X}$ being discrete, the probabilistic graphical model introduced in Section 13.2 is called a Bayesian network.

If the variable X_i has r_i possible values $x_i^1, \ldots, x_i^{r_i}$, then the local distribution, $p(x_i|\boldsymbol{pa}_i^{j,S}, \boldsymbol{\theta}_i)$, is an unrestricted discrete distribution:

$$p(x_i^k|\boldsymbol{pa}_i^{j,S}, \boldsymbol{\theta}_i) = \theta_{x_i^k|\boldsymbol{pa}_i^j} \equiv \theta_{ijk} \tag{13.3}$$

where $\boldsymbol{pa}_i^{1,S}, \ldots, \boldsymbol{pa}_i^{q_i,S}$ denotes the values of $\boldsymbol{Pa}_i^S$, the set of parents of the variable X_i in the structure S. The term q_i denotes the number of possible different instances of the parent variables of X_i. Thus, $q_i = \prod_{X_g \in \boldsymbol{Pa}_i} r_g$.

The local parameters are given by $\boldsymbol{\theta}_i = ((\theta_{ijk})_{k=1}^{r_i})_{j=1}^{q_i})$. In other words, the parameter θ_{ijk} represents the conditional probability of variable X_i being in its kth value, knowing that the set of its parent variables is in its jth value.

The graphical representation is given by a directed acyclic graph where we put edges from X_i's parents ($\boldsymbol{Pa}_i$) to X_i – see Figure 13.1. As we can see in this figure, there is a reduction in the number of parameters we need to determine to obtain the joint distribution over the five variables. In concrete terms, with the use of the Bayesian network in Figure 13.1, this reduction is from 31 parameters to 10.

(a)

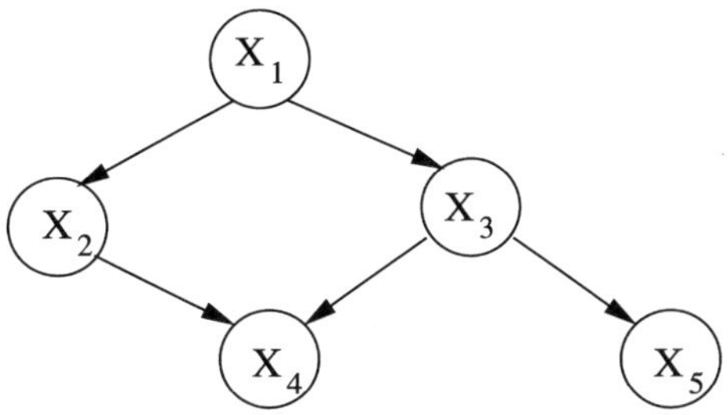

(b)

$$p(X_1 = 0) = 0.20$$
$$p(X_2 = 0|X_1 = 0) = 0.80$$
$$p(X_2 = 0|X_1 = 1) = 0.80$$
$$p(X_3 = 0|X_1 = 0) = 0.20$$
$$p(X_3 = 0|X_1 = 1) = 0.05$$
$$p(X_4 = 0|X_2 = 0, X_3 = 0) = 0.80$$
$$p(X_4 = 0|X_2 = 1, X_3 = 0) = 0.80$$
$$p(X_4 = 0|X_2 = 0, X_3 = 1) = 0.80$$
$$p(X_4 = 0|X_2 = 1, X_3 = 1) = 0.05$$
$$p(X_5 = 0|X_3 = 0) = 0.80$$

Figure 13.1 Joint probability factorization achieved with the Bayesian network attached: $p(x_1, x_2, x_3, x_4, x_5) = p(x_1) \cdot p(x_2|x_1) \cdot p(x_3|x_1) \cdot p(x_4|x_2, x_3) \cdot p(x_5|x_3)$. Note that the reduction in the number of parameters is from $2^5 - 1$ to 10. (a) Bayesian network structure; (b) parameters

In order to understand the introduced notation, we obtain from Figure 13.1 the values expressed in Table 13.1.

Model induction

Once the Bayesian network is built, it constitutes an efficient device to perform probabilistic inference. Nevertheless, the problem of building such a network remains. The structure and conditional probabilities necessary for characterizing the Bayesian network can be provided either externally by experts – time consuming and subject to mistakes – or by automatic learning from a database of cases. On the other hand, the learning task can be separated into two subtasks: *structure learning*, that is, to identify the topology of the Bayesian network, and *parametric learning*, the numerical parameters (conditional probabilities) for a given network topology.

The easier accessibility to huge databases during recent years has led to a large number of model learning algorithms being proposed. We classify the different approaches to Bayesian network model induction according to the nature of the modelling (detecting conditional (in)dependences versus score + search methods) used.

Table 13.1 Variables (X_i); number of possible values of variables (r_i); set of variable parents of a variable ($\boldsymbol{Pa}_i$); number of possible instantiations of the parent variables (q_i)

X_i	r_i	$\boldsymbol{Pa}_i$	q_i
X_1	2	$\emptyset$	0
X_2	2	$\{X_1\}$	2
X_3	2	$\{X_1\}$	2
X_4	2	$\{X_2, X_3\}$	4
X_5	2	$\{X_3\}$	2

The reader can consult some good reviews on model induction in Bayesian networks in the work of Heckerman (1995) and Buntine (1996).

Detecting conditional (in)dependences

Every algorithm that tries to recover the structure of a Bayesian network by detecting (in)dependences has some conditional (in)dependence relations between some subset of variables of the model as input, and a directed acyclic graph that represents a large percentage (and even all of them if possible) of these relations as output. Once the structure has been learnt, the conditional probability distributions required to completely specify the model are estimated from the database – using some of the different approaches to parameter learning – or are given by an expert.

The input information for the algorithms belonging to this category can have one of the following forms:

- a database from which, with the help of some statistical tests (Kreiner, 1989), it is possible to determine the correctness of some conditional (in)dependence relationships,
- an n-dimensional probability distribution where it is possible to test the veracity of the conditional (in)dependence relationships or
- a list containing relations of conditional dependence and independence between triplets of variables.

The most popular algorithm belonging to this category is the PC algorithm (Spirtes, Glymour and Scheines, 1991). As with almost all recovery algorithms based on independence detection, the PC algorithm starts by forming the complete undirected graph, then 'thins' that graph by removing edges with zero-order conditional independence relations, 'thins' again with first-order conditional independence relations and so on.

Score+search methods

Although the approach to model elicitation based on detecting conditional (in)dependences is quite appealing due to its closeness to the semantics of Bayesian networks, a large percentage of structure learning algorithms developed belongs to the category of score + search methods.

To use this learning approach, we need to define a metric that measures the goodness of every candidate Bayesian network with respect to a datafile of cases. In addition, we also need a procedure to move in an intelligent way through the space of possible networks.

In the majority of the score + search approaches, the search is performed in the space of directed acyclic graphs that represents feasible Bayesian network structures.

The number of possible structures for a domain with n variables is given by the following recursive formula obtained by Robinson (1977):

$$f(n) = \sum_{i=1}^{n} (-1)^{i+1} \binom{n}{i} 2^{i(n-i)} f(n-i); \qquad f(0) = 1; \quad f(1) = 1. \tag{13.4}$$

Other possibilities include searching in the space of equivalence classes of Bayesian networks Chickering (1996) – when a score that verifies the likelihood equivalence property is used – or in the space of orderings of the variables (Larrañaga *et al.*, 1996).

The problem of finding the best network according to some criterion from the set of all networks in which each node has no more than K parents ($K > 1$) is NP-hard (Chickering, Geiger and Heckerman, 1994). This result gives a good opportunity for using different heuristic search algorithms.

These heuristic search methods can be more efficient when the model selection criterion, $C(S, D)$, is separable, that is, when the model selection criterion can be written as a product of variable-specific criteria, such as:

$$C(S, D) = \prod_{i=1}^{n} c(X_i, \boldsymbol{Pa}_i, D^{X_i \cup \boldsymbol{Pa}_i}) \tag{13.5}$$

where $D^{X_i \cup \boldsymbol{Pa}_i}$ denotes the dataset D restricted to the variables X_i and $\boldsymbol{Pa}_i$.

Among all heuristic search strategies used to find good models in the space of Bayesian network structures, we have different alternatives: greedy search, simulated annealing, tabu search, genetic algorithms, evolutionary programming etc.

In the following we will review some scoring metrics that have been used in the learning of Bayesian networks from data.

- *Penalized maximum likelihood.* Given a database D with N cases, $D = \{\boldsymbol{x}_1, \ldots, \boldsymbol{x}_N\}$, one might calculate for any structure S the maximum likelihood estimate, $\widehat{\boldsymbol{\theta}}$, for the parameters $\boldsymbol{\theta}$ and the associated maximized log likelihood, $\log p(D|S, \widehat{\boldsymbol{\theta}})$. This can be used as a crude measure of the success of the structure S to describe the observed data D. It seems appropriate to score each structure by means of its associated maximized log likelihood and thus to seek out (using an appropriate search strategy) the structure that maximizes $\log p(D|S, \widehat{\boldsymbol{\theta}})$.

  Using the notation introduced in Section 13.2 we obtain

$$\begin{aligned} \log p(D|S, \boldsymbol{\theta}) &= \log \prod_{w=1}^{N} p(\boldsymbol{x}_w|S, \boldsymbol{\theta}) = \log \prod_{w=1}^{N} \prod_{i=1}^{n} p(x_{w,i}|\boldsymbol{pa}_i^S, \boldsymbol{\theta}_i) \\ &= \sum_{i=1}^{n} \sum_{j=1}^{q_i} \sum_{k=1}^{r_i} \log(\theta_{ijk})^{N_{ijk}} \end{aligned} \tag{13.6}$$

  where N_{ijk} denotes the number of cases in D in which the variable X_i has the value x_i^k and $\boldsymbol{Pa}_i$ has its jth value. Let $N_{ij} = \sum_{k=1}^{r_i} N_{ijk}$.

Taking into account that the maximum likelihood estimate for θ_{ijk} is given by $\widehat{\theta_{ijk}} = \frac{N_{ijk}}{N_{ij}}$, we obtain

$$\log p(D|S, \widehat{\boldsymbol{\theta}}) = \sum_{i=1}^{n} \sum_{j=1}^{q_i} \sum_{k=1}^{r_i} N_{ijk} \log \frac{N_{ijk}}{N_{ij}}. \tag{13.7}$$

When the model is complex, the sampling error associated with the maximum likelihood estimator implies that the maximum likelihood estimate is not really a believable value for the parameter – even when sample sizes appear large. Also, the monotonicity of the likelihood with respect to the complexity of the structure usually leads the search through complete networks. A common response to these difficulties is to incorporate some form of penalty for model complexity into the maximized likelihood.

There is a wide range of suggested penalty functions. A general formula for a penalized maximum likelihood score is as follows:

$$\sum_{i=1}^{n} \sum_{j=1}^{q_i} \sum_{k=1}^{r_i} N_{ijk} \log \frac{N_{ijk}}{N_{ij}} - f(N)\dim(S) \tag{13.8}$$

where $\dim(S)$ is the dimension – number of parameters needed to specify the model – of the Bayesian network with a structure given by S, i.e. $\dim(S) = \sum_{i=1}^{n} q_i(r_i - 1)$. $f(N)$ is a non-negative penalization function. Some examples for $f(N)$ are the Akaike information criterion (AIC) (Akaike 1974) where $f(N) = 1$, and the Jeffreys–Schwarz criterion, sometimes called the Bayesian information criterion (BIC) (Schwarz, 1978), where $f(N) = \frac{1}{2}\log N$.

- *Bayesian scores. Marginal likelihood.* In the Bayesian approach to Bayesian network model induction from data, we express our uncertainty of the model (structure and parameters) by defining a variable whose states correspond to the possible network structure hypothesis S^h and assessing the probability $p(S^h)$.

After this is done, given a random sample $D = \{\boldsymbol{x}_1, \ldots, \boldsymbol{x}_N\}$ from the physical probability distribution for $\boldsymbol{X}$, we compute the posterior distribution of this structure given the database, $p(S^h|D)$, and the posterior distribution of the parameters given the structure and the database, $p(\boldsymbol{\theta}_S|D, S^h)$. By making use of these distributions, the expectations of interest can be computed.

Using Bayes' rule, we have

$$p(S^h|D) = \frac{p(S^h)p(D|S^h)}{\sum_S p(S)p(D|S)} \tag{13.9}$$

$$p(\boldsymbol{\theta}_S|D, S^h) = \frac{p(\boldsymbol{\theta}_S|S^h)p(D|\boldsymbol{\theta}_S, S^h)}{p(D|S^h)} \tag{13.10}$$

where $p(D|S^h) = \int p(D|\boldsymbol{\theta}_S, S^h)p(\boldsymbol{\theta}_S|S^h)d\boldsymbol{\theta}_S$.

In the *Bayesian model averaging* approach we estimate the joint distribution for $\boldsymbol{X}$, $p(\boldsymbol{x})$, by averaging over all possible models and their parameters:

$$p(\boldsymbol{x}) = \sum_{S} p(S|D) \int p(\boldsymbol{x}|\boldsymbol{\theta}_S, S)p(\boldsymbol{\theta}_S|D, S)\mathrm{d}\boldsymbol{\theta}_S. \tag{13.11}$$

If we try to apply this Bayesian model averaging approach to the induction of Bayesian networks, we must sum up all possible structures which results in an intractable approach. Two common approximations to the former equation are used instead. The first is known as *selective model averaging* (Madigan and Raftery, 1994), where only a reduced number of promising structures S is taken into account and the previous equation is approximated in the following way:

$$p(\boldsymbol{x}) \approx \sum_{S \in \mathcal{S}} p(S|D) \int p(\boldsymbol{x}|\boldsymbol{\theta}_S, S)p(\boldsymbol{\theta}_S|D, S)\mathrm{d}\boldsymbol{\theta}_S. \tag{13.12}$$

In the second approximation, known as *Bayesian model selection*, we select a single 'good' model S^h and estimate the joint distribution for $\boldsymbol{X}$ using

$$p(\boldsymbol{x}|D, S^h) = \int p(\boldsymbol{x}|\boldsymbol{\theta}_{S^h}, S^h)p(\boldsymbol{\theta}_{S^h}|D, S^h)\mathrm{d}\boldsymbol{\theta}_S^h. \tag{13.13}$$

A score commonly used in Bayesian model selection is the *logarithm of the relative posterior probability* of the model:

$$\log p(S|D) \propto \log p(S, D) = \log p(S) + \log p(D|S). \tag{13.14}$$

Under the assumption that the prior distribution over the structure is uniform, an equivalent criterion is the *log marginal likelihood* of the data given the structure.

It is possible – see the work of Cooper and Herskovits (1992) and Heckerman, Geiger and Chickering (1995) for details – to compute the marginal likelihood efficiently and in closed form under some general asumptions. Given a Bayesian network model, if the cases occur independently, there are no missing values, and the density of the parameters given the structure is uniform, then the previous authors show that

$$p(D|S) = \prod_{i=1}^{n} \prod_{j=1}^{q_i} \frac{(r_i - 1)!}{(N_{ij} + r_i - 1)!} \prod_{k=1}^{r_i} N_{ijk}!. \tag{13.15}$$

Cooper and Herskovits (1992) proposed the K2 algorithm – see Figure 13.2 – to carry out the search in the space of DAGs. The K2 algorithm assumes that an ordering on the variables is available and that, a priori, all structures are equally likely. It searches, for every node, the set of parent nodes that maximizes the following function:

$$g(i, \boldsymbol{Pa}_i) = \prod_{j=1}^{q_i} \frac{(r_i - 1)!}{(N_{ij} + r_i - 1)!} \prod_{k=1}^{r_i} N_{ijk}!. \tag{13.16}$$

```
Algorithm K2

INPUT: A set of n nodes, an ordering on the nodes, an upper bound u on the
number of parents a node may have, and a database D containing N cases.
OUTPUT: For each node, a printout of the parents of the node.
BEGIN K2
  FOR i := 1 TO n DO
      BEGIN
          Pa_i := ∅;
          p_old := g(i, Pa_i):
          OKToProceed := TRUE
          WHILE OKToProceed AND |Pa_i| < u DO
              BEGIN
                  Let Z be the node in Pred(X_i)\Pa_i that
                  maximizes g(i, Pa_i ∪ {Z});
                  pnew := g(i, Pa_i ∪ {Z});
                  IF pnew > p_old THEN
                      BEGIN
                          p_old := pnew;
                          Pa_i := Pa_i ∪ {Z}
                      END
                  ELSE OKToProceed := FALSE;
              END;
          WRITE('Node:', X_i, 'Parents of this node:', Pa_i)
      END;
END K2.
```

Figure 13.2 The K2 algorithm

The K2 algorithm is a *greedy* heuristic. It starts by assuming that a node does not have parents, then in each step it adds incrementally that parent whose addition most increases the probability of the resulting structure. The K2 algorithm stops adding parents to the nodes when the addition of a single parent cannot increase this probability. Obviously, this approach does not guarantee obtaining the structure with the highest probability.

Gaussian networks

Notation

The other particular case of probabilistic graphical models to be considered in this chapter is when each variable $X_i \in \boldsymbol{X}$ is continuous and each local density function is the linear-regression model:

$$f(x_i|\boldsymbol{pa}_i^S, \boldsymbol{\theta}_i) \equiv \mathcal{N}\left(x_i; m_i + \sum_{x_j \in \boldsymbol{pa}_i} b_{ji}(x_j - m_j), v_i\right) \tag{13.17}$$

Structure

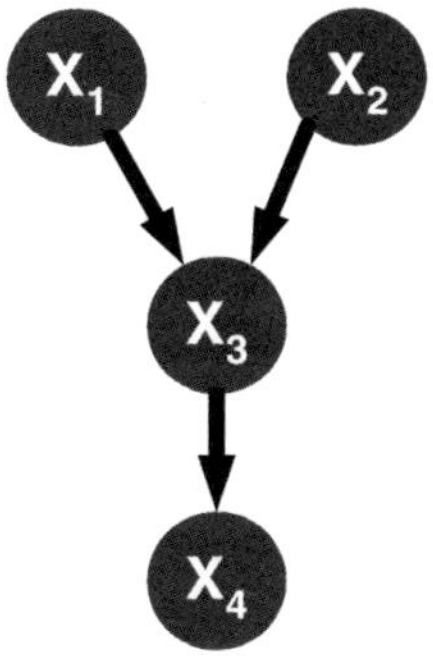

Local densities

$$\theta_1 = (m_1, -, \sigma_1^2)\, f_{X_1} \rightsquigarrow \mathrm{N}\,(x_1; m_1, \sigma_1^2)$$
$$\theta_2 = (m_2, -, \sigma_2^2)\, f_{X_2} \rightsquigarrow \mathrm{N}\,(x_2; m_2, \sigma_2^2)$$
$$\theta_3 = (m_3, b_3, \sigma_3^2)\, f_{X_3|X_1=x_1, X_2=x_2} \rightsquigarrow \mathrm{N}\,(x_3; m_3 + b_{13}(x_1 - m_1) + b_{23}(x_2 - m_2), \sigma_3^2)$$
$$b_3 = (b_{13}, b_{23})$$
$$\theta_4 = (m_4, b_4, \sigma_4^2)\, f_{X_4|X_3=x_3} \rightsquigarrow \mathrm{N}\,(x_4; m_4 + b_{34}(x_3 - m_3), \sigma_4^2)$$
$$b_4 = (b_{34})$$

Factorization of the joint density function

$$f(x_1, x_2, x_3, x_4) = f(x_1) f(x_2) f(x_3 \mid x_1, x_2) f(x_4 \mid x_3)$$
$$= \frac{1}{\sqrt{2\pi}\sigma_1} e^{-\frac{1}{2}\left(\frac{x_1 - m_1}{\sigma_1}\right)^2} \frac{1}{\sqrt{2\pi}\sigma_2} e^{-\frac{1}{2}\left(\frac{x_2 - m_2}{\sigma_2}\right)^2} \frac{1}{\sqrt{2\pi}\sigma_3} e^{-\frac{1}{2}\left(\frac{x_3 - (m_3 + b_{13}(x_1 - m_1) + b_{23}(x_2 - m_2))}{\sigma_3}\right)^2} \frac{1}{\sqrt{2\pi}\sigma_4} e^{-\frac{1}{2}\left(\frac{x_4 - (m_4 + b_{34}(x_3 - m_3))}{\sigma_4}\right)^2}$$

Figure 13.3 Structure, local densities and resulting factorization for a Gaussian network with four variables

where $\mathcal{N}(x; \mu, \sigma^2)$ is a univariate normal distribution with mean μ and variance σ^2. Given this form, a missing arc from X_j to X_i implies that $b_{ji} = 0$ in the former linear-regression model. The local parameters are given by $\boldsymbol{\theta}_i = (m_i, \boldsymbol{b}_i, v_i)$, where $\boldsymbol{b}_i = (b_{1i}, \ldots, b_{i-1i})^{\mathrm{t}}$ is a column vector. We call a probabilistic graphical model constructed with these local density functions a Gaussian network after Shachter and Kenley (1989).

Interpretation of the components of the local parameters is as follows: m_i is the unconditional mean of X_i, v_i is the conditional variance of X_i given $\boldsymbol{Pa}_i$ and b_{ji} is a linear coefficient reflecting the strength of the relationship between X_j and X_i. Figure 13.3 is an example of a Gaussian network in a four-dimensional space.

In order to see the relation between Gaussian networks and multivariate normal densities, we consider that the joint probability density function of the continuous n-dimensional variable $\boldsymbol{X}$ is a multivariate normal distribution if

$$f(\boldsymbol{x}) \equiv \mathcal{N}(\boldsymbol{x}; \boldsymbol{\mu}, \Sigma) \equiv (2\pi)^{-\frac{n}{2}} |\Sigma|^{-\frac{1}{2}} \mathrm{e}^{-\frac{1}{2}(\boldsymbol{x}-\boldsymbol{\mu})' \Sigma^{-1} (\boldsymbol{x}-\boldsymbol{\mu})} \tag{13.18}$$

where $\boldsymbol{\mu}$ is the vector of means, Σ is an $n \times n$ covariance matrix and $|\Sigma|$ denotes the determinant of Σ. The inverse of this matrix, $W = \Sigma^{-1}$, whose elements are denoted by w_{ij}, is referred to as the precision matrix.

This density can be written as a product of n conditional densities, namely

$$f(\boldsymbol{x}) = \prod_{i=1}^{n} f(x_i | x_1, \ldots, x_{i-1}) = \prod_{i=1}^{n} \mathcal{N}\left(x_i; \mu_i + \sum_{j=1}^{i-1} b_{ji}(x_j - \mu_j), v_i\right) \tag{13.19}$$

where μ_i is the unconditional mean of X_i, v_i is the variance of X_i given $X_1, \ldots, X_{i-1}$ and b_{ji} is a linear coefficient reflecting the strength of the relationship between variables X_j and X_i (DeGroot, 1970). This notation gives us the possibility of interpreting a multivariate normal distribution as a Gaussian network where there is an arc from X_j to X_i whenever $b_{ji} \neq 0$ with $j < i$.

The Gaussian network representation of a multivariate normal distribution is better suited to model elicitation and understanding than the standard representation in which one needs to guarantee that the assessed covariance matrix is positive-definite.

Model induction

In this section we present three different approaches to induce Gaussian networks from data. While the first of them is based on edge exclusion tests, the other two belong to score + search methods. As in the section devoted to Bayesian networks, one score corresponds to a penalized maximum likelihood metric and the other is a Bayesian score.

Edge exclusion tests

Dempster (1972) introduced graphical Gaussian models where the structure of the precision matrix is modelled, rather than the variance matrix itself. The idea of this modelling is to simplify the joint n-dimensional normal density by testing whether a particular element w_{ij} with $i = 1, \ldots, n-1$ and $j > i$ of the $n \times n$ precision matrix W can be set to zero. Wermuth (1976) shows that fitting these models is equivalent to testing for conditional independence between the corresponding elements of the n-dimensional variable $\boldsymbol{X}$. Speed and Kiiveri (1986) show that these tests correspond to testing whether the edge connecting the vertices corresponding to X_i and X_j in the conditional independence graph can be eliminated. Hence, such tests are known as edge exclusion tests. Many graphical model selection procedures start by making the $\binom{n}{2}$ single edge exclusion tests – excluding the edge connecting X_i and X_j corresponds to accepting the null hypothesis $H_0 : w_{ij} = 0$, with the alternative hypothesis $H_A : w_{ij}$ unspecified, evaluating the likelihood ratio statistic and comparing it to a χ^2 distribution. However, the use of this distribution is only asymptotically correct. Smith and Whittaker (1998) introduce one alternative to these tests based on the likelihood ratio test.

The likelihood ratio test statistic to exclude the edge between X_i and X_j from a graphical Gaussian model is $T_{lik} = -n \log(1 - r^2_{ij|rest})$ where $r_{ij|rest}$ is the sample partial correlation of X_i and X_j adjusted for the remainder variables. This can be expressed (Whittaker, 1990) in terms of the maximum likelihood estimates of the elements of the precision matrix as $r_{ij|rest} = -\hat{w}_{ij}(\hat{w}_{ii}\hat{w}_{jj})^{-\frac{1}{2}}$.

Score + search methods

- *Penalized maximum likelihood.* Denoting by $L(D|S, \boldsymbol{\theta})$ the likelihood of the database $D = \{\boldsymbol{x}_1, \ldots, \boldsymbol{x}_N\}$ given the Gaussian network model $M = (S, \boldsymbol{\theta})$, we have that

$$L(D|S, \boldsymbol{\theta}) = \prod_{r=1}^{N} \prod_{i=1}^{n} \frac{1}{\sqrt{2\pi v_i}} \mathrm{e}^{-\frac{1}{2v_i}(x_{ir} - mi - \sum_{x_j \in \boldsymbol{pa}_i} b_{ji}(x_{jr} - m_j))^2}. \tag{13.20}$$

The number of parameters, $\dim(S)$, needed to specify a Gaussian network model with a structure given by S can be obtained using the following formula:

$$\dim(S) = 2n + \sum_{i=1}^{n} |\boldsymbol{Pa}_i|. \tag{13.21}$$

In fact, for each variable, X_i, we need to specify its mean, μ_i, its conditional variance, v_i, and its regression coefficients, b_{ji}.

- *Bayesian scores.* In the work of Geiger and Heckerman (1994) the so called BGe (Bayesian Gaussian equivalence) metric is obtained. This metric verifies the interesting property of score equivalence. This means that two Gaussian networks that are isomorphic – represent the same conditional independence and dependence assertions – receive the same score.

The metric is based upon the fact that the normal-Wishart distribution is conjugate with respect to the multivariate normal. This fact allows us to obtain a closed formula for the computation of the marginal likelihood of the data given the structure.

13.4 Inferring Genetic Networks by Means of Probabilistic Graphical Models

In order to represent genetic networks using probabilistic graphical models, we associate each gene – or entity in the molecular system under study – with a random variable. The values of this random variable are determined by the expression level of the gene. Although it is a common practice to discretize it into three values this expression level is originally continuous valued. Depending whether the original information is used or not, probabilistic graphical models for continuous or discrete variables will be of interest. The random variables can include observed variables as well as not observed (latent) or hidden variables – for instance the cluster assignment of a particular gene. Depending on the nature of the studied problem and the properties of the available data, variables in probabilistic graphical models can represent mRNA concentrations, protein modifications or complexes, metabolites or other small molecules, experimental conditions, genotypic information or conclusions such as diagnosis or prognosis.

The following *advantages* of probabilistic graphical models for modelling genetic networks can be listed.

- They are based on probability theory, a scientific discipline with sound mathematical development. Probability theory could be used as a framework to deal with the uncertainty and noise underlying biological domains.

- The graphical component of these models – the structure – allows the representation of the interrelations between the genes – variables – in an interpretable way. The conditional independence between triplets of variables gives a clear semantic. The quantitative part of the models – the conditional probabilities – permits us to establish the strength of the interdependences between the variables.

- Inference algorithms – exact and approximated – developed in these models give us a way to make different types of reasoning inside the model.

- There are already algorithms based on well understood principles in statistics for searching probabilistic graphical models from observational data. These algorithms also permit the inclusion of hidden variables which are not observable in reality. It is also possible to combine multiple local models into a joint global model.

- The declarative nature of the probabilistic graphical models constitutes an advantage to the modelling process by taking additional aspects into account, such as the existence of some edges in the model based on previous knowledge.

- The models are biologically interpretable and can be scored rigorously against observational data.

However, not all the characteristics of the probabilistic graphical models are appropriate for this task. Some of the *disadvantages* are as follows.

- Very little work has been done in the development of learning algorithms able to represent causality between variables (Spirtes, Glymour and Scheines, 1993; Glymour and Cooper, 1999; Pearl, 2000). The description of causal connections among gene expression rates is a matter of special importance for obtaining biological insights about the underlying mechanisms in the cell.

- The characteristics of the analysed databases with very few cases, of the order of dozens, and a very large number of variables, of the order of thousands, make it necessary to adapt the developed learning algorithms. In this way, learning algorithms able to carry out the modelling of subnetworks and at the same time provide robustness in the obtained graphical structure should be of interest (Pe'er *et al.*, 2001).

- The inclusion of hidden variables – where and how many – is a difficult problem when learning probabilistic graphical models from data.

Static and dynamic probabilistic graphical models have been suggested in the literature to reconstruct gene expression networks from microarray data. We first review some works that use *static Bayesian networks* for modelling genetic networks.

Friedman *et al.* (2000) present, for the first time, an application of Bayesian network learning algorithms to the discovery of gene interactions using the *S. cerevisae* cell-cycle measurements of Spellman *et al.* (1998). The authors use a score + search approach to the learning of Bayesian network structure with a Bayesian scoring metric. The so-called sparse candidate algorithm (Friedman, Nachman and Pe'er, 1999) is used to efficiently search in the space of DAGs. The main idea of this algorithm is that it is possible to identify a relatively small number of candidate parents for each gene based on simple local statistics. The search is restricted to Bayesian networks in which only the candidate parents of a gene can be its final parents, resulting in a much smaller search space to be considered. Another interesting aspect is the statistical estimation of the confidence in the edges of the Bayesian network structure. The motivation of this question is due to the very small number of cases stored in the current microarray databases. An effective and relatively simple approach for estimating confidence in the edges is the booststrap method. Generating perturbed versions of the original data set and learning from them, it is possible to collect a set of structures that are fairly reasonable models of the data. Using the frequency of each edge in the different structures, a confidence interval for each edge can be obtained.

Spirtes *et al.* (2000) review current techniques for searching causal relations between variables. They also describe algorithms and data gathering obstacles to applying these techniques to gene expression levels, describing the prospects for overcoming these obstacles. Among the main difficulties pointed out in the work the small sample size, which prevents the building of high-accuracy models, is noted. The other discussed difficulty is the error and noise in the measurement of the expression levels, which do not permit us to be confident about the conditional independences learned from the data. The authors also discuss the necessity of extending the modelling to other types of probability distribution apart from the multinomial (Bayesian networks) or normal (Gaussian networks).

Pe'er *et al.* (2001), based on the work of Friedman *et al.* (2000), define and learn new features that denote the role of a specific gene in the context of the studied gene interactions: mediator, activator and inhibitor. These new features are used to construct subnetworks of strong statistical significance.

Hartemink *et al.* (2001) present an approach to genetic regulatory networks based on Bayesian networks and extensions that allows the inclusion of latent (hidden) variables. The authors use a Bayesian approach to learn the model from data. The graph semantic is also extended to permit annotated edges, able to score models describing relationships of a finer degree of specification, and representing additional information about the type of dependence relationship between the variables. The authors demonstrate the potential of their approach with 52 genomes of Affymetric GeneChip expression data.

Hwang *et al.* (2001) present an Bayesian network application in a supervised classification problem related to cancer diagnosis over the leukaemia data set benchmark. Due to the small sample size of the database, the authors propose a discretization into just two values to avoid unreliable estimation of the parameters of the Bayesian network. In order to induce the Bayesian network, only four genes are selected in a filter way using mutual information and *P*-metric measures. Using these four genes and the class variable – cancer diagnosis – an exhaustive search of the Bayesian network structure with the highest Bayesian score value is done.

Segal *et al.* (2001) introduce a new type of probabilistic graphical model able to express context-specific conditional independences, that is, relationships between triplets of genes that only exist over a subset of the cases of the data set. They demonstrate the power of their approach in two real-world gene expression data sets in yeast.

Chang, Hwang and Zhang (2002) present an application of Bayesian network learning for the dependence analysis over the NCI60 data set. Starting with a database with 890 variables and using a dimensionality reduction technique based on the election of the centroid genes obtained with a cluster technique, the authors finally work with two data sets of 40 and 12 gene-cluster prototypes. The difficulty of reaching biological interpretations of the results obtained with this method is noted by the authors. In the book by Pasanen *et al.* (2003), the chapter devoted to gene regulatory networks is approached with Bayesian networks. The authors present a score + search approach to learn Bayesian network structures from data, using the penalized maximum likelihood score as a metric.

Markowetz and Spang (2003) pay attention to the effects of small sample size and the stability of the solution. Sampling from a Bayesian network model with five variables with three states, the effects of different sample sizes and data perturbation on the reconstruction of the original network topology are evaluated. The authors conclude that active learning from knock-out experiments seems to offer a better approach than learning from passive observations in reconstructing network structure.

Husmeier (2003) provides a brief introduction to learning Bayesian networks from gene-expression data. The method, based on a score + search approach, is contrasted with other approaches to the reverse engineering of biochemical networks. The evaluation of the method is performed by sampling data from a known Bayesian network and trying to recover the structure by MCMC (Markov chain Monte Carlo method).

The author also presents different ROC (receiver operating characteristic) curves for each edge of the Bayesian network as a function of the training set size.

Peña (2004) selects multiple locally optimal models of the data and reports the best of them. The confidence of the final model is reported by studying all the different locally optimal models obtained in the learning phase. Experiments are performed in *Saccharomyces cerevisiae*, obtaining reliable results.

Static Gaussian networks are also used for inferring genetic regulatory networks.

Imoto *et al.* (2003) propose the combination of microarray data and biological knowledge, including protein–protein interactions, protein–DNA interactions, binding site information, existing literature and so on, aimed at estimating a gene network by using a Gaussian network model – the gene expression values are not discretized. A Bayesian scoring is used to measure the goodness of the model. This Bayesian score allows the incorporation of the biological knowledge into the prior probability of the network.

Wu and Ye, Subramanian (2003) propose the use of Gaussian networks to discover gene interactions. The modelling phase defines the independence graph by a set of conditional independence relationships that determine the structure of the graphical model. The partial correlation of two genes by controlling by a third gene is used in obtaining the independence graph. The authors test their methodology using yeast-based microarray data. The obtained results reveal some previously unknown interactions that have solid biological explanations.

Friedman (2004) presents a review paper where the advantages of probabilistic graphical models with respect to clustering methods for inferring cellular networks are discussed. Some learning algorithms for three types of probabilistic graphical model – Bayesian networks, Markov networks and chain graphs – are introduced, illustrating the methodology by several applications to gene expression data.

Dynamic Bayesian networks are able to show how genes regulate each other over time in the complex workings of regulatory pathways. Analysis of time-series data potentially allows us to determine regulatory pathways across time, rather than just associating genes that are regulated together. Two works that use dynamic Bayesian networks for inferring regulatory pathways are reviewed in the next paragraph.

Murphy and Mian (1999) show that most of the proposed discrete models for regulatory networks – including the Boolean network model (Kauffman, 1993; Somogyi and Sniegoski, 1996), the linear model (D'haeseleer *et al.*, 1999) and the nonlinear model (Weaver, Workman and Stormo, 1999) – are special cases of a general class of models called dynamic Bayesian network. The type of dynamic Bayesian network considered by the authors verifies the first-order Markov property, which states that the future is independent of the past given the present. In this work a review of techniques for learning discrete time dynamic Bayesian networks with discrete and continuous states and hidden variables are presented. Unfortunately no empirical evidence is shown.

Ong and Page (2001) introduce an approach for determining transcriptional regulatory pathways by applying dynamic Bayesian networks to time-series gene expression data from DNA microarray hybridization experiments. In their approach an initial dynamic Bayesian network that exploits background knowledge of operons – a sequence of genes that are transcribed together – and their associated genes is built. The authors use a previously published operon map that maps every known and putative gene in the *E. coli* genome into its most probable operon. The structural EM algorithm (Friedman, 1998) is used to infer the remaining structure of the dynamic Bayesian network from *E. coli* gene expression data. This work constitutes the first application of dynamic Bayesian networks to time series gene expression microarray data. The main conclusions of the work are that background knowledge about an organism's genome (in this case, an operon map) can be used to construct the initial, core structure of the dynamic Bayesian network, and that the experimental results provide additional insights into the organism's regulatory network.

13.5 Conclusions

Through this paper, we have presented an overview of the methodology of inferring genetic regulatory networks by probabilistic graphical models. After reviewing classical methods for modelling the nature of regulatory networks – Boolean rules, differential equations and stochastic models – we have introduced methods to learn Bayesian and Gaussian networks, discussing the appropriateness of these probabilistic graphical models to carry out this task.

Among the advantages of using probabilistic graphical models we note that they are based on probability theory – an adequate framework to deal with the uncertainty and noise underlying to biological domains; the graphical components of these models also permit the representation of the relationship between genes, allowing for interpretation from a biological point of view. In this way the research community has a set of solid inference and learning algorithms based on well understood principles in statistics that can be used for reasoning inside the graphical structure and searching these probabilistic models from observational data respectively.

As pointed out by Friedman (2004), it is expected in the near future to see an explosion in the quantity and diversity of high-throughput data sets, including new experimental assays, new experimental designs and examinations of systems at the levels of a single cell, a composite organ, a whole organism and a society. The use of computational analysis methods will be critical for gleaning biological insight from these data sets. To cope with these challenges, the field of computational biology should develop new methodologies as well as adapting existing ones, taking the characteristics of the analysed domains into account. The declarative semantics of probabilistic graphical models is well suited for composing different submodels in a principled and understandable manner.

Acknowledgements

This work was partially supported by the Spanish Ministry of Science and Technology (TIC2001-2973-C05-03), the Spanish Ministry of Health (PI021020), the Basque Government (ETORTEK-GENMODIS and ETORTEK-BIOLAN) and the University of the Basque Country (9/UPV 00140.226-15334/2003).

References

Abkowitz, J. L., Catlin, S. N. and Guttorp, P. (1996) Evidence that hematopoiesis may be a stochastic process in vivo. *Nature Med*, **2**, 190–197.

Akaike, H. (1974) New look at the statistical model identification. *IEEE Trans Automatic Control*, **19** (6), 716–723.

Arkin, A., Ross, J. and McAdams, H. H. (1998) Stochastic kinetic analysis of developmental pathway bifurcation in phage lambda-infected *Escherichia coli* cells. *Genetics*, **149** (4), 1633–1648.

Becquet, C., Blachon, S., Jeudy, B., Boulicaut, J.-F. and Grandrillon, O. (2002) Strong-association-rule mining for large-scale gene-expression data analysis: a case study on human sage data. *Genome Biol*, **3** (12), 1–16.

Becskei, A. and Serrano, L. (2000) Engineering stability in gene networks by autoregulation. *Nature*, **405**, 590–593.

Buntine, W. (1996) A guide to the literature on learning probabilistic networks from data. *IEEE Trans Knowledge Data Eng*, **8** (2), 195–210.

Castillo, E., Gutiérrez, J. M. and Hadi, A. S. (1997) *Expert Systems and Probabilistic Network Models*. Springer, New York.

Chang, J.-H., Hwang, K.-B. and Zhang, B.-T. (2002) Analysis of gene expression profiles and drug activity patterns by clustering and Bayesian network learning. In *Methods of Microarray Data Analysis II*. Kluwer, Dordrecht, 169–184.

Chickering, D. M., Geiger, D. and Heckerman, D. (1994) *Learning Bayesian Networks is NP-Hard*, technical report, Microsoft Research, Redmond, WA.

Chickering, M. (1996) Learning equivalence classes of Bayesian networks structures. In *Proceedings of the Twelfth Conference on Uncertainty in Artificial Intelligence*, Kaufmann, Portland, OR, 150–157.

Cooper, G. F. (1990) The computational complexity of probabilistic interference using belief networks. *Artific Intell*, **42**, 393–405.

Cooper, G. F. and Herskovits, E. A. (1992) A Bayesian method for the induction of probabilistic networks from data. *Machine Learning*, **9**, 309–347.

Cowell, R. G., Dawid, A. P., Lauritzen, S. L. and Spiegelhalter, D. J. (1999) *Probabilistic Networks and Expert Systems*. Springer, New York.

Creighton, C. and Hanash, S. (2003) Mining gene expression databases for association rules. *Bioinformatics*, **19** (1), 79–86.

Dawid, A. P. (1979) Conditional independence in statistical theory. *J R Stat Soc Series B*, **41**, 1–31.

DeGroot, M. (1970) *Optimal Statistical Decisions*. McGraw-Hill, New York.

Dempster, A. P. (1972) Covariance selection. *Biometrika*, **32**, 95–108.

D'haeseleer, P., Liang, S. and Somogyi, R. (2000) Genetic network inference: from co-expression clustering to reverse engineering. *Bioinformatics*, **16**, 707–726.

D'haeseleer, P., Wen, X. Fuhrman, S. and Somogyi, R. (1999) Linear modelling of mRNA expression levels during CNS development and injury. In *Proceedings of the Pacific Symposium on Biocomputing*.

Friedman, N. (1998) The Bayesian structural EM algorithm. In *Proceedings of the Fourteenth Conference on Uncertainty in Artificial Intelligence*, Kaufmann, Madison, WI. 129–138.

Friedman, N. (2004) Inferring cellular networks using probabilistic graphical models. *Science*, **303**, 799–805.

Friedman, N., Linial, M., Nachman, I. and Pe'er, D. (2000) Using Bayesian networks to analyze expression data. *Computational Biol*, **7**, 601–620.

Friedman, N., Nachman, I. and Pe'er, D. (1999) Learning Bayesian network structure from massive datasets: the sparse candidate algorithm. In *Proceedings of the Uncertainty in Artificial Intelligence Conference*, Kaufmann, Stockholm, 206–215.

Geiger, D. and Heckerman, D. (1994) *Learning Gaussian Networks*, technical report, Microsoft Advanced Technology Division, Microsoft Corporation, Seattle, WA.

Glass, L. and Kauffman, S. A. (1973) The logical analysis of continuous, non-linear biochemical control networks. *J Theor Biol*, **39**, 103–129.

Glymour, C. and Cooper, C. (1999) *Computation, Causation and Discovery*, MIT Press, Cambridge, MA.

Hartemink, A. J., Gifford, D. K., Jaakkola, T. S. and Young, R. A. (2001) Using graphical models and genomic expression data to statistically validate models of genetic regulatory networks. In *Pacific Symposium on Biocomputation* 6, 422–433.

Heckerman, D. (1995) *A Tutorial on Learning with Bayesian Networks*, technical report, Microsoft Advanced Technology Division, Microsoft Corporation, Seattle, WA.

Heckerman, D., Geiger, D. and Chickering, D. M. (1995) Learning Bayesian networks: the combination of knowledge and statistical data. *Machine Learning*, **20**, 197–243.

Herrero, J., Díaz-Uriarte, R. and Dopazo, J. (2003) An approach to inferring transcriptional regulation among genes from large-scale expression data. *Comparative Functional Genomics*, **4**, 148–154.

Hill, C. C., Tomasi, J. M. and Sethna, J. P. From stochastic to deterministic descriptions of gene expression. In preparation.

Husmeier, D. (2003) Reverse engineering of genetic networks with Bayesian networks. *Biochem Soc Trans*, **31** (6), 1516–1518.

Hwang, K.-B., Cho, D.-Y., Park, S.-W., Kim, S.-D. and Zhang, B.-T. (2001) Applying machine learning techniques to analysis of gene expression data: cancer diagnosis. In *Methods of Microarray Data Analysis*. Kluwer, Dordrecht, 167–182.

Imoto, S., Higuchi, T., Goto, T., Tashiro, K., Kuhara, S. and Miyano, S. (2003) Using Bayesian networks for estimating gene networks from microarrays and biological knowledge. In *Proceedings of the European Conference on Computational Biology.*

Jensen, F. V. (1996) *An Introduction to Bayesian Networks*. University College of London, London.

Jensen, F. V. (2001) *Bayesian Networks and Decision Graphs*. Springer, New York.

Jensen, F. V., Olesen, K. G. and Andersen, S. K. (1990) An algebra of Bayesian belief universe for knowledge based systems. *Networks*, **28** (5), 637–659.

Kauffman, S. A. (1971) Differentiation of malignant to benign cells. *J Theor Biol*, **31**, 429–451.

Kauffman, S. A. (1993) *The Origins of Order, Self-Organization and Selection in Evolution*. Oxford University Press, Oxford.

Kreiner, S. (1989) *On Tests of Conditional Independence*, technical report, Statistical Research Unit, University of Copenhagen.

Lander, E. (1999) Array of hope. *Nature Genetics*, **21** (3), 3–4.

Larrañaga, P., Kuijpers, C. M. H., Murga, R. H. and Yurramendi, Y. (1996) Searching for the best ordering in the structure learning of Bayesian networks. *IEEE Trans Systems, Man Cybernetics*, **41** (4), 487–493.

Lauritzen, S. L. (1996) *Graphical Models*. Oxford University Press, Oxford.

Lauritzen, S. L. and Spiegelhalter, D. J. (1988) Local computations with probabilitis on graphical structures and their application to expert systems. *J R Stat Soc Series B*, **50** (2), 157–224.

Liang, S., Fuhrman, S. and Somogyi, R. (1998) REVEAL, a general reverse engineering algorithm for inference of genetic network architectures. In *Proceedings of the Pacific Symposium on Biocomputing*, 18–29.

Madigan, D. and Raftery, A. (1994) Model selection and accounting for model uncertainty in graphical models using Occams window. *J Am Stat Assoc*, **89**, 1535–1546.

Markowetz, F. and Spang, R. (2003) Reconstructing gene regulation networks from passive observations and active interventions. In *Proceedings of the European Conference on Compuational Biology.*

McAdams, H. H. and Arkin, A. (1997) Stochastic mechanisms in gene expression. *PNAS*, **94** (3), 814.

Murphy, K. and Mian, S. (1999) *Modelling Gene Expression Data Using Dynamic Bayesian Networks*, technical report, Department of Computer Science, University of California at Berkeley.

Neapolitan, E. (1990) *Probabilistic Reasoning in Expert Systems*. Wiley, New York.

Neapolitan, E. (2003) *Learning Bayesian Networks*. Prentice-Hall, Upper Saddle River, NJ.

Ong, I. M. and Page, D. (2001) *Inferring Regulatory Pathways in E. coli Using Dynamic Bayesian Networks*, Technical Report 1426, Computer Sciences, University of Wisconsin – Madison.

Pasanen, T., J. Saarela, I., Saarikko, T. T. M. T. M. V. and Wong, G. (2003) *DNA Microarray. Data Analysis*. CSC – Scientific Computing, Helsinki.

Pearl, J. (1998) *Probabilistic Reasoning in Intelligent Systems: Networks of Plausible Inference*. Kaufmann, San Mateo, CA.

Pearl, J. (2000) *Causaliy, Models, Reasoning, and Inference*. Cambridge University Press, Cambridge.

Pe'er, D., Regev, A., Elidan, G. and Friedman, N. (2001) Inferring subnetworks from perturbed expression profiles. *Bioinformatics*, **17**, 215–224.

Peña, J. (2004) Learning and validating Bayesian network models of genetic regulatory networks. In *Proceedings of the Workshop on Probabilistic Graphical Models*. 161–168.

Robinson, R. W. (1977) Counting unlabeled acyclic digraphs. In *Lecture Notes in Statistics* 622. Springer, New York, 28–43.

Schwarz, G. (1978) Estimating the dimension of a model. *Ann Stat*, **7** (2), 461–464.

Segal, E., Taskar, B., Gasch, A., Friedman, N. and Koller, D. (2001) Rich probabilistic models for gene expression. *Bioinformatics*, **17** (1), 243–252.

Shachter, R. and Kenley, C. (1989) Gaussian influence diagrams. *Management Sci*, **35**, 527–550.

Shafer, G. R. (1996) *Probabilistic Expert Systems*. Society for Industrial and Applied Mathematics, Philadelphia, PA.

Smith, P. W. and Whittaker, J. (1998) Edge exclusion tests for graphical Gaussian models. In *Learning in Graphical Models*. Kluwer, Dordrecht, 555–574.

Somogyi, R. and Sniegoski, C. A. (1996) Modeling the complexity of genetic networks: understanding multigenic and pleiotropic regulation. *Complexity*, **1** (6), 45–63.

Speed, T. P. and Kiiveri, H. (1986) Gaussian Markov distributions over finite graphs. *Ann Stat*, **14**, 138–150.

Spellman, P., Sherlock, G., Zhang, M., Iyer, V., Anders, K., Eisen, M., Brown, P., Botstein, D. and Futcher, D. (1998) Comprehensive identification of cell cycle-regulated genes of the yeast *Saccharomyces cerevisiae* by microarray hybridization. *Mol Biol Cell*, **9**, 3273–3297.

Spirtes, P., Glymour, C. and Scheines, R. (1991) An algorithm for fast recovery of sparse causal graphs. *Soc Sci Comput Rev*, **9**, 62–72.

Spirtes, P., Glymour, C. and Scheines, R. (1993) *Causation, Prediction, and Search*, Lecture Notes in Statistics 81. Springer, New York.

Spirtes, P., Glymour, C., Scheines, R., Kauffman, S., Animale, V. and Wimberly, F. (2000) Constructing Bayesian networks models of gene expression networks from microarray data. In *Proceedings of the Atlantic Symposium on Computational Biology.*

Szallasi, Z. (2001) *Genetic Networks Analysis*, technical report, Uniformed Services University of the Health Sciences.

Weaver, D. C., Workman, C. T. and Stormo, G. D. (1999) Modelling regulatory networks with weight matrices. In *Proceedings of the Pacific Symposium on Biocomputing.*

Wermuth, N. (1976) Model search among multiplicative models. *Biometrics*, **32**, 253–263.

Whittaker, J. (1990) *Graphical Models in Applied Multivariate Statistics*. Wiley, New York.

Wu, X., Barbara, D., Zhang, L. and Ye, Y. (2003) Gene interaction analysis using *k*-way interaction loglinear model: a case study on yeast data. In *ICML03 Workshop on Machine Learning in Bioinformatics.*

Wu, X., Ye, Y. and Subramanian, K. R. (2003) Interactive analysis of gene interactions using graphical Gaussian model. In *BIOKDD03: 3rd ACM SIGKDD Workshop on Data Mining in Bioinformatics*, 63–69.

14

Integrative Models for the Prediction and Understanding of Protein Structure Patterns

Inge Jonassen

Abstract

Protein structures are complex objects that can be described and classified in different ways. In this chapter we give a brief introduction to protein domains, structure and domain classification, structure comparison and prediction. We discuss structural patterns and applications of such patterns to the mentioned problems. It is shown that structure patterns can be useful both for uncovering relationships between different structures and for structure prediction

Keywords

protein structure, classification, structure comparison, alignment, pattern discovery, model evaluation, structure prediction

14.1 Introduction

Proteins are often represented as amino acid sequences, but their evolution and function cannot be properly understood without considering the three-dimensional structure. Proteins frequently have modular architectures, where the modules have different evolutionary histories and may fold more or less independently (domains). Therefore, when analysing protein structure and evolution, it is useful to break individual structures into domains. Furthermore, it is useful to decompose the universe of protein structures into groups and sub-groups of whole structures or of

Data Analysis and Visualization in Genomics and Proteomics Edited by Francisco Azuaje and Joaquin Dopazo

domains. Having performed such a decomposition and classification, one may study the classes of similarly folding and evolving protein domains and utilize information about all of these to understand the relationships between sequence, structure and function. However, it is not straightforward to break structures into modules (or domains), and even less so to obtain a rigorous classification of the resulting modules.

Approaches for exploring the relationships between protein structures are traditionally based on methods for comparing pairs of structures. Such methods are appropriate for comparing structures that are relatively similar, but are less suited for discovery of patterns found in very diverse sets of proteins. To identify such patterns, it is more appropriate to apply methods that do not rely on pairwise comparisons and instead utilize information from several structures at the same time. Patterns or motifs may be found across classes of similarly folding domains and proteins and can be seen to introduce an additional dimension in a classification of the protein structure universe (see Figure 14.1).

One great challenge is how to predict the structure of a protein given its sequence. This is an extremely challenging problem, where both advances in computational power and new inventive algorithmic techniques may advance the field. The approach described above – breaking proteins into building blocks and identifying recurring motifs – is also of great value in this context. First, *ab initio* structure prediction methods could use commonly used building blocks as templates. Second, it may be possible to identify relationships between building blocks, their combinations and sequence information.

In this chapter we give an overview of approaches to the problem of protein classification and comparison, motif discovery and protein structure prediction. The structure of the chapter is as follows. In Section 14.2 we discuss methods for structure

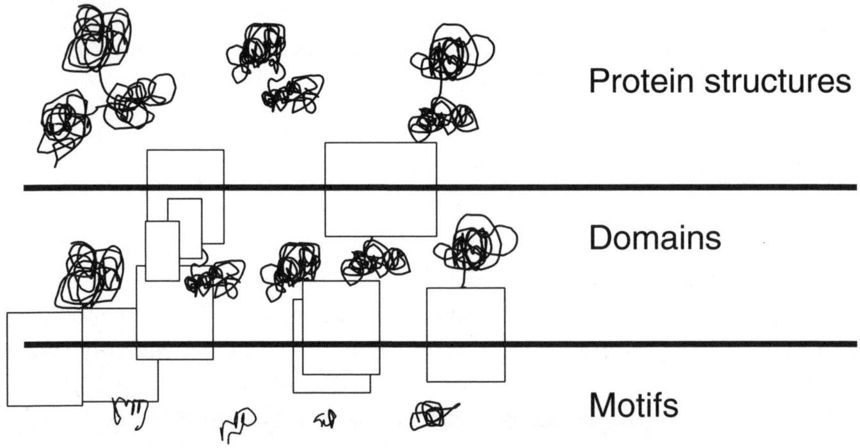

Figure 14.1 Schematic illustration of how protein structures can be broken into domains (a domain is an independently folding unit). Domains can be classified according to sequence and structure similarity in different ways (not shown)

prediction and in Section 14.3 we discuss protein structure comparison and classification. Finally, in Section 14.4 we describe methods for discovery of recurring packing patterns (motifs) in protein structures and an approach to use packing patterns in the evaluation of structural models.

14.2 Structure Prediction

Protein structure prediction can be defined as the process of predicting structural features from the sequence of a protein. The features may be the location and type of secondary structures along the backbone of the protein (referred to as secondary structure prediction), prediction for each residue of whether it is exposed or not or the three-dimensional coordinates of each residue in the protein.

Secondary structure prediction

A large number of methods have been developed for secondary structure prediction and relatively reliable predictions are produced. The methods are based on sequence local properties. For instance, when predicting the secondary structure type of one residue, the amino acid types of the residue itself as well as its nearest neighbours in both directions along the sequence are taken into account. The most successful methods learn patterns in such short sequences associated with each possible secondary structure type and use for instance artificial neural networks to learn and represent the patterns. It has also been shown that if one includes not only one sequence but also evolutionarily related ones (homologues) in the form of alignments, the prediction accuracy is improved. Naturally, for this to succeed, one needs to use alignments both for training the neural network classifier and when performing the actual prediction. For example methods see, e.g., the work of Rost (1996) and Jones (1999).

Ab initio tertiary structure prediction

Analogous ideas have been used for *ab initio* tertiary structure prediction. In this case, one needs to associate sequence features with tertiary structure features. The simplest is to find sequence patterns associated with the local structure of a backbone fragment. For instance, in the I-site approach of Bystroff and Baker, geometrically similar backbone fragments are collected, and sequence patterns are derived from the corresponding sequence alignments (Bystroff and Baker, 1997). When predicting the structure of a protein sequence, a match to one of these patterns biases a randomized algorithm to select the fragment associated with the pattern to the model for this part of the sequence. See Figure 14.2 for an illustration. This idea is one of the elements in

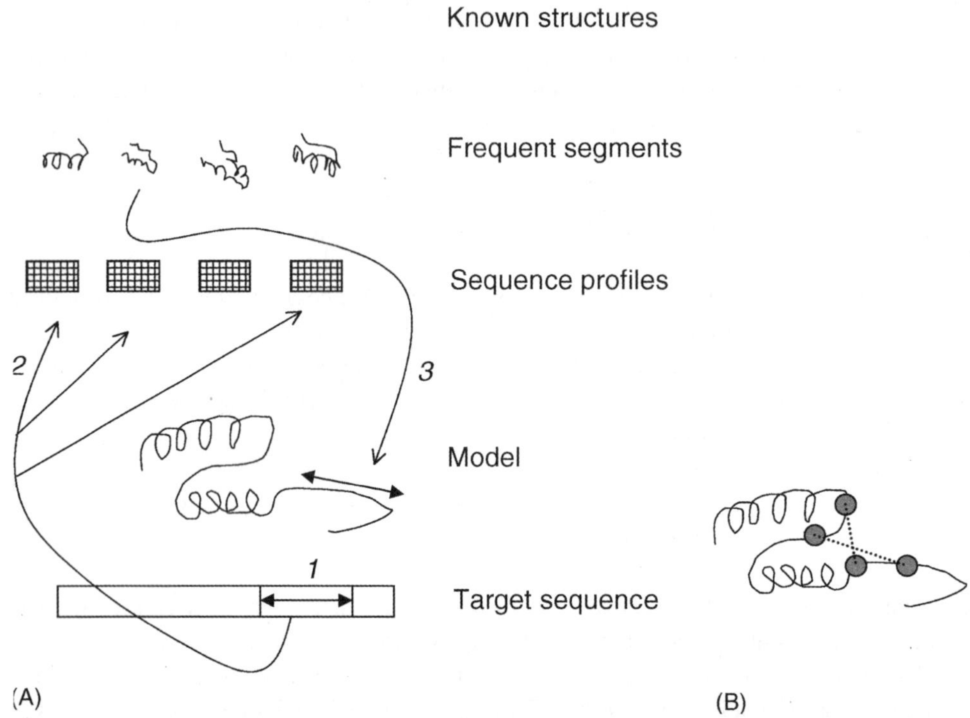

Figure 14.2 (A) Schematic illustration of the Rosetta algorithm. (B) Illustration of additional constraints imposed by an assumed zinc-ion-coordinating cluster where the four coordinating residues should form a tetrahedron. This is an example of a tertiary long-range interaction that can be used either in the evaluation of models or in the generation of models

the Rosetta method of Baker that obtained the best results in the most recent CASP – critical assessment of structure prediction methods (CASP 5, 2003). A library (I-sites) of (sequence profile, structure segment) is generated from the Protein Data Bank (PDB). A model is made for a target sequence through a series of steps (iterations) where in each step a segment in the model is chosen randomly, the corresponding sequence fragment is matched against the I-site sequence profiles and a structure fragment is chosen randomly with the segments associated with profiles obtaining a high score having higher probabilities of being selected. The fragment is then inserted into the model, the model is evaluated and a new step performed.

A limitation of the I-site and Rosetta approach is that the structure patterns used are local along the sequence, and tertiary interactions (e.g. hydrogen bonding between residues dispersed along the backbone) are not captured in this way. If one has hypotheses or experimental evidence about tertiary interactions, it would be advantageous to utilize these in structure prediction. However, this makes the structure space exploration more complicated. For example, an MCMC (Markov chain Monte-Carlo) approach can be used quite easily with the Rosetta approach, where one at each step randomly changes the local conformation of a sequence fragment using the library of sequence-fragment patterns. Analogous steps combining sequence local and tertiary structure patterns are more challenging to realize. One approach in this direction is the GADGET method proposed by Petersen and Taylor (2003), where tertiary interactions (zinc-coordinating residues) are used to constrain the space of possible solutions (models) and distance geometry is used to impose these constraints while sequence local conformational changes are introduced in an MCMC fashion.

Homology modelling

Homology modelling is the most reliable method for protein structure prediction. It builds a structural model of a *target* sequence based on the three-dimensional structure of a *template*. The template is most often chosen by homology, i.e. the template and the target are evolutionarily related. Since structure evolves less rapidly than sequence, it can often be assumed that the three-dimensional structures of two homologous proteins will be similar. As for secondary structure prediction, one can obtain better models if one includes a set of sequences homologous to the target in the form of an accurate alignment. In this case it is possible to use information from all of the sequences when modelling. Patterns of conservation and patterns of substitutions are informative about the constraints imposed by the structure. Naturally, if the alignment is inaccurate, the information will be misleading. See Figure 14.3. The target sequence is searched against a sequence database using for example BLAST. A set of homologous sequences is found and these or a subset of these are aligned to the

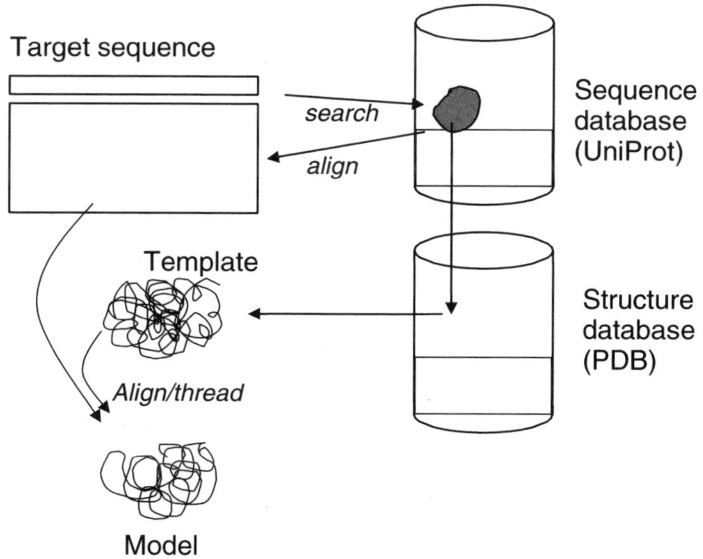

Figure 14.3 Schematic overview of a simple homology modelling method

target sequence. If there is a homologous protein with known structure, this structure can be used as a template and the target sequence alignment can be aligned with the template – a process often referred to as threading. In order to produce a complete model, one may need to re-model loops and optimize placing of side-chains.

A critical step when performing the homology modelling is the alignment between the target and the template. One needs a scoring scheme assessing each alignment and a method to search for the highest scoring alignments. Scoring schemes may be based on energy calculations so that the highest score is assigned to the alignment giving the model with the lowest energy. However, this is computationally demanding and simpler heuristic methods are often used. One example is to use knowledge-based potentials by scoring each pairwise interaction (pair of residues close in space) in the model by how frequently one finds interacting residues with the same amino acid types in true protein structures. A weakness of this approach is that only pairwise interactions are considered. An alternative approach utilizing packing patterns, enabling consideration of multiple interactions, is described below (Section 14.4).

14.3 Classifications of Structures

Protein structures can be classified into groups so that each group contain structures with some common characteristics. Groupings may be nested so that a hierarchy is formed. This is an approach that has been used by several groups. One natural limitation of such an approach is that the hierarchy and the classification will only contain already solved structures and may not represent the full spectrum of structures in nature. Also, one should note that the solved structures are biased by the choice of crystallographers depositing structures into the Protein Data Bank (PDB – Berman *et al.,* 2000) and naturally by which structures are amenable to structure determination. The structural genomic projects will contribute to the number of solved structures and probably obtain a more representative set of structures as these projects aim to 'fill holes' and pick targets for structural studies that allow homology modelling of proteins that currently cannot be modelled in this way.

The following databases classify proteins by their structures. Since proteins are often made up of several domains – each domain being an independently folding unit – it is natural to classify the individual domains rather than complete proteins. See Table 14.1

Table 14.1 Protein classification databases

Database/tool	Reference	URL
CATH	Orengo *et al.*, 1997	http://www.biochem.ucl.ac.uk/bsm/cath/
SCOP	Murzin *et al.*, 1995	http://scop.mrc-lmb.cam.ac.uk/scop/
DALI	Dietman and Holm, 2001	http://www.ebi.ac.uk/dali/domain/
Periodic table	Taylor, 2002	

for a summary of references and web sites for the classifications described, and also refer to Chapter 2 in this book.

CATH

The CATH database comes from University College London from the groups of Christine Orengo and Janet Thornton (Orengo *et al.*, 1997; Pearl *et al.*, 2000). The database acronym indicates the four levels of the hierarchy of CATH – class, architecture, topology, and homologous superfamily. The groups at the top level of the hierarchy – the classes – are defined by the secondary structure content of the structures (mainly alpha, mainly beta, mixed alpha/beta, few secondary structures) and are assigned automatically for most proteins. A group at the architecture level contains proteins with roughly the same spatial arrangements of secondary structure elements, where the way the backbone connects the secondary structure elements may differ. Toplogy classes are assigned to architecture groups manually. Two proteins are said to have the same toplogy if they have the same architecture and in addition they have the corresponding secondary structures in the same order along their backbones. The homologous superfamily level groups together proteins that are evolutionarily related at a level that can be detected on the sequence level. Proteins are assigned to toplogy and homologous superfamily classes automatically using sequence and structure alignment programs.

SCOP

The SCOP (Structural Classification Of Proteins) database is constructed and maintained by Alexey Murzin and colleagues (Murzin *et al.*, 1995; Andreeva *et al.*, 2004). It is a hierarchy where at the top one discriminates between the major classes of proteins based on their secondary structure content (e.g. alpha, beta, alpha/beta, alpha+beta proteins); the next level contain folds having a certain secondary structure architecture and constraints on the topology. Further, the protein domains are grouped into superfamilies and families – both contain protein domains that the SCOP authors believe are evolutionarily related. The SCOP database was constructed and is maintained largely manually and so depends on the biological expertise of the group. While this has some advantages, it means that the database is one subjective view of the protein universe and that it is hard to accurately represent the logic or rules underlying the classifications.

DALI

The DALI database is generated fully automatically by use of the Dali tool for structure alignment (Holm and Sander, 1996) and contains four levels referred to as fold space attractors, fold types functional families and sequence families. Proteins are broken into domains prior to the classification using criteria of compactness and recurrence, the domains are clustered to form a hierarchy and subtrees in the hierarchy are used to define the groupings at the different levels in the DALI classification. A definite advantage with DALI is that it is fully automated, which makes it self-consistent and also enables it to be up to date to a greater extent than classifications that require manual labour. Dietmann and Holm (2001) show that the classification performed by DALI closely resembles that of the SCOP database.

The periodic table of protein structures

Taylor (2002) proposed an alternative approach to classification of protein structure in the form of a periodic table. The table contains protein 'ideal forms' based on principles of protein folding and idealized geometrical shapes. This overcomes several weaknesses of the approaches described above. First, ideal forms can also be generated for protein folds that have not been seen so far. Second, by matching a protein to all ideal forms the matches between forms and substructures can be used to help define the domains. In construction of SCOP and CATH, domain boundaries are defined prior to the construction of the classification. Third, the classification produced using the periodic table is completely automated, and as the classification involves no manual steps or human judgment it is objective.

14.4 Comparing Protein Structures

There exists a large number of methods for comparing protein structures. The methods are designed to uncover different kinds of similarities/patterns. Some are aimed at identifying the core shared by a set of structures – such a core may be described as a set of secondary structure elements. Others are aimed at identifying patterns of individual residues associated with a binding or an active site.

Protein structure descriptions and patterns

A complex object such as a protein structure can be described at a number of different levels with emphasis on features of interest in a particular analysis. Some of the more common representations are

- three-dimensional coordinates of all atoms – this is the most complete representation and the one used in PDB (Berman *et al.*, 2000; http://www.rcsb.org/pdb/) and is referred to as tertiary structure (or quaternary if there are multiple chains)

- three-dimensional coordinates of one or two representative atoms (or pseudo-atoms – e.g. carbon alpha or mean side-chain) per residue – these points may be ordered (by the backbone) or un-ordered

- vector representation of the secondary structures, either using coordinates or relative positions; direction and order of secondary structures along the backbone may or may not be included in the representation

- location of secondary structures along the backbone (secondary structure)

- sequence of amino acids (primary structure).

Methods to compare protein structures typically take two or more descriptions on one of the forms described above and aim to find corresponding (matching) elements in the descriptions. If the descriptions specify an order of the elements (e.g. residues in the order along the protein backbone) and the correspondence is required to match up elements in the same order, the correspondence is an alignment.

Alignment methods

In a similar way as two or more sequences can be aligned, one can align protein structures. An alignment of two structures can be described as a list of residue pairs, each pair containing one residue from each structure, so that the paired residues come in the same order along the proteins' backbones in both structures. A good alignment allows for the residues of one structure to be superposed (e.g. by a transformation and a translation) onto the corresponding residues in the other structures so that the coordinates of paired residues come close in space. Naturally, one also wants to align a large number of residues. The quality of the alignment is often described by a measure of the geometric fit (root mean square deviation – RMSD) together with the number of residues involved. In this formulation we treat the structures as rigid bodies and do not allow for any modification of the geometry of either structure (rigid-body transformation).

For alignment of sequences, it is common to use procedures based on dynamic programming (DP). For standard scoring schemes, the DP algorithm produces mathematically optimal results, and, for alignment of two sequences, it can be efficiently computed. For sequence alignment, this is possible since the scores assigned to different sub-alignments are independent of each other. When an optimal solution for an alignment of a prefix pair has been computed, it is never necessary to adjust this alignment later in the process. For structure alignment, this property does not hold. A decision to add (or remove) one or several residue pairs to (from) the alignment may change the transformation needed to obtain the best RMSD, and this will change the scoring (and potentially the optimality) of all parts of the alignment.

Given a fixed superposition (transformation of one structure on top of the other), it is possible to perform an alignment using DP, i.e. to find a set of co-linear residue pairs close together in space under the given superposition. Conversely, given an alignment, one can calculate a superposition based on the alignment. Hence, given some starting point (initial alignment), one can perform an iterative process alternating between alignment and transformation. This procedure has been explored and implemented by a number of groups since the early 1970s (Rao and Rossman, 1973).

A weakness of the approach is that in each iteration one needs to select one specific alignment and the results depend critically on this one alignment. Taylor and Orengo (1989) proposed a method where one could explore and use several different alignments and utilize the results from all to form one alignment. This method was called SAP – Structure Alignment Program – and the algorithm referred to as double dynamic programming (DDP), since it performs DP on two levels. In the original version of SAP, a DP matrix is used for alignment for each residue pair (i, j) where i comes from structure 1 and j from structure 2. The alignment for pair (i, j) is performed using a scoring scheme derived from a superposition with respect to alignment of i and j (using a neighbourhood around i and j to obtain the superposition). The scores of the best path through each lower-level DP matrix are propagated to an upper-level (summary) matrix, where a final DP is performed to produce the final structure alignment. This final alignment sums up contributions from all alignments considered by the lower-level DP steps so that the lower-level DP matrices that obtained the highest scores have more influence on the high-level DP.

In a later version of SAP (Taylor, 1999), lower-level DP is only performed for a relatively small number of residue pairs. Additionally, the procedure is iterated where the output from one iteration biases the selection of pairs for which to perform lower-level DP in the next iteration. Since only a small number of iterations is necessary (fewer than 10), and also a small proportion of all pairs are subjected to lower-level DP, this iterative version is typically much faster and also turns out to be more accurate.

Geometric methods

Geometric hashing (GH) is another approach that has been explored for the comparison of two and multiple structures. GH is a method developed in the computer vision community for alignment and comparison of geometric shapes. The method can be applied in different ways to perform various analyses. For example one can consider all possible reference frames in each structure (every set of three non-collinear points defines a reference frame), and representing all atoms (residues) in the structure with respect to each reference frame. The representation is done through the use of a hash table so that one can very efficiently identify all reference frames with an atom in a particular location. If similar substructures are found in multiple structures, they will have similar configurations of atoms with respect to at least some reference frames. Such similar substructures can be identified rapidly using the hash tables. After an initial identification of similar substructures, these are clustered, e.g. by similarity of transformations, and extended. Refinements of the method have been suggested both to allow for matching on the secondary structure level (Alesker, Nussinov and Wolfsson, 1996) and to allow for alignment of proteins with flexibility (hinge movements) (see, e.g., Shatsky, Nussinov and Wolfsson, 2002). For an introduction and overview, see Wolfsson and Rigoutsos (1997).

Graph-based methods

A number of groups have proposed the use of graph methods for identification of similarities between protein structures, and in particular similarities on the secondary structure level. These methods typically represent each structure as a labelled graph where nodes represent secondary structure elements and edges are included between elements in geometrical proximity. Both nodes and edges may be labelled with properties associated with the corresponding structural elements or relationships. Similarities between structures can now be identified as subgraph isomorphisms. The comparison of graphs to identify shared subgraphs is a computationally hard problem, but since the graphs are relatively small and reasonable heuristics can be applied the methods can be competitive in their performance. Examples of methods are given by Koch and Lengauer (1997) and Artymiuk *et al.* (1994).

14.5 Methods for the Discovery of Structure Motifs

Similar substructures or motifs can be identified by analysing alignments obtained using methods for structure alignment. However, in many cases it is not trivial to obtain the best structure alignment and it may be advantageous to search for recurring patterns without having to perform an alignment first. In this case, one can use methods for pattern or motif discovery. These can take as input a set of protein structure descriptions and produce patterns matching all structures, an unexpected number of structures or at least some minimum number of structures.

Patterns may be represented as generalizations over structure description formalisms, so that a structure is said to match the pattern if its description can be specialized from the pattern. The matching may be done approximately, in which case one may give an upper bound on the distance between a specialization of the pattern and the structure description. The distance may be the root means square between the coordinates of the pattern and those of the potentially matching (sub-) structure or it may be one defined between protein sequences (e.g. number of mismatches).

The SPratt method

We describe in some more detail the SPratt method that we have developed (Jonassen, Eidhammer and Taylor, 1999; Jonassen *et al.*, 2002). This method is focused on the discovery of patterns consisting of individual conserved residues coming close together in space. The residues may be involved in binding sites or may be central to the protein folding and need not be close in sequence. The method works by focusing on sets of residues coming close together in space and comparing such residue sets in terms of their amino acid types and in terms of geometric similarity, while at the same time requiring that the order of the residues along the proteins' backbones is conserved. The reason for the last requirement is that SPratt is intended to identify packing patterns conserved through evolution between evolutionarily related proteins.

Algorithmically, the method works by constructing for each residue i in each structure a neighbour string containing the amino acid types of all residues within a maximum distance d from residue i. The residue i is also included and referred to as the central residue of the string. The amino acids are written down in the order in which they appear along the backbone of the protein. Internally, SPratt records the correspondence between each position in the neighbour string and the structure residues. Given n proteins of average length m, we obtain nm neighbour strings. In order to identify structure patterns matching at least k out of the n structures, we search for matching sets of neighbour strings so that the set contains neighbour strings from at least k structures. To reach this goal we analyse every neighbour string from each of the $n - k + 1$ smallest (fewest residues) structures and treat each of these neighbour strings as a *seed*. Each seed is analysed to find whether it can be generalized to a pattern that matches at least k of the structures. The seed can be generalized in a large number of different ways by removing from it any set of residues effectively forming subsequences of the seed string. The only constraint is that the central residue should never be removed and thus be contained in all generalizations. The set of possible generalizations is explored by a depth first search, where the starting point for the search is the pattern consisting simply of the central residue of the seed. All neighbour strings having a central residue of the same amino acid type as the seed match this initial pattern. The search is performed recursively by in each step extending the pattern under consideration in all possible ways by adding one seed element to the left or one to the right of the ones already included. When a pattern P is extended, e.g. to P-A, all matches to P are examined to see whether they can be extended to match P-A. If a pattern P does not match at least k structures, then no extension of P will either, and we do not explore extensions of P, effectively *pruning* the search. When a pattern contains four or more elements, the geometric conformation of the involved residues is compared with those of the residues of the seed. If the RMSD is too high, the match is discarded. In this way SPratt identifies patterns matching a minimum of k structures and where the structural similarity of the matches can be limited by defining an upper limit on the allowed RMSD. See Figure 14.4 for an illustration of the method. For each input structure, a neighbour string is constructed for each residue. Part (b) illustrates how the string is constructed. In the next step (search), SPratt searches for patterns that match neighbour strings from at least k structures. The patterns specify amino acid type and coordinates for a set of residues.

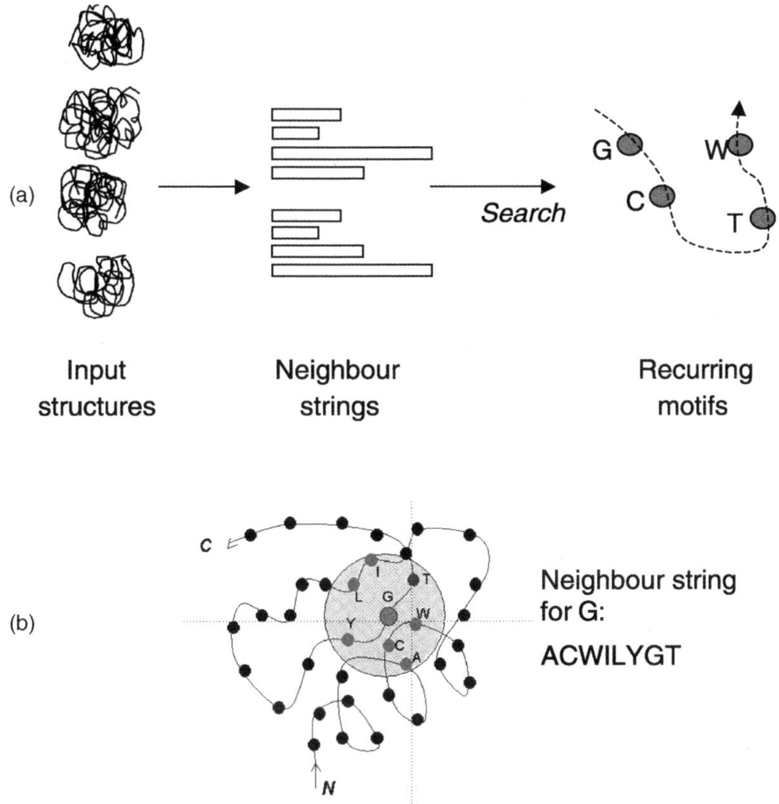

Figure 14.4 Illustration of the SPratt method

The algorithm has also been extended to work with amino acid match sets instead of single amino acids. In this case, we use a pre-defined set of allowed amino acid match sets. When exploring generalizations of a seed we also generate all possible generalizations of the amino acid match sets of the seed. For example, if a pattern contains an I and [ILV] and [IVF] are allowed amino acid sets, then the search will consider all three variants. If one finds a set of patterns all having the same set of matches, all will be reported by SPratt, but the one with the most constrained amino acid match sets will have the highest score. A natural extension is to allow each structure to be associated with a number of homologous sequences represented by an alignment so that each position in the structure can be associated with a column in the alignment and its amino acid set. In this way, one may constrain the search to patterns conserved in the respective sequence alignments. Naturally, one should be careful to ensure that the sequence alignments are accurate.

A weakness with the SPratt approach is that it will normally only identify patterns consisting of a small number of residues and the reported RMSD calculated for this small number of residues is not easy to assess. Therefore, it may be reasonable to try to extend the alignment to include more than the residues described by the pattern. A structure alignment program can be used for this purpose and we have evaluated both the use of an iterative procedure alternating alignment and superposition and using the SAP program (Taylor, 1999) to extend the alignment. Both methods work well and a comparative assessment will be published elsewhere. An advantage of using SPratt in combination with a pairwise method such as SAP is that SPratt is able to take into account information from more than two structures at the time, whereas SAP can only utilize information from two structures at a time. In this way, SPratt can identify elements shared by many structures and SAP can be used to if possible extend the alignment. Thinking about this in multiple alignment terms, SPratt identifies shared blocks (motifs – that would correspond to conserved vertical columns in a typical alignment representation) while SAP extends these horizontally to include larger parts of the structures.

The Sprek method

Recently, we have also shown that packing patterns can also be used to assess structural models (Taylor and Jonassen, 2004). For this application of packing patterns we extended the packing patterns to include amino acid match sets derived from homologous sequences aligned to the protein structure under analysis and to include secondary structure type in addition to amino acid match set for each residue. We compiled a library of patterns found in native structures by using a combination of the CAMPASS and HOMSTRAD databases (Sowdhamini *et al.*, 1998; Mizuguchi *et al.*, 1998) to define a representative set of structures, each with a set of aligned homologous sequences. We used this for evaluation of structural models, where each model was built not for a single sequence but for an alignment of homologous sequences. For each such model, we constructed packing patterns using the procedure used to generate the pattern library. The resulting patterns were matched against the library to obtain the number of library patterns matching each packing pattern from the model. Extensive evaluation comparing this method, named Sprek, with more sophisticated methods reveals that our method produce competitive results. Further work will include refining our initial implementation of the method, a project that is expected to improve the results and make the method even more competitive.

14.6 Discussion and Conclusions

In this chapter we have given a taste of approaches to the analysis and prediction of protein structures. This chapter cannot give an exhaustive overview of all problems nor of all approaches or methods. For further reading, a number of excellent reviews and books can be consulted, e.g. The work of Eidhammer, Jonassen and Taylor (2004), Holm and Sander (1999) and Gibrat, Madej and Bryant (1996).

We have described several approaches to breaking down the complexity of the universe of protein structures both by breaking down individual structures into 'building blocks' – domains – and by constructing classifications of these domains. Such classifications are currently being constructed semi-manually. An alternative approach – a 'periodic table' of protein structures – was described that allows one to simultaneously decompose a structure into domains and at the same time classify the resulting domains. This approach, building on principles of biophysics, gives an objective way to classify proteins by their architectures and provides an excellent complement to existing protein structure classifications such as CATH and SCOP.

We have presented some examples of methods for predicting protein structures and for comparing structures. In particular, we described in more detail the SPratt method that allows for the automatic and efficient discovery of local packing patterns in large sets of structures – without requiring laborious alignment of the structures under analysis. We have also described the Sprek method, where packing patterns such as those used in SPratt can be applied to evaluate structure models. The method, even though it has not been much optimized, shows performance competitive with more advanced and refined methods. This illustrates that alternative approaches sometimes allow for representation of features not easily captured by conventional approaches and that this may result in methods that may supplement and compete with the more traditional ones.

Understanding protein structure and its relationships to function is critically important in order to understand how a cell or an organism works. The number of structures that have been solved experimentally is increasing, and methods to compare, classify and identify recurring patterns can help to better understand underlying principles and relationships to evolution and function. An understanding of the proteins' structure, interactions and dynamics will be a major component in the understanding of biological systems and will therefore play a central role in the field of systems biology.

Protein structure prediction methods exploit data on known structures either directly as in homology-based prediction methods or indirectly for example through neural networks (or structural pattern libraries) trained on (or derived from) known structures. Currently, the most successful *ab initio* prediction methods (e.g. Rosetta) use elements of known structures as building blocks. We believe that methods able to utilize different types of building block will be able to achieve even better predictions. Building blocks found in proteins of known structure can be described as structural patterns. Patterns of the form used in the SPratt and Sprek methods are able to capture information useful for evaluation of structural models. Different forms of patterns capture different aspects of protein structure and may be used in combination in the building or evaluation of models.

Given a protein structure or model, it is far from trivial to predict the function of the protein. In high-throughput structure determination or model building projects, one needs accurate methods for predicting various aspects of protein function from structure. This is an active field of research, and one that can be coupled with high-throughput functional experiments such as screens for protein interactions or gene expression measurements.

References

Alesker, V., Nussinov, R. and Wolfsson, H. J. (1996) Detection of non-topological motifs in protein structures. *Protein Eng*, **9**, 1103–1119.

Andreeva, A., Howorth, D., Brenner S. E., Hubbard, T. J. P., Chothia, C. and Murzin, A. G. (2004) SCOP database in 2004: refinements integrate structure and sequence family data. *Nucleic Acids Res*, **32**, D226–D229.

Artymiuk, P. J., Poirrette, A. R., Grindley H. M., Rice, D. W. and Willett, P. (1994) A graph-theoretic approach to the identification of three-dimensional patterns of amino acid side-chains in protein structures. *J Mol Biol*, **243**, 327–344.

Berman, H. M., Westbrook, J., Feng, Z., Gilliland, G., Bhat, T. N., Weissig, H., Shindyalov, I. N. and Bourne, P. E. (2000) The Protein Data Bank. *Nucleic Acids Res*, **28**, 235–242.

Bystroff, C. and Baker, D. (1997) Blind predictions of local protein structure in CASP2 targets using the I-sites library. *Proteins Suppl*, **1**, 167–71.

CASP-5. (2003) *Proteins: Proteins, Structure, Function and Genetics*, **53**, S6. Available from http://predictioncenter.llnl.gov/casp5.

Dietmann, S., Holm, L. (2001) Identification of homology in protein structure classification. *Nature Struct Biol* **8**, 953–957.

Eidhammer, I., Jonassen, I., Taylor, W. R. (2004) *Protein Bioinformatics, an Algorithmic Approach to Sequence and Structure Analysis*. Wiley, New York.

Gibrat, J. F., Madej, T., Bryant and S. H. (1996) Surprising similarities in structure comparison. *Curr Opin Struct Biol,* **6**, 377–385.

Holm, L. Sander, C. (1996) Mapping the protein universe. *Science*, **273**, 595–603.

Holm, L, Sander, C. (1999) Protein folds and families: sequence and structure alignments. *Nucleic Acids Res*, **27**, 244–247.

Koch, I., Lengauer, T. (1997) Detection of distant structural similarities in a set of proteins using a fast graph-based method. In *ISMB97*, AAAS Press, 167–187.

Jonassen, I., Eidhammer, I., Conklin, D., Taylor, W. R. (2002) Structure motif discovery and mining the PDB. *Bioinformatics*, **18**, 362–367.

Jonassen, I., Eidhammer, I., Taylor, W. R. (1999) Discovery of local packing motifs in protein structures. *Proteins*, **34**, 206–219.

Jones, D. T. (1999) Protein secondary structure prediction based on position-specific scoring matrices. *J Mol Biol*, **292**, 195–202.

Mizuguchi K., Deane C.M., Blundell T. L., Overington J. P. (1998) HOMSTRAD: a database of protein structure alignments for homologous families. *Protein Sci*, **7**, 2469–2471.

Murzin A. G., Brenner S. E., Hubbard T., Chothia C. (1995) SCOP: a structural classification of proteins database for the investigation of sequences and structures. *J Mol Biol*, **247**, 536–540.

Orengo, C. A., Michie, A. D., Jones, S., Jones, D. T., Swindells, M. B., and Thornton, J. M. (1997) CATH – a hierarchic classification of protein domain structures. *Structure*, **5** (8), 1093–1108.

Pearl, F. M. G, Lee, D., Bray, J. E, Sillitoe, I., Todd, A. E., Harrison, A. P., Thornton, J. M. and Orengo, C. A. (2000) Assigning genomic sequences to CATH. *Nucleic Acids Res.*, **28**, 277–282.

Petersen, K., Taylor, W. R. (2003) Modelling zinc-binding proteins with GADGET: Genetic Algorithm and Distance Geometry for Exploring Topology. *J Mol Biol*, **325**, 1039–1059.

Rao, S. T., Rossmann and M. G. (1973) Comparison of super-secondary structures in proteins. *J Mol Biol*, **76**, 241–256.

Rost, B. (1996) PhD: predicting one-dimensional protein structure by profile based neural networks. *Methods Enzymol*, **266**, 525–539.

Shatsky, M., Nussinov, R., Wolfsson, H. J. (2002) Flexible protein alignment and hinge detection. *Proteins*, **48**, 242–256.

Taylor, W. R. (2002) A 'periodic table' for protein structures. *Nature*, **416** (6881), 657–660.

Sowdhamini R., Burke D. F., Huang J.-F., Mizuguchi, K., Nagarajaram H. A., Srinivasan N., Steward R. E. and Blundell T. L. (1998) CAMPASS: A database of structurally aligned protein superfamilies. *Structure*, **6**, 1087–1094.

Taylor, W. R. (1999) Protein structure comparison using iterated double dynamic programming. *Protein Sci*, **8**, 654–665.

Taylor, W. R., Jonassen, I. 2004. A structural pattern-based method for protein fold recognition. *Proteins: Structure, Function, Bioinformatics*, **56**, 222–234.

Taylor, W. R. and Ovengo, C. A. (1989) Protein structure alignment. *J Mol Biol*, **208** (1), 1–22.

Wolfsson, H. J., Rigoutsos, I. 1997. Geometric hashing: an overview. *IEEE Comput Sci Eng*, **4** (4), 10-21.

Index

Note: Figures and Tables are indicated by *italic page numbers*, footnotes by suffix 'n'

ab initio structure prediction methods 240, 241–3, 253
AD-ORFeome 1.0 library 119–20
AD-wormcDNA library 119
adaptive quality-based clustering 163, *171*
agglomerative (hierarchical) clustering 34, 157–9, 177
AIC-based model averaging 199
Akaike information criterion (AIC) 224
alignment methods/tools
  for protein sequence analysis 144–5
  for protein structure comparison 247–8
Ambrosia 3D structure viewer 148, *149*
annotation databases 44, 48–9
  see also sequence databases
annotations 13, 49, 88, 103
antagonistic relationships 199
applied ontologies 101–2
Arabidopsis thaliana, phenotypic characteristics 86
ArrayExpress database 21, 23, *26*

bacterial rhodopsin, sequence analysis *142*
Basic Local Alignment Search Tool (BLAST) 141, 143, 144
Bayesian Gaussian equivalence (BGe) metric 229
Bayesian information criterion (BIC) 224
Bayesian model averaging approach 199, 225
Bayesian model selection 225
Bayesian multiple imputation method (for missing values) 75
Bayesian networks 220–6
  application to gene interactions 231
  dynamic, genetic networks represented using 233–4
  genetic networks represented using 231–3, 233–4
  integrative models based on 32, 71–3, 180–8
  model induction in 221–6
    conditional (in)dependences detection approach 222
    score + search methods 222–6, 232
  notation 220–1
  static, genetic networks represented using 231–3
Bayesian scores 224, 229, 233
benchmarking 52
  lack of in cluster analysis 168, 170
bias, observational studies affected by 206
bibliographic databases 17, *24*, 47–8
biclustering algorithms 166–8, *171*, 178
biobanks 85
BioCreative contest 50, 56
bioinformatics, origins 139
biological data analysis, limitations of traditional approach 4, 33
biological databases 11–23, 47–9
  bibliographic databases 17, *24*, 47–8
  clustering databases 19–20, *25–6*
  enzyme databases 22–3, *26*
  expression databases 21, *26*, 176, 177
  gene databases 19, *25*, 48–9
  interaction databases 22, *26*, *52*, 176
  listed *24–6*, *48*
  pathway databases 23, *26*
  prediction-of-genomic-annotation databases 19, *25*

Data Analysis and Visualization in Genomics and Proteomics Edited by Francisco Azuaje and Joaquin Dopazo

biological databases (*Continued*)
  protein classification databases 20, *25–6*
  sequence databases 17–19, *24–5*
  structure databases 21, *26*
  taxonomy databases 17, *24*
  2D-PAGE databases 21–2, *26*
biological functions, protein–protein interactions studied by 66
biological information visualization, limitations of traditional approach 4–5, 33
biological relevance 205
biologists–statisticians collaboration 207–8
Biomolecular Interaction Network Database (BIND) 22, *26*, 176
bio-ontologies 101–3
  and text mining 49
  see also Gene Ontology (GO)
BIOSIS databases 17, *24*
Biotechnology and Biological Sciences Research Council (BBSRC, UK) 91
biplot 200
Bonferroni correction 105, 107
Boolean rules, for genetic networks 217
bootstrap methods 169, 204, 231
'borrowing' approach to class comparison 196
bovine rhodopsin, sequence analysis *149*
Braunschweig Enzyme Database (BRENDA) 22, *26*
Brenner, S. 113

Caenorhabditis briggsae 116
Caenorhabditis elegans 113–14
  gene expression map 124
  genetics analysis 126
    and protein–protein interaction 126–7
  genome 114, 116
  genomic-context information integrated with protein interaction datasets 128
  interactome
    biological properties 123
    cross-talk with other genomics and post-genomics datasets 123–8
    and gene expression data 123–6
    initial version 114–15
    overlap with gene expression data 123–6
    overlap with phenotype datasets 126–7
    two-hybrid screens to map 118–20
    visualization of 121
  interactome–phenome relationships 127
  interactome–phenome–transcriptome relationships 128
  interactome–transcriptome relationships 32, 124
  orthologues with *S. cerevisiae* 128
  protein localization studies 127–8
Cambridge Structural Database (CSD) 21, *25*
chemical–genetic profiles 90
Class, Architecture, Topology, Homologous (CATH) database 20, *25*, *244*, 245
class comparison 194–8
class discovery 194
  see also unsupervised analysis
class prediction 194, 198–201
  biological interpretation 200
classification, of protein structures 244–6
classification level, integrative approach applied at 37
classification methods, for class prediction 198–9
cluster analysis 36, 153–73
  by biclustering algorithms 166–8, *171*, 178
  distance metrics for 156–7
  by hierarchical clustering 34, 157–9, 177, *186*
  by *k*-means clustering 34, 159, *186*
  limitations 170–1
  model-based methods 163–6, *171*, 177
  by quality-based clustering algorithms 162–3
  by self-organizing maps 34, 159–60, 177, *186*
  by SOTA 108, *109*, 161, *162*, *171*
  by two-dimensional clustering 178
cluster coherence, testing 168
cluster quality, assessing 168–70
clustering algorithms
  first-generation 154, 157–60
    limitations 160–1
  second-generation 154, 161–8, 177, 178
  unsupervised analysis of microarray data using 177–8
clustering based integrative prediction framework 34–5
  tasks and tools for *35*
clustering based studies, evaluation methods required 6
clustering databases 19–20, *25–6*
clusters of genes

functional analysis of
examples 108, *109*
FatiGo used 106–7
other tools 107–8
Clusters of Orthologous Groups (COGs) database 20, *25*
CluSTr database 19–20, 23, *25*
Colour Interactive Editor for Multiple Assignments (CINEMA) 148, *149*
complementary information integration 37
computational biology 216
computer desktop environment 146
computer technology, advances 137–8
conditional density function 219
conditional (in)dependences, detection of 222
conditional mass probability 219
confounding, observational studies affected by 206–7
controlled vocabulary 13, 49, 102
see also Gene Ontology; ontologies
corpus development *45*, 50
Critical Assessment of Information Extraction in Biology *see* BioCreative contest
cross-community information transfer 139
cross-validation 168, 204
curated repositories 101
see also Database of Interacting Proteins (DIP); InterPro database; Kyoto Encyclopedia of Genes and Genomes (KEGG) database
Cytoscape visualization tool 121

DALI database *244*, 245–6
data, meaning of term 100
data analysis
limitations of traditional approach 4, 33
meaning of term 4
data integration 12–16
of data in different formats 13–14
for function prediction 179–80
identification of common database objects and concepts 12–13
various approaches listed *15*
data management, difficulties 100
data mining 7, 89
see also information mining
data visualization
in class prediction 200
integrative tools 33
limitations of traditional approach 4–5, 33
meaning of term 4
data warehousing 14–16
Database of Interacting Proteins (DIP) 22, *26*, *52*, 101, 176
databases 17–26, 47–9
integration of phenotypic data in 93–5
limitations for text mining 47
listed *24–6*, *48*, *52*, *94*, *244*
see also biological databases
Dayhoff, M.O. 139
decision trees 32, 37, 69–71
phenotype predictions using 88, *89*
dendrograms 157, 158
density function 219
design principles 5–8
diagonal linear discriminant analysis (DLDA) *198*
differential equations, for genetic networks 217–18
differential expression
class comparison 194–8
ROC curves for evaluating 201–3
directed acyclic graph (DAG) 13, 102, 219
in probabilistic graphical models 219, 222
directed network 76
DiscoveryLink system *15*, 16
distance-based clustering methods 156, 157–61
distance metrics for 156–7
Distributed Annotation Server (DAS) system 14
DNA damage repair (DDR), proteins involved in 127
domains, in proteins *140*, *240*, 253
double dynamic programming (DDP), protein structure alignment using 248
Drosophila melanogaster, protein interaction map 115, 129
drug-resistance phenotype 90
dynamic Bayesian networks, genetic networks represented using 233–4
dynamic programming (DP), protein structure alignment using 247

edge exclusion tests, Gaussian networks induced from data 228
electron crystallography, protein interactions studied by *51*

EMBL/GenBank/DDBJ nucleotide sequence database 12, 18, 23, *24*
empirical Bayes methods 196
Ensembl tool 5, 19, *25*, 33
EnsMart tool *15*, 16, 33
entropy (in information theory) 70
ENZYME database 22, *26*
enzyme databases 22–3, *26*
Escherichia coli bacteriophage T7, protein–protein interaction map 114
essentiality, protein interactions studied by 66, 91
Euclidean distance 157
European Molecular Biology Laboratory (EMBL), databases 12, 18, 23, *24*, 48
expectation maximization (EM) algorithm 74–5
  in cluster analysis 164, 165, 166, 167
external validity indices 36

factor analysis, mixture of 164–6
false discovery rate (FDR) method 105, 106, 195
false positives (FP) 184
family wise error rate (FWER) method 105–6, 195
Fast Assignment and Transference of Information using Gene Ontology (FatiGO) 106–7
FastA software 143
feature diversity 7
feature redundancy 7
feature selection 7, 37, 205
figure-of-merit (FOM) method, cluster quality assessed using 168–9
filter-based (feature selection) methods 7
filtering, data preprocessing by 155–6
flucanzole-sensitive genes 90
FlyBase database 19, *25*, *48*, 49
forward genetics 84, 85–7
Free Software Foundation 209
free-text processing 101, 138
  see also text mining
FunAssociate 107–8
function prediction
  assessment of accuracy 34, 184–5
  data integration for 179–80
functional analysis, integrated data analysis techniques for 34–6
functional annotations
  predictions of phenotype from 88–9
  unsupervised analysis 177–8
functional categories, enrichment of, cluster quality assessed by 169–70
functional composition 87
functional genomics
  bio-ontologies in 101–3
  data analysis and prediction methods for 5
  data sources 175–6
  goals 176–7
  information mining in 99–100
  ontologies and 99–112
functional similarity, between proteins 66

GADGET method 243
Gateway recombinational cloning *117*, 118
Gaussian networks 226–9
  genetic networks represented using 233
  model induction in 228–9
    edge exclusion tests 228
    score + search methods 229
  notation 226–8
gene databases 19, *25*, 48–9
gene–drug interaction 90
gene expression analysis 30
gene expression correlation
  relationship with interactome data sets 30, 31–2
  and text mining 55–6
gene expression data
  and *C. elegans* interactome 123–6
  obstacles to utility 217
gene expression databases 21, *26*, 176, 177
gene expression microarray data
  functional annotation by unsupervised analysis 177–8
  NCBI database 176, 177
gene expression signatures 200
gene function, identification of 88
gene interactions, Bayesian network learning algorithms application 231
Gene Ontology (GO) 13, 49, 102–3, 184
  annotations 88, 102, 103, 184
    evaluation of gene function prediction using 184
  distribution of terms 104
    comparison between clusters of genes 106–7
    significance testing 104–6
    sliding window for comparison 110

Fast Assignment and Transference of Information using 106–7
functional genomics experimental results translated using 103–4
term extraction tools 103–4, *187*
and text mining 49, 55
types of terms 49, 102, 184
gene shaving method 166
GeneMerge 107
General Repository for Interaction Datasets (GRID) *52*, 176, 181
generalized conditional probability distribution 219
generalized probability distribution 218
genetic networks 216–18
probabilistic graphical models used to represent 229–34
advantages 230
disadvantages 230–1
dynamic Bayesian networks used 233–4
static Bayesian networks used 231–3
static Gaussian networks used 233
genetic redundancy 87
genetics *see* forward genetics; reverse genetics
GENIA corpus 50
genomic annotation
errors 44
see also annotation databases
genomic data analysis 185–8
Genomic Diversity and Phenotype Connection (GDPC) database 93–4
geometric hashing, structure comparison using 248
GEPAS tools 5, 108
Gibbs sampling, application to clustering 167, *171*
gold-standard datasets 67
GOStat 108
graphics hardware, advances 138
GRID *see* General Repository for Interaction Datasets
Grid approach to data integration *15*, 16
growth in bioinformation 138

heat map 158
heuristic search methods 223
hierarchical clustering 34, 157–9, 177
compared with MAGIC *186*
distance measures between clusters 158
visualization of results 158–9
hierarchical network 121, *122*
high-throughput data 138, 176
increase in amount 234
probabilistic framework for integrated analysis of 180–8
homology modelling, protein structure prediction by 243–4
human–computer interaction 146
Human Genome Organization (HUGO), *Genew* database 19, *25*
human protein interaction datasets 129

I-site approach (for protein structure prediction) 241–2
I-view visualization tool 121
ID3 algorithm 70–1
ImMunoGeneTics (IMGT) databases 18–19, *24*
inclusive analysis 106, *107*
induced graph, in network analysis 79
information, meaning of term 100
information extraction (IE) *45*, 46–7, 100–1
information gain 70
information mining
in genome-wide functional analysis 99–100
see also data mining
information retrieval (IR) *45*, 46
information sources, free text vs curated repositories 100–1
information visualization
integrative tools 33
limitations of traditional approach 4–5, 33
input representation level, integrative approach applied at 37
instability problem (in wrapper method) 7
IntAct database 22, *26*, *52*
integrated databases 23, *24*
integration
of annotation 14
of data analysis techniques for supporting functional analysis 34–6
of data in different formats 13–14
of informational views and complexity 31–4

integrative approach to data analysis and visualization 3–5, 29–31
  application at classification level 37
  application at feature pre-processing level 37
  application at input representation level 37
  challenges and opportunities 145–7
  complementary information integration approaches 37
  computational categories 30–1
  multiple data types 31–4, 41–133
  multiple prediction models and methods 34–6, 135–255
  redundant information integration approaches 37
IntEnz database 22, *26*
interaction databases 22, *26*, *52*, 176
interaction maps, intersection of 179
interactome 31
  C. elegans
    biological properties 123
    initial version 114–15
    two-hybrid screens to map 118–20
    visualization of 121
interface, computer 146
International Protein Index (IPI) database 20, *25*
interologues, *C. elegans* 120
InterPro database 20, 23, *25*, 101, 141–2
  graphical output *141*
  typical entry *142*
Interviewer visualization tool 121

'jack-of-all-trades' approach 145
jackknife correlation 157, 168
Jeffreys–Schwarz criterion 224
joint density function 219
  factorization for Gaussian network *227*
joint generalized probability distribution 218–19
joint probability mass 219

k-means clustering 34, 159, *186*
k nearest neighbours (KNN) method *198*
  for missing values 74
knockout mutants 87
knowledge discovery 7
Knowledge Discovery and Data Mining (KDD) Challenge Cup 56, 89
Kyoto Encyclopedia of Genes and Genomes (KEGG) Pathway Database 23, *26*, 101

laboratory robots, phenotype growth experiments 90
latent class methods 196, 201
learning
  meaning of term 68
  see also machine learning; supervised learning
leave-one-out cross-validation 168
lexicons, and text mining 49–50
likelihood ratio 63, 72
  in mRNA expression data 63, *64*
likelihood ratio test, Gaussian networks induced from data 228

machine learning
  in class prediction 198, 199
  for gene interactions 218
  meaning of term 68
  for protein–protein interactions 67–73
MAGIC *see* Multisource Association of Genes by Integration of Clusters
marginal likelihood, in Bayesian network model induction 224
Markov chain Monte Carlo (MCMC) method 232, 242
mass probability 219
mass spectrometry, protein interactions studied by *51*
mean substitution method (for missing values) 74
Medical Subject Headings (MeSH) terms 49
MEDLINE database 17, *24*, 47–8
Mendel, Gregor 83–4
Mendelian Inheritance in Man (MIM) database 19, *25*
metabolic models, phenotypic behavior analysed using 93
microarray data
  cluster analysis of 153–73, 177–8
    assessing cluster quality 168–70, 178
    by biclustering algorithms 166–8, 178
    distance metrics for 156–7
    by distance-based clustering methods 156, 157–61
    external standards 178

by hierarchical clustering 157–9, 177, *186*
internal standards 178
by *k*-means clustering 159, *186*
by model-based methods 156, 163–6, 177
by quality-based clustering algorithms 162–3
by self-organizing maps 159–60, 177, *186*
by Self-Organizing Tree Algorithm 161
preprocessing of 155–6
by filtering 155–6
by missing value replacement 155
by nonlinear transformation 155
by normalization 155
by rescaling 156
by standardization 156
microarray databases 21, 23, *26*
Microarray Gene Expression Data (MGED) Society 13
data format 13
Ontology 102
MicroArray Gene Expression Markup Language (MAGE-ML) 8, 14
Minimal Information About a Microarray Experiment (MIAME) standard 13–14
minimum description length method 166
minP step-down method 105, 107
Missing At Random (MAR) mechanism 73–4
Missing Completely At Random (MCAR) mechanism 73
missing value mechanisms 73–4
missing value problem 73–5, 80
missing value replacement, data preprocessing by 155
mitogen-activated protein kinase (MAP kinase) pathways, in yeast signalling network 77
mixture of factor analysis model, clustering using 164–6
mixture models, clustering using 163–6
model averaging 199
model over-fitting 6, 204
model-based clustering 156, 163–6, *171*, 177
mixture of factor analysis 164–6
mixture model of normal distributions 163–4
modular analysis of networks 77, 79
modular network 121, *122*
molecular biology databases 17–23, 47–9
listed *24–6*, *48*
see also biological databases
Molecular Interactions Database 176
molecular markers 86
molecular signatures 200
motifs
in proteins *140*, *240*
pattern discovery methods 249–52
mRNA expression microarrays, protein interactions studied by *51*, 63, *64*
multiple classification techniques, integrative framework for 34
multiple testing 105–6, 195–6
Multisource Association of Genes by Integration of Clusters (MAGIC) 34, 180–8
as annotation aid 187–8
application to *S. cerevisiae* data 185–8
Bayesian network architecture *182*, 183
combination of clustering methods 183
compared with optimized clustering methods 185–6
evaluation of 184–5
input format 181
method of constructing Bayesian network 183
output 186
quality control role 188
system design 181–4
Munich Information Center for Protein Sequences (MIPS)
complexes database 67, 120
Comprehensive Yeast Genome Database (CYGD) *24*, *48*, 93, 176

naïve Bayes classifier 71–3
named entity recognition (NER) *45*, 46
National Center for Biotechnology Information (NCBI)
bibliographic database 17, *24*, 47–8
Gene Expression Omnibus database 176, 177
Map Viewer 19, *25*, 33
taxonomy database 17, *24*
natural language processing (NLP) 45–7
future developments 56
see also text mining
nearest neighbours methods 74, *198*

network analysis
  future challenges 80
  of protein interactions 75–9
network clustering method 77, 79
network modularity 77, 79
network topology 75–7, 121–3
  average connectivity/degree 75, 122
  clustering coefficient 75, 123
  degree exponential of power-law distribution 122–3
  distribution degree 122
  path lengths 75, 123
network visualization 77, *78*
  C. elegans interactome 121
neural networks 32, 37, 241
NEWT taxonomy database 17, *24*
nomenclature/ontology databases 13, *24*
non-additive relationships 199
nonlinear transformation procedures, data preprocessing by 155
normal distributions, mixture model of 163–4
normalization procedures, data preprocessing by 155
nuclear magnetic resonance (NMR) spectroscopy, protein interactions studied by *51*, 128
null hypothesis tests 36

observational studies 205–7
'omic' data sets, relationships between 31–2, 84
Onto-Express 107
ontologies 49, 101–4
  factors affecting success 102–3
  and functional genomics 99–112
  and text mining 49
  see also Gene Ontology (GO)
Open Bioinformatics Foundation 209
Open Biological Ontologies (OBO) initiative 13, 49, 102
open reading frames (ORFs)
  C. elegans genome 116
  see also ORFeome
Open Source Initiative 209
ORFeome, *C. elegans* 116–17
Osprey visualization tool 121

parametric learning 221
part-of-speech (POS) tagging *45*, 46
partial least squares (PLS) methods *198*
PATHFINDER network for pathology diagnosis 183
pathway databases 23, *26*, 101
pattern discovery 249–52
PC algorithm 222
Pearson correlation 63, 124, 156–7
penalized maximum likelihood score, in probabilistic graphical models 223–4, 229
penalized regression models 197n5, *198*
'periodic table' of protein structures *244*, 246, 253
PharmGKB database 94
phenotype 83, 84
  forward genetics 84, 85–7, 126
  prediction from genomic data 87–8
  prediction from other data sources 88–90
  reverse genetics 84, 87–8, 126
phenotype databases, listed *94*
phenotypic data 83–4
  integration in databases 93–5, 126–7
  integration with phylogenetic data 85
  integration with systems biology 90–3
phenotypic tests, *C. elegans* 119
Phred score 120
plaid model 166, 178, 201
polyacrylamide gel electrophoresis *see* two-dimensional polyacrylamide gel electrophoresis
portal-based approaches 141–2, 151
  limitations of monolithic approach 142–3
  in UTOPIA system 147, *148*
positive-predictive value (PPV) 201n12
post-genomics data integration 123–8, 129
prediction-of-genomic-annotation databases 19, *25*
predictive generalization 6
predictive value negative (PVN) 201n12
predictive value positive (PVP) 201n12
predictors
  error rate 203–4
  evaluation of, ROC curves used 201–3
pre-specified groups of genes, ranking of genes 196–8
principal component analysis (PCA) 80
probabilistic clustering algorithms 178
probabilistic graphical models 216, 218–29
  advantages 230, 234
  Bayesian networks 220–6
  disadvantages 230–1

Gaussian networks 226–9
genetic networks represented using 229–34
notation 218–19
semantics 218–19
probabilistic integration of data 180–8
advantage of approach 180–1
prognostic prediction 194, 198–201
PROPER system 46
protein arrays, protein–protein interactions studied by *51*
protein classification 244–6
databases 20, *25–6*, 66, 244–6
text-mining tools 55
Protein Data Bank (PDB) 20, 21, *25*, 242, 244
collaborating organizations 21
structure representation in 246
protein essentiality 66
protein functions
experimental assessment of 44
similarity 66
protein–protein interaction networks 75–9
biological properties 123
C. elegans
two-hybrid screens to map 118–20
visualization of 121
see also interactome
protein–protein interactions
databases 22, *26*, 50, *52*, 101
experimental techniques 50, *51*
machine learning on 67–73
network analysis of 75–9
prediction of 32–3, 50–2
by genome-based methods 51, 61–81
genomic features 63–7
by physical docking algorithms 52
by sequence-based methods 51–2
text mining of 50–5
see also interaction
protein sequence analysis 139
alignment methods/tools 144–5
'gap' concept 150–1
integrated approach 146, 147–9, 151
methods and databases 139–41
portal-based approach 141–2, 151
limitations of monolithic approach 142–3
tool-based approach 143–5, 151
protein sequence databases 18, *24*, *243*
protein structure analysis 139, 239–55
protein structure comparison 246–9
alignment methods 247–8
geometric methods 248–9
graph-based methods 249
structure descriptions/representations 246–7
protein structure databases 242, *243*, 244
protein structure motifs *140*, *240*
pattern discovery methods 249–52
protein structure prediction 241–4, 253
ab initio tertiary structure prediction 241–3, 253
homology modelling 243–4
secondary structure prediction 241
protein tagging, difficulties 46
Proteome Analysis Database 20
PubMed database 17, *24*, 44, 47
information retrieval system 46
text mining using 47, 101

QT_Clust procedure 162, *171*
quality-based clustering algorithms 162–3
quantitative trait loci (QTL) analysis 84, 86–7
limitations 86

random forests method *198*
model averaging by 199
ranking of genes 194–8
gene-by-gene approach 195–6
prespecified groups of genes 196–8
Rashomon effect 205
receiver operating characteristic (ROC) curves 185, *186*, 200–3, 233
see also ROC-based statistics
redundant information integration 37
RefSeq sequence database 18, *24*
regression imputation method (for missing values) 74
reinvention of methods 204–5
rescaling procedures, data preprocessing by 156
RESID database 21, *25*
resubstitution rate 204
REVerse Engineering ALgorithm (REVEAL) 217
reverse engineering process 216
reverse genetics 84, 87–8
Robot Scientist 90–1, *92*

ROC-based statistics
  diagnostic utility of medical tests evaluated using 203
  ranking of genes using 196, 203
  see also receiver operating characteristic (ROC) curves
Rosetta approach (for protein structure prediction) 242

Saccharomyces cerevisiae
  application of MAGIC to functional genomic data 185–8
  diauxic shift 108, *109*
  Genome Database (SGD) *25*, 34, *48*, 49, 93, 176
  interactome 114, 129
  metabolic network 93
  multiple functional databases *24*, *25*, 34, *48*, 49, 93, 176
  orthologues with *C. elegans* 128
  phenome–interactome relationships 127
  Promoter Database (SCPD) 181
  signalling network 77
  transcriptome–interactome relationships 124
sample size
  Bayesian network applications affected by 232
  observational studies affected by 206
scale-free networks 76, 121, *122*, 129
scientific article databases 17, *24*, 47–8
Search Tool for the Retrieval of Interacting Genes/Proteins (STRING) 20, *52*
secondary structure prediction, proteins 241
selective model averaging 225
self-organizing maps (SOM) 34, 159–60, 177, *186*
Self-Organizing Tree Algorithm (SOTA) 161, *162*, *171*
  example of use 108, *109*
  iterative procedure 161, *162*
semantic obstacles 101, 138
sensitivity 201
  Bayesian network approach 72–3, 185
sensitivity analysis, cluster quality assessed using 169
sensivity/specificity calculations 185
  ROC curves constructed using 202–3
sequence analysis methods and databases 139–41
sequence databases 17–19, *24–5*, *140*
Sequence Retrieval System (SRS) 15–16
SGD database *25*, 34, *48*, 49, 93, 176
Sidak correction 105, 107
significance testing 104–6
single-gene mutants 87
singular value decomposition (SVD) 74
small-world networks 76, 123
software, expectations from 208
source code (of software), availability 209
sparse candidate algorithm 231
specificity 185, 201
SPratt pattern discovery method 249–52, 253
  in combination with SAP 252
  weakness of approach 251–2
Sprek pattern discovery method 252, 253
spurious associations 100, 104
  statistical testing for 104–6
squared Pearson correlation 157
stacking 199
standardization procedures, data preprocessing by 156
Stanford Microarray Database (SMD) 21, *26*
static Bayesian networks, genetic networks represented using 231–3
static Gaussian networks, genetic networks represented using 233
statisticians
  advice needed from during experimental design 207–8
  typical questions asked by *208*
stochastic models, inter-gene regulation represented using 216, 218
Structural Classification of Proteins (SCOP) database *25*, *244*, 245
Structure Alignment Program (SAP) 248
  in combination with SPratt method 252
Structure Assignment With Text Description (SAWTED) system 55
structure comparison, for proteins 246–9
structure databases 21, *26*
structure learning 221
structure prediction, for proteins 139, 241–4
structured vocabulary *see* ontologies
SUISEKI protein interaction discovery tool 53–4
supervised analysis 193–214
supervised learning 5, 68
  compared with unsupervised learning 68–9

support vector machines (SVMs) 46, 55, 67, 90, 180, *198*
survival data methods (for class prediction) *198*
SwissProt database *24*, 48–9
synergistic relationships 199
System for Information Extraction of Interactions *see* SUISEKI
systems biology
  integrated models required 3, 62, 91
  integration with phenotype data 90–3

taxonomy databases 17, *24*
tertiary structure prediction, proteins 241–3
text mining 45–7, 100–1
  advantages and drawbacks 101
  databases and resources for 47–50
  lexicons/thesauri and 49–50
  meaning of term 45
  ontologies and 49
  other applications in genomics 55–6
  of protein–protein interactions 50–5
therapies, and protein interaction data 129
thesauri, and text mining 49
threading (in protein structure prediction) *243*, 244
time-shifted approach, for genetic networks 218
toolbox-based approaches 143–5, 151
  in UTOPIA system 148–50
TopNet (network topological analysis) tool 76–7, *78*
topological analysis of networks 75–7
transcriptome 31
transcriptome–interactome relationships 32
true positives (TP) 185
two-dimensional polyacrylamide gel electrophoresis (2D-PAGE), databases 21–2, 23, *26*
two-hybrid screening, *see also* yeast two-hybrid (Y2H) approach

ubiquitin-dependent protein catabolism, gene cluster for 187–8
UK Biobank 85
undirected network 76
Unified Medical Language System (UMLS) 49–50, 102
UniGene database 20, *25*
UniProt database 12–13, 18, *24*, 48
UniRef database 20, 23, *25*
University of California, Santa Cruz (UCSC), Genome Browser 19, *25*, 33
unsupervised analysis, functional annotation by 177–8
unsupervised learning 68–9
User-friendly Tools for Operating Informatics Applications (UTOPIA) 147–50
  architecture *148*
  basic approach 146
  interoperability in 148–50
  sequence alignment editor 148, *149*
  3D structure viewer 148, *149*
  virtual filing system (UFS) 147

validation of cluster output 168–9
validation dataset 6
VisAnt tool 33

Ward's method (in hierarchical clustering) 158
worldwide Protein Data Bank (wwPDB) 20, 21, *25*
  see also Protein Data Bank (PDB)
Worm Interactome v. 5 (WI5) dataset 120, 127
  integration with *C. elegans* transcriptome and phenome datasets 124–6
WormBase database *25*, *48*, 49
wrapper methods (for feature selection) 7

X-ray crystallography
  protein interactions *51*, 128
  protein structure 21
XML (markup language) 14

yeast
  metabolic network 93
  multiple functional databases *24*, *25*, 34, *48*, 49, 93, 176
  signalling pathways 77
  see also Saccharomyces cerevisiae
yeast two-hybrid (Y2H) screens, protein–protein interactions studied by *51*, 114, 118–20